THE SPANISH ARCADIA

Sheep Herding, Pastoral Discourse, and Ethnicity in Early Modern Spain

The Spanish Arcadia

Sheep Herding, Pastoral Discourse, and Ethnicity in Early Modern Spain

JAVIER IRIGOYEN-GARCÍA

UNIVERSITY OF TORONTO PRESS
Toronto Buffalo London

Toronto Buffalo London
www.utppublishing.com

ISBN 978-1-4426-4727-5

Library and Archives Canada Cataloguing in Publication

Irigoyen-García, Javier, 1975 –, author
The Spanish Arcadia: sheep herding, pastoral discourse, and ethnicity in early modern Spain / Javier Irigoyen-García

Includes bibliographical references and index.
ISBN 978-1-4426-4727-5 (bound)

1. Shepherds – Spain – Social life and customs. 2. Sheepherding – Spain – History. 3. Pastoral sytems – Spain – History. 4. Pastoral literature, Spanish – History and criticism. 5. Spanish literature – Classical period, 1500–1700 – History and criticism. 6. Shepherds in literature. 7. Ethnology – Spain – History. 8. Spain – Historiography. I. Title

GN585.S7G37 2014 306.094609'03 C2013-902883-8

University of Toronto Press acknowledges the financial assistance to its publishing program of the Canada Council for the Arts and the Ontario Arts Council.

University of Toronto Press acknowledges the financial support of the Government of Canada through the Canada Book Fund for its publishing activities.

Contents

List of Illustrations vii

Acknowledgments ix

Introduction: A Country of Shepherds 3
Race, Religion, and Culture in Early Modern Spain 5
The Figure of the Shepherd in Early Modern Spain 11
Pastoral Literature and Cultural Supersession 15

Part One: Sheep Herding and Ethnocentrism in Early Modern Spain

1 Sheep Herding and Discourses on Race 35
Writing the History of Spanish Sheep 36
Sheep Herding and Racial Terminology 38
The Good Shepherd and *limpieza de sangre* 43
Representing Jacob's Livestock 59

2 Rustic Culture and the Invention of the Spanish People 75
The Adoration of the Shepherds 78
Apparitions to Shepherds 85
Performing Rustic Culture 89
Dressing the Shepherd 94
Rustic Speech as Relic 108

3 In the Land of Pan: Pastoral Classicism and Historiography 112
Ancient Place Names and Pastoral Cartographies 113
Inhabiting the Past with Shepherds 121
Etymology and Sheep Herding 129

Paganism and Ethnic Identity 134
The God Pan and the Name of Spain 142

Part Two: Contesting Ethnocentrism within the Arcadia

4 The Moor in Arcadia 151
El Abencerraje in Montemayor's *La Diana* 152
Reluctant Shepherds: Gaspar Mercader's *El prado de Valencia* 161
The Limits of Cultural Cleansing in Cervantes's Pastoral 168

5 Imagining the Spanish Arcadia after 1609 186
Jews and Gypsies in Lope de Vega's *Pastores de Belén* 187
Nostalgia for the Moor: Jacinto de Espinel Adorno's *El premio de la constancia* 197
Pastoral Hierarchies: Juan de Barrionuevo y Moya's *Primera parte de la soledad entretenida* 209

Conclusion: Pan's Labyrinth 219
The Pastoral Habitus: Early Modern to Present 220
From Blood Purity to Whiteness 225
Pastoral and Ethnocentrism: Future Directions 234

Notes 239

Works Cited 281

Index 321

Illustrations

1 Juan Pantoja de la Cruz, *The Birth of Christ with the Royal Family* 56
2 Pedro Orrente, *Jacob Watering Laban's Sheep before Peeled Branches* 64
3 Bartolomé Esteban Murillo, *Jacob Laying Peeled Rods before the Flocks of Laban* 65
4 Pedro Orrente, *Laban Catches up with Jacob* 68
5 Pedro Orrente, *Laban Catches up with Jacob* (detail) 69
6 Bartolomé Esteban Murillo, *Laban Searching for His Stolen Household Gods* 70
7 Bartolomé Esteban Murillo, *Laban Searching for His Stolen Household Gods* (detail) 71
8 Jusepe Ribera, *Jacob among the Sheep* 72
9 Jusepe Ribera, *Jacob with the Herd of Laban* 73
10 Juan de Borgoña, *Nativity* 80
11 Luis de Morales, *Adoration of the Shepherds* 81
12 Bartolomé Esteban Murillo, *The Adoration of the Shepherds* 82
13 Bernardo Asturiano, *Pilgrimage to the Virgin of Cabeza* 87
14 *Libro en que se qüentan los amores de Viraldo y Florindo* 98
15 Christoph Weiditz, "Das ist ain kastilianischer Hiertt" (This is a Castilian Shepherd) 106
16 *Amanece que no es poco* 231

Acknowledgments

I was lucky enough to be able to work with Barbara Fuchs, who wisely guided me when I was writing my dissertation and has inspired me and supported me in so many ways during all these years. Julio Baena, besides participating as a reader for this project, is the ultimate person responsible for nurturing my passion for early modern Spanish literature while I was a graduate student at the University of Colorado at Boulder. Yolanda Martínez-San Miguel revised the initial dissertation thoroughly and gave me useful orientations on the theoretical frame that have proved critical in writing this book. There are many people who have contributed with their comments to improve this book: Abel Alves, Frederick de Armas, Michael Armstrong-Roche, María Judith Feliciano Chaves, Georgina Dopico Black, Edward H. Friedman, Rosilie Hernández-Pecoraro, Marcus Keller, Margaret Marek, Miguel Martínez, Natalio Ohanna, Juan José Pastor Comín, Michael Solomon, John Slater, Laura Zorrilla, and Anne J. Cruz, as well as my colleagues and graduate students at the University of Illinois. I would also like to thank all the anonymous readers in this and other related projects for the time and energy they have invested in providing me with so much valuable feedback. At the University of Toronto Press, I would like to thank Suzanne Rancourt for her assistance throughout the process, and to Miriam Skey for her meticulous copy editing. A special mention goes to Linde Brocato; without her help and advice I do not think I would have been able to complete this project.

Some of the materials included in this book are extensively revised versions of articles that have been previously published elsewhere. Parts of my discussion on clothing in chapter 3 appeared in "'La música ha sido hereje': Pastoral Performance and Cultural Hybridity in *Los Baños de Argel*" (*Bulletin of the Comediantes* 62.2 (2010): 45–62), and my discussion on Cervantes's pastoral in chapter 4 is a rewriting of "'¡Qué si destas diferencias de música resuena la de los albogues!': lo pastoril y lo morisco en Cervantes" (*Cervantes: Bulletin of the Cervantes Society of America* 28.2 (2008): 119–46). I am thankful to Edward H. Friedman and Bruce R. Burningham for permission to reproduce these articles.

The archival research and the time involved in writing this book was made possible with the support of several grants from the Department of Spanish, Italian, and Portuguese, the Campus Research Board, and the Hewlett International Grants Program at the University of Illinois at Urbana-Champaign, as well as the School of Arts and Sciences at the University of Pennsylvania and the Program for Cultural Cooperation between Spain's Ministry of Culture and United States' Universities.

Of course, I cannot forget *os meus carrabouxos* ...

THE SPANISH ARCADIA

Sheep Herding, Pastoral Discourse, and Ethnicity in Early Modern Spain

Introduction: A Country of Shepherds

Después de 1895 di en fijarme en las diferencias que imprime
en los pueblos que éstos hubieran sido en sus mocedades, y durante siglos,
labradores o pastores, y cuánto puede sacarse para la psicología del castellano
el hecho de que su espíritu sea de ganadero más que de labrador.

[After 1895 I began to notice the differences that are imprinted on people
by the fact that they had been, for centuries in their youth, plowmen or shepherds,
and I noticed how much could be deduced about the psychology of the Castilian
by the fact that his spirit belongs to the herdsman rather than to the plowman.][1]

Miguel de Unamuno, *En torno al casticismo*

Miguel de Unamuno locates the initial inspiration for his *En torno al casticismo* (1902) [*On Castilian National Essence*] in this reflection on how the importance of being a country of herdsmen affects the essence of Spanish national identity. While the prologue states that his reflection over the state of Spain preceded the Spanish-American war of 1898, and that, therefore, the book is not influenced by it, it is significant that he is recalling precisely that particular memory in the wake of the moment in which the Spanish empire was dissolved with the loss of Cuba, Puerto Rico, and the Philippines. Unamuno, when looking for a stable sign of national identity from time immemorial in a time of crisis, construes Spain as a country of shepherds.

To be sure, there is an economic explanation for the Spanish attachment to sheep herding. The production of wool and its exportation to the rest of Europe were the main economic activities in Castile from the central Middle Ages well into the eighteenth century. The privileging of sheep herding over other economic activities was favoured by the royal interest in short-term revenues with the thirteenth-century creation of the Mesta, a powerful guild that watched over the protection of pastures and sheep tracks. However, the pressure of population growth, with the subsequent development of agriculture, and the loss of Spain's monopoly over the Merino breed, led to the decay of sheep herding that was already marked in the nineteenth century (Phillips and Phillips 73–96).[2] When Unamuno wrote the prologue to *En torno al casticismo*, sheep herding had yielded its prominence to other economic activities, but it still maintained some of the cultural prestige it enjoyed in the past, and could therefore be used as the point of departure in a reflection about national identity.

Unamuno resorts to the image of the Castilian shepherd not only because it is traditional, but also because it allows him to develop a system of inclusions and exclusions in its construction of Spanishness. On the one hand, Unamuno conflates professional and cultural characteristics when he states that the Castilian belongs to a "raza de ganaderos, de pastores" [race of livestock owners, of shepherds] (18), collapsing any distinction between the broader term and the more specific. Thus in his account the common activity of sheep-herding identity subsumes class difference between livestock owners and dispossessed shepherds, as if they belonged to the same racial community.[3] On the other hand, the mention of the "race of livestock owners and shepherds" triggers the evocation, in the very same passage, of the "race" that opposes and therefore defines Spaniards: "El odio mismo del castellano al morisco no creo arrancara de otra razón: era el odio de los hijos de Abel a los de Caín, porque también los abelinos odian y envidian" [I don't believe the very hatred that the Castilian felt for the Morisco stems from any other reason: it was the hatred that the descendants of Abel felt for those of Cain, because Abelians also hate and envy] (18). Unamuno inscribes the opposition between Moriscos and Old Christians within an eschatological paradigm that is simultaneously religious, racial, and cultural. The exclusion is twofold: at the same time that he excludes Moriscos from the category "Castilian" as if Castilian were a synonym of Christian, he also excludes the rest of Spain's traditional regions from "Spanish" as if Castile were the unique core of Spanish identity.

The reason why Unamuno is evoking this stark opposition between the Castilian and the Morisco in 1898 is beyond the scope of this book, although it may be related to the anxieties about Spanish identity and the fragile imperial projection in North Africa.[4] What is intriguing to me is why the mention of sheep herding brings to the fore the allusion to an ethno-religious Other that had been excised from Spain almost three centuries earlier. Is Unamuno establishing a particular connection between sheep herding, the construction of Spanishness, and the exclusion of the Islamic legacy? As this book tries to demonstrate, he is rather reproducing an imaginary of difference that was bequeathed to him from tradition. But before exploring the role of sheep herding in this construction of identity, we first need to look at the long history of Spanish ethnocentrism for which Unamuno is offering an extremely simplified view.

Race, Religion, and Culture in Early Modern Spain

The *convivencia* of Christianity, Judaism, and Islam in medieval Iberia, along with its unsettling at the beginning of the early modern period, has been one of the most prominent objects of study in Hispanic studies. The Black Death prompted violence against Jews throughout Europe; in the Iberian Peninsula, pogroms against Jewish communities began at the end of the fourteenth century, and continued until their expulsion in 1492 (Nirenberg, *Communities* 18–19); in 1482, the Catholic Monarchs established the Inquisition in Spain in order to oversee the orthodoxy of converted Jews, usually called *conversos* or *cristianos nuevos*.

Whereas Muslims under Christian rule, also known as *mudéjares*, had been granted religious freedom during the Middle Ages, the fall of the Nasrid kingdom of Granada in 1492 soon turned to a new royal policy promoting cultural and religious homogenization, even in the face of protests by ecclesiastics engaged in conversion efforts. Castilian Muslims were forced to convert to Christianity in 1502, and those of Aragon in 1525. Simultaneously, the monarchs passed successive laws that tried to eradicate cultural practices that were considered to be related to Islam. In 1526 Charles V banned Morisco cultural practices, such as the use of certain kinds of garments and the Arabic language, but the prohibition was lifted with a forty-year moratorium when the Moriscos agreed to pay a levy. With the end of the moratorium in 1566, Philip II refused to renew it, a decision that led to the uprising of the Alpujarras in 1568. By 1571 the uprising was repressed and the Moriscos of the kingdom were deported from Granada and dispersed throughout the rest of Castile. This period was marked by rising ethnic tensions that ended in 1609 with Philip III's decree of the expulsion of all Moriscos from Spain.[5]

The conversion of Jews in 1391 and 1492, and that of Muslims in 1502 and 1525, did not necessarily entail their full integration into Christian society. In fact, although those who converted were labelled *cristianos nuevos* (literally, New Christians), in opposition to the *cristianos viejos* (Old Christians), those who boasted of a lineage that had been Christian for many generations, early modern segregationist discourse continued to justify itself through religious difference. The integration of New Christians was increasingly hindered from the fifteenth century on with the emergence of "estatutos de limpieza de sangre" [statutes of blood purity], which required genealogical proof that an individual was "limpio de sangre" [of pure blood] or had no "raça de moro o judío" [taint of Moor or Jew], in which case they were not permitted to enter various private and ecclesiastical institutions, or to travel to the New World.[6] This required a redefinition of the ethnic boundaries constituted by religious differences in the Middle Ages that were reshaped in the early modern period around genealogy. Scholars who hold that the early modern conflict was purely religious essentially reproduce early modern conflation of genealogy and religious orthodoxy underlined in the concept of blood purity, which was justified with the argument that New Christians were prone to heresy.[7]

The main methodological problem with this view that the conflict was only religious is that it reflects only the perception charted by Old Christians, which transmutes the traditional religious boundaries into genealogical, ethno-religious determinism, without considering that many New Christians were sincere Christians and followed orthodox belief and practice (Tueller).[8] The misleading use of terms such as *converso* and *cristiano nuevo* reveals the underlying genealogical logic, since they were applied not only to the individuals who converted to Christianity, but also to their descendants, as Fray Lope de Barrientos complained in the middle of the fifteenth century: "quando yo pienso en mí qué razón hay para poder llamar conuersos aquellos que son hijos de nietos de conuertidos, quedo muy espantado; porque aquellos que nacieron cristianos no sauen cosa alguna de los judaicos usos" [I am horrified when I ponder the reason for naming *conversos* those who are great-grandsons of people who converted, because those who were born Christians know nothing about Jewish customs] (191).

For the defenders of the religious interpretation, terms such as "racism" or "ethnocentrism" constitute anachronisms that blur the differences between past and present systems of social segregation. The question of whether Moriscos, Jews, and Old Christians were different "races" has different answers depending on both the theoretical view adopted and the historical period considered. From the current point of view in popular Western culture, which defines "race" as a taxonomic category based on alleged phenotypic differences, there was no racism in the doctrine of purity of blood in early modern Spain, since it was not dependent on physical traits.[9] Consequently, Antonio Domínguez Ortiz, following Claudio Sánchez Albornoz, denies the existence of racism in early modern Spain, arguing that the descendants of Christians, Jews, and Muslims shared the same physical traits (*La clase social* 141–3). This however confuses our modern definition of race with that of early modern Spaniards, or, to put it in anthropological terminology, it confuses the etic and the emic perspectives, the views of the researcher and the informant (Harris 31–48).[10]

This book follows instead the line of inquiry opened by scholars who explore the analogies between early modern and modern racism, such as Christiane Stallaert, José María Perceval, Henri Méchoulan, Albert Sicroff, David Nirenberg, or María Elena Martínez. At first sight, the perspective of social actors in the early modern period and our own seems to be completely different, since the former is expressed through a mix of genealogical and religious language while the latter is based mainly on biological difference. However, as Stallaert shows, even though Naziism's ethnocentrism resorted to a pseudoscientific language, the term "race" was used in the popular sense, as equivalent to a people bound by kinship and lineage (*Ni una gota* 60). Stallaert identifies a similar divorce between theory and practice in early modern Spain, where religion played the same discursive role that science had in Nazi Germany, pointing out that the criteria for racial classification was in both cases genealogical (*Ni una gota* 219).[11] The paradigm of "scientific racism," based on alleged biological difference, is a manipulation of scientific discourse in order to reinforce social criteria for segregation, much as the defenders of blood purity appropriated the religious language of Christianity in order to justify the exclusion of New Christians.

One consequence of strictly differentiating past and present ethnocentrism is the sanitizing of Spanish history by denying any equivalence with such overcharged ideologies as racism or Naziism. Another outcome of such an approach is that it precludes the identification of modern manifestations of racism as such when not openly articulated through biological criteria. Such arguments against the existence of an early modern racism, as Nirenberg suggests, ultimately derive from a naturalization of modern racism ("Race" 74).[12] While the parallels between early modern and modern racisms and ethnocentrisms are clear, they should not be collapsed. Both Stallaert and Nirenberg, even when defending the analytical analogy between past and present racism, are careful to delimit the differences, refusing any easy continuity between them.[13]

The issues dealt with here necessarily raise the question of the distinction between the concepts of race and ethnicity. An ethnicity is usually defined as a community formed by criteria that can be cultural, linguistic, religious, genealogical, or even economic, whereas the term "race" is preserved for communities constituted upon a biological basis. However, since the "visibility" of phenotype is only meaningful when it attains a value for social hierarchization, its perception is culturally mediated and becomes thus a sign of ethnicity.[14] Inversely, "ethnic differences" tend to be conceived by social participants as conveying hereditary and genealogical notions, reinscribing them into similar processes of biologization and naturalization. It is this malleable conception of any racial system that Peter Wade identifies as the "strategic equivocation between nature and culture" (14).[15] Since purely phenotypic criteria are not operative in early modern Spanish policies of exclusion regarding *limpieza de sangre*, this book deals with "ethnocentrism" instead of "racism," and "ethnicity" instead of "race."

This option does not aim to erase the term "race" from my descriptive vocabulary, avoiding it as a problem. Rather, I aim to categorize "race" as one of the specific tags that has been historically used to conceptualize collective constructions of community and difference, the content of which has changed over time and in the different societies that have adopted it.[16] What we now think of as traditional racism, based on alleged biological differences, anthropologists point out, has continued to evolve its criteria and its descriptive language, without altering its segregationist objectives.[17] The advantage of the concept of ethnicity is that it allows one to better describe historical and local variations of social exclusion, acknowledging the strategic and situational features of collective identity. It also provides a theoretical frame that explains the mutability of the boundaries of collective identity and why the selection of group markers evolves along with the relation between the different social groups (Stallaert, *Etnogénesis* 13–15). In this kind of strategic and situational selection of group markers, early modern discourses on blood purity amalgamate biological, genealogical, religious, and cultural factors to the ever-changing social and historical circumstances, seldom distinguishing them.[18]

The implementation of statutes of blood purity at the level of the individual and the quest for ethno-religious homogeneity were accompanied by the construction of collective cultural "purity." In order to develop (or rather invent) the timeless features of an essential Spanish identity, humanists developed a cultural archaeology privileging the prehistoric, Graeco-Latin, and Visigothic historical periods. Concomitantly they tried to identify the "Moorish" elements of Spanish culture that had to be erased.[19] A note on terminology is needed here: throughout this book, I use "Islamic" and "Muslim" when it refers specifically to religion; "Morisco" when referring to the descendants of Iberian Muslims (regardless of whether they are actually Christian or crypto-Muslim); "Arabic" is used for the language; and the inhabitants of North Africa are described as "North African." Yet all these concepts are conflated, even today, under the problematic term *moro* [Moor], a highly diffuse category that can be used to refer, alternatively, to race, culture, or religion. When referring to individuals, I use the categories referred to above, using "Moor" only when translating Spanish texts that use the word *moro*, or when I refer to the idealized, gallant image of the Muslim, especially when dealing with literary texts in chapters 4 and 5. However, when analysing cultural practices, it is necessary to preserve the equivocal term "Moorish" because it best reflects the imaginary of early modern Spaniards regarding a wide array of quotidian cultural practices and objects that are perceived to derive from prolonged contact with Islam (independently of whether this is their real origin or not). This is why José Antonio González Alcantud considers "lo moro" as a cultural trope or cultural figure (115), and Barbara Fuchs analyses "Moorishness" as a complex set of perceptions and strategies taking place in both the literary realm and the regulation of material culture within the formation of national identity (*Exotic Nation* 2–7).

The conceptualization of discrete unitary cultures is an ideological binarism that imposes a fictitious representation of difference, but in the social discourse of early modern Spain there was certainly an imaginary of cultural difference that, like the imaginary of genealogical difference, was operative and continues to be up to the present. The construction of Spain as an essential identity was contested by Américo Castro in his *España en su historia: cristianos, moros y judíos* (1948), where he suggested that Spanish identity is not an ahistorical product that could be identified with an essentially Christian identity, but that it was marked by the *convivencia* between different cultures and religions. Sánchez Albornoz reacted against Castro's thesis in his *España: un enigma histórico* (1956), stating that the Islamic period was an interference of the natural historical development of a fundamentally Christian Roman-Visigothic identity in which the Semitic elements were only incidental. The controversy between these two intellectuals, both exiled because of the Spanish Civil War, has framed for decades the academic debate over Spanish identity. Thus, while authors like María Rosa Menocal posit Iberian *convivencia* as a historical model for current ethnic and religious conflict (266–81), others such as Serafín Fanjul categorically exclude the Islamic period from exerting any influence over Spanish culture. The line of inquiry initiated by Castro was subjected to recent revisions that question many of the assumptions inherited from the same tradition against which he was writing. Paul Julian Smith, while acknowledging Castro's deconstruction of the tenets of Spanish nationalism based on a timeless identity, points out that Castro himself rearticulated this essentialism in his conception of the three ethno-religious groups as a conflict between "castas" (55). Similarly, Nirenberg also criticizes his deterministic view between genealogy and cultural and intellectual production ("Race" 83–4).[20]

Yet imaginaries are real. As Étienne Balibar states, reformulating Benedict Anderson's concept of "imagined communities," even though collective identities are largely imaginary, "imagined communities" are real in the sense that their fictitious nature does not preclude them from structuring social discourses and practices (93). The early modern process of cultural cleansing constituted as a double dialectical move of denial and of search disguised as a "recuperation" of the national origin, inventing a tradition in which Moriscos and Jews are elided. It is important to emphasize the interrelatedness of these imaginaries of difference so that the parallel processes of "de-Orientalization" and "re-Latinization" of early modern Spain are not naturalized as the unavoidable outcomes of the so-called *reconquest*.[21]

By invoking "Spain" in the early modern period I raise the related question of nationalism along with race and culture. The increasing pressure towards religious homogenization and the ensuing ethnic unrest of early modern Spain coincides with the unification of the different kingdoms of the Iberian Peninsula under a single monarchy. Castile and Aragon came under one monarchy after the marriage of the Catholic Monarchs in 1469; the Nasrid kingdom of Granada fell in 1492; and Ferdinand of Aragon annexed Navarre in 1512. In 1580, Philip II claimed the crown of Portugal after the death of King Sebastião in the battle of Alcazarquivir (1578), achieving the union of all the Iberian kingdoms within one monarchy until 1640, when Portugal rebelled against the centralizing projects of the Count-Duke of Olivares. Subsequent rebellions in Catalonia, Aragon, and Andalusia were repressed in the 1640s. The union of all Iberian kingdoms under one monarchy fuelled a teleological vision of Spain and the rising of a sentiment of national identity articulated around ethnic and religious homogeneity (Wulff 16–20).[22]

The concept of national identity applied to the early modern period and particularly to Spain faces objections similar to those already discussed in relation to the refusal to accept the existence of "racism" in the early modern period, arguably because both concepts are tightly related, as Balibar shows (37–67). Methodological caution is well justified: it would be anachronistic to speak of early modern Spain as if it were exactly the same concept of the nation which emerged in the Enlightenment, disseminated through an educational and institutional apparatus centralized by the state. In early modern Spain, we find nothing of the sort.[23] There was simply no universal education nor were there centralized institutions other than the monarchy and the Inquisition. The Spanish monarchy was composed of different realms and kingdoms with distinct political and cultural identities: Castile, Aragon, Navarre, and Portugal had different legal systems, and the monarchy had only limited control over most of the territory that it ruled.[24]

Yet, as Mateo Ballester Rodríguez demonstrates, the existence of a sentiment of Spanish identity is amply attested in medieval and early modern Iberia, at least in the elite classes (81–157). Identifying the presence of a Spanish identity in the early modern period does not necessarily entail supporting the view that Spain has an eternal ontological existence as a nation. Indeed, I agree with Tamar Herzog when she proposes "to abandon the quest for 'identity' and examine instead processes of 'identification,' that is, the processes through which people claimed to be or were identified as members of the community" (6). This book underscores the contingency and historicity of Spanish identity, analysing how its definition changed content and scope over time. Throughout these pages, I refer to "Spain" when talking about the early modern period because it is the term most commonly employed by early modern Spaniards, even if it was hardly a political reality. While I understand the methodological reasons of many scholars who prefer to use the term "Iberian Peninsula," I keep this term to refer to this territory from prehistoric times to the medieval period – a period for which early modern Spaniards also used "Spain" as a way to identify with an essential immemorial identity.

The Figure of the Shepherd in Early Modern Spain

This historical and theoretical overview explains how, many years later, Unamuno replicates this essentialized opposition between the Castilian and the Morisco in his *En torno al casticismo*, even though the latter had been expelled from Spain nearly three centuries earlier. Yet what is the importance of sheep herding in this opposition?

When stressing the preeminence of sheep herding as the central feature of an essential Spanish identity in which the Other is not included, Unamuno is not inventing anything new. Even up to the present, most historiography on the Moriscos, when dealing with their trades, tends to emphasize that they worked mainly as peasants, muleteers, and craftsmen, mentioning only rarely their participation in sheep herding.[25] In 1991, Bernard Vincent, in his "Les morisques et l'élevage," corrected this error, criticizing the historiographical silence about Morisco sheep herding (155). Further, even though the participation of the Moriscos in early modern sheep herding and the ensuing wool trade has been noted repeatedly in local studies, general accounts of the history of the Moriscos still tend to perpetuate this imaginary division of labour determined by ethnicity that was largely crafted, as we shall see, by the early modern Spanish imaginary, in which all shepherds are Christians and no Moor nor Morisco is a shepherd.[26]

The early modern Christian sources most commonly used in research about the Moriscos systematically avoid representing them as shepherds, making this division of labour one of the criteria for the construction of their stereotype.[27] In a 1587 letter issued to Philip II, Martín de Salvatierra, bishop of Segorbe, maintained that the Moriscos used to work only in "officios vajos y mecanicos ... como son hortelanos, aguaderos, bodegoneros, panaderos, carpinteros, alvañiles y otros semejantes" (qtd. in Boronat y Barrachina 1:626) [despicable mechanical trades, such as peasants, water boys, storekeepers, bakers, carpenters, masons, and other similar trades], and a similar list is found in the 1588 report on the Moriscos by Alonso Gutiérrez (Boronat y Barrachina 1:635). For Pedro Aznar Cardona's *Expulsión justificada de los Moriscos Españoles* (1612), the Moriscos "eran dados a officios de poco trabajo, texedores, sastres, sogueros, esparteñeros, olleros, çapateros, albeytares, colchoneros, hortelanos, recueros, y revendedores" (2:34v) [they were used to trades of little effort, such as weavers, tailors, ropemakers, *esparto* craftsmen, potters, shoemakers, veterinarians, mattress makers, peasants, muleteers, and street vendors], excluding them from raising livestock, perhaps because he considers it one of the country's beneficial trades (2:64v). Similarly, another apologetic text on the expulsion, Jaime Bleda's *Corónica de los Moros de España* (1618) reproduces Aznar Cardona's list almost verbatim, equally omitting sheep herding (1023–5).

Conversely, when Moriscos do appear as shepherds it is significantly in texts that in one way or another aim at defending their place within the nation. Thus Pedro de Valencia's *Tratado acerca de los Moriscos de España* (1606) (*Treatise on the Moriscos of Spain*) asserts that the Moriscos are for the most part "cavadores, segadores, pastores, hortelanos, correos de a pie, recueros, herreros y de otros oficios de trabajo" (87) [diggers, harvesters, shepherds, gardeners, postmen, muleteers, blacksmiths, and other trades of great effort]. In opposition to the negative stereotypes analysed above, Pedro de Valencia emphasizes that Moriscos are devoted to "oficios de trabajo," and includes trades that other writers considered proper only for Christians, such as sheep herding. In case this inclusion is overlooked in his aforementioned list of trades, he explicitly praises the Islamic interdiction against wine, relating it to the sheep herding of earlier Islam:

> Eran Árabes y pastores, conveníales mudar regiones a menudo buscando comodidad de pasto para sus ganados en ymbierno, y verano. Si ponían viñas obligábanse a ocupación perpetua de cultivarlas, no querrían alejarse mucho de ellas. Beber vino a los pastores demás de ser costoso tiene inconvenientes, que por irlo a buscar a los pueblos dexen el ganado y se entretengan en las tabernas, y se duerman y no estén alertas y tan en sí como requiere el oficio. (88)
>
> [They were Arabs and shepherds, and they had to move to different regions seeking pasture for their livestock in winter and summer. If they planted vineyards, they would need to take care of them perpetually, and they would not have wanted to move away from them. Drinking wine, for shepherds, is expensive, and it causes many problems, since they leave the herds behind in order to look for wine in towns, and they linger in taverns, and fall asleep and are not as alert as their job requires.]

The acknowledgment of the Islamic – and, by extension, Morisco – abilities with regard to sheep herding is inscribed within Pedro de Valencia's larger project of defending the Moriscos against the rising voices of early seventeenth-century Spain that were clamouring for their expulsion.[28]

As we can see, Unamuno's discovery is the repetition of an identity, with its inclusions and exclusions, formed in the late medieval and early modern articulations of a Spanish imaginary. This book is primarily about how Old Christians represented themselves as an idealized community by appropriating the figure of the shepherd (not about the Jewish or Muslim communities nor their descendants in early modern Spain), and how this reshaping served to recreate a symbolic difference that opposed an equally formulaic stereotype of New Christians, by resorting to religious, cultural, and biological features. Paradoxically, the longevity of the figure of the shepherd as an ethnocentric metaphor has remained possible because, as can be seen in my conclusion, social actors rarely make this connection explicit. It belongs to what Pierre Bourdieu defines as the *habitus*, a system of tropes and practices that do not require conscious acceptance of its ideological significance (53). What I call the "pastoral habitus" can only be reconstructed by looking at the tacit projections and denials of the discourses – fictional and non-fictional, canonical and non-canonical – that use sheep herding and the figure of the shepherd to reshape an image of the Spanish people.

Arguably the reason why the shepherd becomes attached to Old Christian – hence Spanish – identity is because, in the early modern period, the figure of the shepherd was ubiquitous, prominent in a wide range of discourses, from religious and political treatises to fictional literature and popular cultural practices, and almost always associated with Christianity in one way or another. Biblical passages reflecting sheep herding were widely quoted and glossed in religious sermons that explored pastoral care, based on Christ's articulation of his role as the Good Shepherd in John 10. By extension, the trope of the Good Shepherd was used as well to refer to the monarch and in general to any kind of ruler.[29] Because of the presence of the shepherds at the birth of Christ (Luke 2:8–20), they became a popular religious figure in the liturgical, literary, and pictorial tradition of the Adoration of the Shepherds, as well as in the legends of Marian apparitions to shepherds that spread throughout the Iberian Peninsula during the Middle Ages. Pastoral drama, which probably developed from the medieval liturgy of the Nativity, constituted the beginning of early modern theatre as developed by Juan del Encina and Lucas Fernández at the end of the fifteenth and the beginning of the sixteenth centuries. At the same time, representations of sheep herding became part of both urban and courtly festivities, which, although secularized, clearly emerged from the Christian tradition.

I am not the only one to connect the figure of the shepherd with Old Christian identity. Several scholars, such as Elaine Wertheimer, Ronald Surtz ("Cardinal" 230–2), and Alfredo Hermenegildo ("Sobre la dimension" 40–5) have seen this association in the early Castilian theatre, and Julián Gállego has suggested it to explain the success of the Adoration in painting (257). Despite these isolated intuitions, which will be discussed throughout this book, the cultural link between shepherds and an exclusivist view of Spanish identity remains to be explored in a systematic way, analysing how it informs other genres and cultural manifestations in less obvious ways. This is the case in the stylized forms of pastoral, cultivated in Garcilaso de la Vega's eclogues and those of his followers, and in the pastoral romances that – after Jacopo Sannazaro's *Arcadia* (1502) – became major editorial best-sellers in the 1560s after the publication of Jorge de Montemayor's *La Diana* (c. 1559).[30]

Certainly, the figure of the shepherd was central to the medieval and early modern European imaginary, and not exclusively in Spain. Both pastoral tropes and genres can ultimately be traced to previous European, classical, and biblical models, which explains why the bibliography on Spanish pastoral literature (romance, poetry, and drama) has traditionally promoted the analysis of literary motifs and the interchange of formalistic structures with the rest of Europe.[31] Yet the overwhelming presence of the figure of the shepherd in every aspect of early modern Spanish culture indicates that pastoral and sheep herding had very specific and differentiated social uses. As Julia Kristeva observes, "the passage from one signifying system to another" is a transposition that alters the original representability of the signifier (59–60); in this case, the shared discursive tradition of pastoral metaphors in Europe was resemanticized when it was inscribed into the semiotic system of early modern Spain, which had its own signifying practices of cultural and social oppositions based on the shepherd as cultural archetype within the ethno-religious pluralism foreclosure in 1492.

Pastoral Literature and Cultural Supersession

The new model of courtly manners was introduced into Spain at a very significant moment of Spanish history, recounted in an anecdote that has had dramatic consequences for Spanish literary scholarship. During the stay of the Venetian ambassador, Andrea Navagero, in the court of the emperor Charles V in Granada in 1526, Navagero suggested that the Catalan poet, Juan Boscán, introduce the poetic forms and themes of the Italian Renaissance in Spain. In his letter to the Duchess of Soma that prefaces the first edition of his works, Boscán gives an account of their meeting, focusing not only on the content of their conversation, but also on the context in which it took place: "estando un día en Granada con el Navagero ... tratando con él en cosas de ingenio y de letras y especialmente en las variedades de muchas lenguas, me dixo por qué no provava sonetos y otras artes de trobas usadas por los buenos authores de Italia" (Boscán 118) [One day in Granada with Navagero, in dealing with him in matters of ingenuity and letters and especially in the varieties of many languages, he said to me, why not try out sonnets and other types of rhymes used by the good authors of Italy]. Yet it seems that it was the sight of the old Nasrid court, the Alhambra, with its characteristic Moorish style, that suggested the "varieties of languages" and was the trigger that made Navagero recommend that the Spaniards become more like the Italians.[32]

Whether or not Navagero's proposal had anything to do with an orientalist view of Granada, Boscán's letter does link poetic change with an awareness of cultural difference. When dealing with the adoption of new poetic verse forms, Boscán argues lack of known origin as one of the reasons to abandon traditional Castilian metric verses in favour of that of the Italians:

> Vi que este verso que usan los castellanos, si un poco acertadamente queremos mirar en ello, no hay quien sepa de dónde tuvo principio. Y si él fuese tan bueno que se pudiese aprovar de suyo, como los otros que hay tan buenos, no havría necesidad de escudriñar quiénes fueron los inventores dél. Porque él se

> trahería su autoridad consigo y no sería menester dársela de aquellos que le inventaron. Pero él agora ni trahe en sí cosa por donde haya de alcançar más onra de la que alcança, que es ser adimitido del vulgo ni nos muestra su principio con la autoridad del cual seamos obligados a hazelle onra. (118–19)

> [I saw that the kind of verse used by the Castilians, if we look at it more correctly, we realize that nobody knows its origin. And if it was so excellent that it could be approved by itself, there would be no need to seek those who invented it, because it would bring its own authority and we would not need to give the credit to those who invented it. But now it brings nothing to attain more honour than it already has, which is the admission of the commons, and it does not show its beginnings with the authority with which we would be obliged to honour it.]

By contrast, Boscán praises the prestige of the Italian poetic forms by tracing them to Homer, thus promoting antiquity and known origins as the valid criteria for legitimizing a new canon of literary taste and civility.

By insisting on the lack of known origin of the Castilian verse forms, what he suggests in the terms "falta de origen conocido," commonly found in texts dealing with inquiries into purity of blood, is not so much a lack of antiquity, but uncertainty about the genealogy of their creators and therefore the kinds of cultural influences they invoked. Taking into account that the debate takes place in Granada in 1526, only two generations after the city's fall to Christians, Boscán's suspicion for Castilian verses is arguably caused by the anxiety about Arabic cultural influence in Spain.[33] Not only the setting but the timing is also essential to the interpretation of this influential letter. Boscán and Navagero are in Granada in 1526 as members of the court who travelled with Charles V. Also in 1526 the emperor celebrated the Congregation of the Royal Chapel, in which he dictated prohibitions against several aspects of Morisco culture in order to force cultural assimilation (Gallego Burín and Gámir Sandoval 198–205). Morisco protest led to a forty-year moratorium regarding implementation of the measure, after the Moriscos of Granada agreed to pay 80,000 *ducados*, only to be finally implemented by Philip II in 1566. Boscán's emphasis on where and when this encounter with Navagero took place situates his literary agenda within a larger project of cultural cleansing upon Spain's inclusion into Charles V's empire.

Ignacio Navarrete (58–72) analyses how the adoption of Italian renaissance norms was conceptualized in Spain within a messianic context that promoted both the idea of empire and cultural cleansing after the fall of

Granada in 1492, and in which Italy's colonial situation encouraged the adoption of Petrarchan poetic forms.[34] Yet, even though the encounter between Navagero and Boscán in 1526 has traditionally been identified as the foundational moment of the adoption of Petrarchism as a model for cultural supersession of an ethno-religiously mixed past, fifteenth-century Castilian humanism had already endeavoured to achieve a similar literary aim, by which "early modern Iberian society strove to verify its own renovation and to show that it had overcome the 'dark,' multicultural Islamic and Jewish past in order to be reborn in the era of the 'renaissance'" (Dangler 30). The earlier and the later moments of cultural cleansing produce waves of poetic transformation which claim a connection with classicism and humanism within the larger process affecting almost every aspect of early modern Spanish culture.[35] Thus the sixteenth-century reintroduction of Renaissance artistic forms, among them a courtly form of pastoral eclogue as well as the model for pastoral romance provided by Sannazaro's *Arcadia* (1504), contributes to the project of cultural supersession.

The Spanish adoption of Petrarchan poetry, which included the cultivation of highly stylized pastoral eclogues, emerges, as we have seen, amid political concerns with cultural heterogeneity. Equally, the editorial history of Spanish pastoral romances shows a striking parallelism with debates concerning Spanish culture and the place of the Moriscos, even though they do not seem to explicitly address this issue. As we shall see, pastoral romances became one of the most representative genres in the negation of Moorish Spain – but served as well as a space for contestation. Their success goes beyond any facile explanation based on literary influence or a formalistic imitation of the prestigious Italian Renaissance models – while Italian influence is undeniable, the uses and connotations of this genre are not necessarily the same when transposed to early modern Spanish society. As Werner Krauss points out, even though other European pastoral romances such as Sannazaro's *Arcadia*, Honoré d'Urfé's *L'Astrée*, or Sidney's *Arcadia* were major literary works in their respective national literatures, they did not constitute a genre per se, as the Spanish pastoral romance did – although certainly pastoral drama and poetry were cultivated to a similar extent in the rest of Europe ("Localización" 363). The fact that more pastoral romances were published in Spain than in the rest of Europe altogether is symptomatic of social and symbolic values that surpass any explanation based on pan-European aesthetic tastes, and this exorbitant success begins with the first Spanish example, *Los siete libros de la Diana*. Against critical opinions such as that of Marcelino Menéndez Pelayo, for whom "[n]inguna razón histórica justificaba la aparición del género bucólico" [no historical reason justified the appearance of the bucolic genre] (185), I pursue a sociohistorical analysis of the genre, such as that of Rosilie Hernández-Pecoraro, who states that we should consider "how this highly formulaic mode takes on and attends to the specific conditions surrounding both its production and consumption" (*Bucolic* 18).

Montemayor's *Los siete libros de la Diana* was published for the first time around 1559,[36] initiating the vogue of the pastoral romance among Spanish readers. Most of the pastoral romances, including the most influential ones, were published between 1560 and 1620: Alonso Pérez's *Segunda parte de la Diana* (1563), Gaspar Gil Polo's *Primera parte de Diana enamorada* (1564), Jerónimo de Arbolanche's *Las Abidas* (1566), Antonio Lofrasso's *Los diez libros de la fortuna de amor* (1573), Luis Gálvez de Montalvo's *El pastor de Fílida* (1582), Miguel de Cervantes's *La Galatea* (1585), Bartolomé López de Enciso's *Desengaño de celos* (1586), Bernardo González de Bobadilla's *Ninfas y pastores de Henares* (1587), Bernardo de la Vega's *El pastor de Iberia* (1591), Jerónimo de Covarrubias Herrera's *La enamorada Elisea* (1594), Lope de Vega's *Arcadia* (1598), Gaspar Mercader's *El prado de Valencia* (1600), Juan de Arce Solorzeno's *Tragedias de amor* (1607), Bernardo de Balbuena's *Siglo de Oro en las selvas de Erífile* (1608), Cristóbal Suárez de Figueroa's *La constante Amarilis* (1609), Jacinto de Espinel Adorno's *El premio de la constancia y pastores de Sierra Bermeja* (1620), Miguel Botelho de Carvalho's *El pastor de Clenarda* (1622), Jerónimo de Tejeda's *Tercera Parte de la Diana* (1627), Gabriel de Corral's *La Cintia de Aranjuez* (1629), Gonzalo de Saavedra's *Los pastores de Betis* (1633), and Juan de Barrionuevo y Moya's *Primera parte de la soledad entretenida* (1638).

To this list, we should consider the inclusion of religious rewritings of the genre, known as pastoral romances "a lo divino," such as Bartolomé Ponce's *Clara Diana a lo divino* (1580), Lope de Vega's *Pastores de Belén* (1612), Francisco Bramón's *Los sirgueros de la Virgen sin original pecado* (1620), and Ana Francisca Abarca de Bolea's *Vigilia y octavario de San Juan Baptista* (1679), as well as other works that, although not properly pastoral, include pastoral episodes in their main plot, such as miscellanies and Byzantine romances (Castillo Martínez, "Introducción" xiii–xviii) and books of chivalry (Martín Romero). Along with the printed editions, we need to include a significant and growing number of manuscripts, some discovered recently, which attest to the vitality of the genre beyond the circuit of print; some are of known authorship, like the recently discovered manuscript by Juan Pérez de Lazárraga (c. 1567), which is one of the first literary expressions in Basque language, or *El teatro pastoril en la ribera del*

Tajo (1582), by Luis Hurtado de Toledo.[37] Other anonymous manuscripts include *Libro en que se qüentan los amores de Viraldo y Florindo* (c. 1541), and *La pastora de Mançanares y desdichas de Pánfilo* (c. 1620).

Beyond texts that are still extant, several pastoral romances are documented but have been lost. By looking at printing privileges issued for pastoral titles that had no success in finding a printer, Jaime Moll has documented their existence, as well as their loss of editorial currency with the genre's decadence ("La narrativa" 34). Moll's discoveries include titles such as Gabriel Hernández's *Tercera parte de la Diana* (1582), Gabriel Fernández de Córdoba's *Tercera y cuarta parte de la Diana* (1601) – arguably the same author trying to publish an extended version of his previous work – Gabriel Pascual de Fresno's *El pastor Rusticio* (1601), and Jerónimo de Covarrubias Herrera's *Segunda parte de la enamorada Elisea* (1603) ("La narrativa" 34–5). These discoveries complement his earlier ones: Francisco Meléndez's *Pastores de Tormes* (1622), Antonio Hurtado Martínez's *Primera parte de Piroeles y Musidoro en Arcadia* (1622), and Gregorio de Lobariñas's *Pastores de Tormes, en prosa y verso* (1626) (Moll, "Aproximaciones" 66–7). There are yet more traces of now-lost pastoral romances. In his prologue to the 1642 reprint of *Diálogo de la verdadera honra militar*, Andrés de Uztarroz mentions that he had also read Jerónimo Jiménez de Urrea's *La famosa Épila*, from which he copies the introductory paragraph (Castillo Martínez, "Introducción" xv). In his short novel "Guzmán el Bravo," Lope de Vega promised *El pastor de Galatea* (*Novelas a Marcia Leonarda* 337), either lost or left unwritten. Another prolific writer, Manuel de Faria e Sousa, claimed to have written two pastoral romances in his youth, which he later burned (140).

The trajectory of the genre as a literary phenomenon clearly extends beyond its editorial viability, but, while there is wide consensus in identifying Montemayor's *La Diana* (c. 1559) as the inaugural work of the genre in Spain, it is less clear when we should date its decline. Its waning during the seventeenth century is commonly attributed to a change in aesthetic taste (Krulls-Hepermann; Castillo Martínez, "De las lágrimas a la risa"; Arredondo), but a look at chronology suggests that socio-political events may have played a crucial role in the sudden apathy of early modern Spanish readers. As we shall see, it is not by chance that interest in pastoral romances dropped dramatically by the 1620s, right after the expulsion of the Moriscos, finalized in 1614 – although propaganda favouring the expulsion made much use of pastoral elements, as will be shown in chapter 5. Most of the titles published after this date have little to do with the vitality of the genre in Spain: Jerónimo de Tejeda's *Tercera parte de la Diana* (1627) was published in Paris and was intended for a French audience where the pastoral romances were still in vogue for different reasons (Solé-Leris 132–3; López Estrada, "Los libros de pastores" 158); Gonzalo de Saavedra's *Los pastores del Betis*, published posthumously by his son in 1633, was probably written between 1603 and 1604 (Solé-Leris 135); Juan de Barrionuevo y Moya's *Primera parte de la soledad entretenida* (1638) is mostly a Byzantine romance of which only the first part is properly pastoral, and something similar could be said of most of the books with pastoral indications published after 1620.

For Juan Bautista Avalle-Arce, the last representative pastoral romance would be *Pastores de Belén* (1612) by Lope de Vega, while later pastoral romances are but courtly miscellaneous romances or divine versions (*La novela pastoril* 272–3). Marcos Morínigo offers an even more limited development of the genre, which he circumscribes to the period between 1560 and 1585, while the remaining pastoral romances would be only proof of "la dilatada agonía del género" (46) [the long death throes of the genre]. Willard King holds the opposite view, showing that the late case of *Vigilia y octavario de San Juan Baptista* (Zaragoza, 1679) by Ana Francisca Abarca de Bolea makes it necessary to extend the life of the genre beyond 1633 (121n16). Paul Carranza (5) and Benjamin Nelson (244–6) use a similar periodization to mine; for them, however, the event that largely conditions the evolution and final disappearance of the Spanish pastoral romance is the rise and decadence of the imperial project.[38]

But even more important than the chronology of the first edition of different titles in considering the life span of the genre is the history of their reprints. Montemayor's *La Diana* was undoubtedly one of the best-sellers in early modern Spain, with some thirty reprints between 1559 and 1624 (Fosalba, *Diana* 14–16; López Estrada, "Prólogo" lxxxvii–xcvi).[39] Yet after the 1624 edition printed in Lisbon, it was not published again until the end of the eighteenth century. We find a similar situation with the most successful continuators of Montemayor: Gaspar Gil Polo's *La Diana Enamorada* had seven reprints between 1563 and 1624, and Alonso Pérez's *Segunda parte de la Diana* is known to have had at least fifteen editions within the very same period (Smieja 717–18). Cervantes's *La Galatea* was highly dependent on its author's fame: after its publication in Alcalá de Henares in 1585, it had only one reprint in Lisbon (1590); after the success of *Don Quixote*, however, it enjoyed five reprints between 1611 and 1618, only to fall into oblivion again after its author's death in 1616; it remained there until the end of the eighteenth century. Luis Gálvez de Montalvo's *El pastor de Fílida* knew five reprints between 1582 and 1613, again with no further reprint until the end of the eighteenth century. Juan de Arce

Solorzeno's *Tragedias de amor* (1607) was strikingly reprinted in 1647, probably because it includes a favourable portrayal of Pedro Fernández de Castro, 7th Count of Lemos, whose remains were translated to Monforte in 1646. The only exceptions to the sudden fall into oblivion of the pastoral romances were those of Lope de Vega: *La Arcadia* had eleven reprints between 1598 and 1612, and ten more during the rest of the seventeenth century; *Pastores de Belén* also had nine reprints between its first publication in 1612 and 1676. But Lope de Vega's case is a clear exception, since the success of his pastoral romances during the seventeenth century was very likely dependent on his personal fame as a playwright and poet, and not on the viability of the genre itself.

From this editorial trajectory, we can trace the rise and decline of Spanish pastoral romances between 1559 and the 1620s. This is also the period when the genealogical and cultural anxieties over national identity reached a new apex, starting with the reinvigoration of the statutes of blood purity in the 1550s and the repression and eventual expulsion of the Moriscos between 1609 and 1614.[40] Examination of how these events overlap with the editorial history of the pastoral romance strongly suggests that they participated in the cultural and national anxieties of early modern Spain. At the same time, the historical momentum of the pastoral novel also keeps pace with that of the rise and decline of maurophile literature, such as Moorish ballads (1575–1600), the influential Moorish novel *El Abencerraje* (c. 1561), and Ginés Pérez de Hita's *Guerras civiles de Granada* (first part 1595, second part 1619). Pastoral romance also declined as a genre at the same time that the debates on the fate of Moriscos waned in the early 1620s subsequent to their expulsion. This coincidence suggests that it is precisely when this object of denial was lost that the genre forfeited its raison d'être in the Spanish imaginary. As Fredric Jameson suggests, "the failure of a particular generic structure ... to reproduce itself ... alerts us to the historical ground, now no longer existent, in which the original structure was meaningful" (146). The symbolic opposition to everything Moorish in Spain was the basis of pastoral's success between the 1560s and 1620s, and, conversely, with its loss, the literary figure of the shepherd relinquished its privileged place in the cultural imaginary, and its function had to be redefined within the changed social system of symbolic oppositions.

A more explicit dialectical opposition between shepherd and Moor is found in the compilations of ballads that became wildly popular during the second half of the sixteenth century. Ballads about historical figures and about the frontier wars with Muslims were composed and circulated in oral form during the Middle Ages (Ruiz Lagos 9–10), and by the middle of the sixteenth century these medieval ballads became of antiquarian interest and were transcribed and compiled into print anthologies. From the Spanish *romance*, meaning "ballad," these compilations were called *romanceros viejos* [anthologies of old ballads]. By contrast, literate authors began to imitate this traditional poetic form in the second half of the sixteenth century, creating the *romancero nuevo* [anthologies of new ballads].

Among the topics favoured in the *romancero nuevo* were historical episodes, but also sentimental stories of shepherds or gallant Moors. In Spanish scholarship, the cycle of ballads dealing with Moorish themes is usually called the *romancero morisco*, although these ballads are not about actual Moriscos, but instead praise the gallant idealized Moors of the frontier wars of the *reconquest*, who were represented in a positive light even though they usually appeared as enemies of Christians. The cycle of Moorish ballads gathered momentum between 1575 and 1585, serving as the colophon of a traditional Spanish attachment to everything Moorish, but its success suddenly declined towards 1600. While Chantal Colonge attributes the lack of interest to a shift in literary taste (147), other critics link the decline of the genre to the socio-historical context marked by anti-Morisco propaganda (García Valdecasas 154–8; Carrasco Urgoiti, "Vituperio" 132–8; Ruiz Lagos 18; Márquez Villanueva, "Lope" 172–5).

In opposition to these popular ballads representing an idealized vision of the Moor, whose alleged descendants the Moriscos were still living among Old Christians, a whole series of parodies appeared in the 1590s and again in 1610–11, at the same time as the most violent discourses against Moriscos were articulated. One of the most celebrated anti-Morisco ballads, "Tanta Zaida y Adalifa," first published in 1593, accused the poets of Moorish ballads of committing treason that was aesthetic, anti-patriotic, and suspect of heresy: "Renegaron de su ley / los romancistas de España, / y ofrecieron a Mahoma / las primicias de sus gracias" (Ruiz Lagos 162) [Spain's writers of ballads reneged on their religion and offered to Muhammad the first fruits of their wits]. As María Soledad Carrasco Urgoiti points out, such parodies suggest that their authors perceived the popularity of Moorish ballads as a threat that praised religious and cultural difference as valid options within Spanish identity ("Vituperio" 121).

These parodies of Moorish ballads reinserted the discussion of proper Spanish identity within the genre itself, usually articulating it as an opposition between the pastoral and the Moorish, two of the most prominent literary themes of the moment.[41] The parodic ballad "Oídme, señor Belardo" [Hear me, Sir Belardo] urged authors to abandon Moorish ballads and embrace instead the shepherds' imagery: "Recoged vuestro gabán, / y echad el zurrón al hombro" (Ruiz Lagos 175) [pick up your peasant tunic and throw your pouch on your shoulder]. Similarly, the ballad "Toquen aprisa a rebato" [Hurry, sound the alarm] called for leaving Moorish themes behind by proposing pastoral as the appropriate alternative: "Entre los toscos pastores, / en el soto y en la vega, / al son de sus instrumentos / puede cantar sus endechas" (Ruiz Lagos 184) [Among the rustic shepherds, in the mountain meadow and the valley, to the sound of their instruments, he can sing his mournful verses]. The ballad "Descúbranse ya esas caras" [Now uncover these faces] criticizes the practice of disguising real people behind Arabic names, proposing to substitute Moorish themes by either the pastoral or the epic celebration of Spanish deeds: "Si importa celar los nombres ... / ¿Por qué no vais a buscarlos / a las selvas y cabañas ... / o a la felice Numancia?" (Ruiz Lagos 168) [If it's important to conceal names, why don't you go seek them in forests and huts ... or in the happy Numancia?]. It is striking that these parodic ballads put more emphasis on pastoral as an alternative for Moorishness than on the epic heroism of the *reconquest*. Gabriel Lasso de la Vega is one of the few poets to propose such an alternative in his ballads "Señor Moro vagabundo" [Sir Vagabond Moor] (28–31) and "Por Dios, señores Poetas" [For God's Sake, Sir Poets] (37–9). One possible explanation is that, as chapter 3 will show, the vague pastoral world effectively negates eight centuries of Islamic presence in the Iberian Peninsula, while the Castilian epic, even though praising its Christian heroes, cannot fail to acknowledge the existence of the Moor.

While in "Descúbranse ya esas caras" the first alternative to leaving the Moorish topic behind is pastoral ("las selvas y cabañas"), the second one refers to a historiographical myth as an equally privileged locus of national literary representation ("la felice Numancia"). Both bucolic and epic were simultaneously cultivated by Cervantes in the 1580s, when he wrote both his pastoral romance *La Galatea* and the epic drama *La Numancia*, a coincidence that reinforces the connection between the two literary projects. Fernando de Herrera also underlined the complementary value of epic and pastoral when he sang of the fight against the infidel, suggesting that it will be celebrated in both genres altogether: "avrá quien cante umilde su tormento, / quien belígero orror i aguda espada, / i quien el dulce i rústico lamento" (583) [some will sing their torment with humility, some with warlike horror and sharp sword, and some with mellow and rustic lament]. According to some literary treatises, such as Francisco Cascales's *Tablas poéticas* (published in 1617, but written in 1604), epic and pastoral were similar genres (132).

Pro-Morisco ballads reacted to the attacks by satirizing the pastoral element. The anonymous ballad "¿Por qué, señores poetas / no volvéis por vuestra fama?" [Why, sir poets, do you not defend your fame?] responds to anti-Morisco accusations by defending the Spanishness of Moorish cultural practices: "¿Qué le aprovecha a Gazul/tirar al otro la lança,/si hoy un ninfo del Leteo/quiere deshacer sus zambras?" (Ruiz Lagos 169) [What's the use of Gazul throwing his spear at another, if nowadays a male nymph from the river Lethe wants to undo his *zambras*?].[42] By calling the anti-Morisco poet "ninfo" [male nymph], the ballad mocks the pastoral mode as a feminizing genre that cannot compete with the martial and courtly values promoted by Moorish themes.

The literary conflict was resolved with the victory of pastoral over Moorishness within the context of the expulsion of the Moriscos, made explicit in the picaresque novel *Alonso, mozo de muchos amos*, by Jerónimo de Alcalá y Yáñez. Although published between 1624 and 1626, after the expulsion of the Moriscos, most of the story is located at the beginning of the seventeenth century. At some point, Alonso recounts that a former master used to write Moorish ballads, but decided to change topics after the expulsion of the Moriscos: "Hízonos Dios merced de que en este tiempo saliese la cédula real del católico rey Don Felipe III, nuestro señor, en que mandaba desterrar los moriscos de España ... y con esta nueva mudó de sujeto, dejando a los devotos del falso profeta por seguir las humildes chozas de los pastores" (1174) [God granted us that in that time the royal decree was passed by the Catholic king Philip III, our lord, banishing all Moriscos from Spain ... and with this news he changed his subject, abandoning the devotees of the false prophet to seek out the humble huts of the shepherds]. The change is corroborated by the compilations of ballads, in which, since the 1590s, pastoral ballads start to outnumber Moorish ones, which in turn tended to disappear by the beginning of the seventeenth century (Graña 53–6).

Similarly, this book contests interpretations that assume that pastoral romances are a stylized literary genre that bears no relation to its social milieu, such as that of Avalle-Arce, for whom they are "fuera del tiempo y del espacio, en el puro reino de las ideas" (*La novela pastoril* 103) [out of time and space, in the pure realm of ideas]. This has been the traditional interpretation of Spanish pastoral romances, as followed by Menéndez Pelayo, Avalle-Arce, Francisco López Estrada, Hugo Rennert, Amadeu Solé-Leris, and Pilar Fernández-Cañadas de Greenwood, to cite only the most representative names. The first studies that questioned the alleged detachment of the pastoral romances from their social reality interpreted them as produced mainly by *conversos*.[43] More recently, Nelson, Carranza, and Leah Middlebrook (103–37) have explored pastoral's connection with discourses on empire, and Hernández-Pecoraro (*Bucolic Metaphors*) has analysed the genre from a feminist and psychoanalytic perspective.[44] As she proposes:

> when we ... examine these texts' symbolic intentions, as well as the fissures and inconsistencies they concurrently manifest, we come closer to a deeper understanding of their historical and social significance. In the process it becomes possible to expose and examine those circumstances that the idyll is meant to negate. (*Bucolic* 18–19)[45]

My work thus participates in the larger project of analysing the social aspects negated by the figure of the shepherd and, more specifically, how it serves as a symbolic negation of the ethnic diversity of early modern Spain (most notably in relation to the legacy of the Islamic period), and as an affirmation of ethnocentrism articulated around the Old Christian community and the rise of Spanish identity.[46]

Written during a period in which herding privileges were contested, most of the books I analyse herein emphasize the essentiality of herding in Spanish history and its existence as a nation, which may well reflect the defence of the interests of the landowning nobility and the ecclesiastical hierarchy. Moreover, in their idealizing of sheep herding, pastoral fictions erase the class difference between the livestock owner and the actual shepherd, signalling the coexistence of different forms of exclusion in the social imaginary. While I argue that early modern Spanish pastoral romance is a genre aimed at opposing the Moorish cultural and genealogical legacy, I shall show that it reflected as well entangled issues of class confrontation. As John Beverley points out, Américo Castro's emphasis on "ethnic confrontation" failed to take into account how class differences interfered with racial classifications and to some extent informed them ("Class" 141–9). For Jaime Contreras Contreras, the discourse on blood purity is but the symbolic capital of hegemonic local classes, a fiction that is instrumentalized beyond any plausible reality for most of the cases, and that ultimately serves as a supplemental rhetoric that tests the real strength of individual social position (87–95).[47] The words of Unamuno with which I began this introduction, which oscillate between defining the Castilian either as shepherd or as livestock owner, reflect perfectly how pastoral serves to promote a homogeneous conception of national identity that subsumes class difference. As such, pastoral exemplifies the main distortion that ideology operates, as Louis Althusser puts it: "What is represented in ideology is ... not the system of the real relations which govern the existence of individuals, but the imaginary relation of those individuals to the real relations in which they live" (111). In this sense, we can consider "pastoral habitus" as an "ideologeme," which Jameson defines as "the imaginary resolution of the objective contradictions to which it thus constitutes an active response" (118). Pastoral, much like national sentiment, obviates class struggle, as well as the conflicts between rural and urban spaces, by construing sheep herding as an essential feature of the "Spanish people" (Hernández-Pecoraro 53).

Since the aim of this book is to contextualize pastoral romance within a social milieu that equates the figure of the shepherd with the Old Christian, and to analyse the genre as an instrument to promote ethnic segregation and cultural cleansing, a wide range of texts is included to document the extent of this ideological conflation. In order to show that it is not a purely literary project, I trace the use of sheep herding and pastoral metaphors in theological, economic, scientific, historiographical, and pictorial documents. Even though the conflation of literary and non-literary texts makes this book fall within cultural studies, in many ways it does not attempt to distance itself from a certain tradition of the study of pastoral romances; rather, this book largely participates in the line of inquiry initiated by López Estrada in his seminal work *Los libros de pastores en la literatura española* (1974), which aimed to find a meaningful global interpretation of the heterogeneous pastoral manifestations in early modern Spain.[48] In fact, as López Estrada points out, the term used by early modern Spaniards was *libros de pastores* [books of shepherds], and not *novelas pastoriles* [pastoral romances], which is anachronistic within the early modern literary system (*Los libros de pastores* 20–1). The genre was thus defined more by its literary motifs, centred around the presence of shepherds, and less by its formal features, and it could comprise both pastoral romances in prose as well as compilations of pastoral poetry and eclogues, and even pastoral drama.

Readers used to early modern pastoral literature will likely notice one significant absence in this book: it barely touches on the concept of the bucolic. It is important to note that "pastoral" has many possible definitions, and this book follows only one of them. There is one broad definition of "pastoral" that equates it to the bucolic, to the extent that they become interchangeable. Such a broad definition of "pastoral" would comprise the idealized description that Christopher Columbus makes of the New World in his diaries, or the sentimental interludes in chivalric and Moorish romances, or almost any description of nature as idyll. There is a second definition, the one used in this book, that limits "pastoral" to texts that deal with shepherds, summed up by Leo Marx in his motto "no shepherd, no pastoral" (45). This might be regarded as an oversimplification, but it is appropriate within the scope of this book, which is to analyse a particular discursive system centred on the figure of the shepherd. Focusing on this restricted notion of pastoral allows exploration of the equivalences between disparate texts that explicitly mention shepherds but that may have nothing to do with the bucolic.

Furthermore, the identification between "pastoral" and the terminology of sheep herding is more immediate in Spanish. English *pastoral* and French *pastorale* refer both to the bucolic literary genre and evangelical care, but, although there is a common awareness that "pastoral" derives from the Latin root of sheep herding, these languages use different words to refer to the actual person who guards sheep – English *shepherd* or French *berger*. Spanish not only has maintained the Latin word *pastor* as the common term to refer to shepherds and ecclesiastic personnel, but has also coined two adjectives with very different connotations. Spanish *pastoral*, used mainly for pastoral care, coexists with the parallel term *pastoril*, used to refer to the material aspects of sheep herding. The adjective *pastoril* was preferred for pastoral fictions, suggesting a stronger semantic bond in Spanish between the genre and the exploits of actual shepherds. In its origins, *pastoril* had a derogative connotation at least until the sixteenth century, as when Barrientos aimed to disqualify the anti-Semitic rioters of mid-fifteenth-century Toledo by labelling them "de baja sangre pastoril" [of low shepherd stock/blood] (184). Thus, even though *pastoral* had a new semantic scope, its automatic application as synonym for bucolic would be anachronistic in the case of early modern Spanish pastoral romances, since the latter retained their lexical bond with the terminology of sheep herding.[49]

Another lexical peculiarity of Spanish is that, even though *pastor* is usually translated as "shepherd," that is, the herdsman who guards sheep, it may have as well a more general scope and can be used to refer to any other kind of herdsman. Therefore we can find expressions such as *pastor de cabras* [goatherd] or *pastor de cerdos* [pigherd], along with the equivalent terms *cabrero* and *porquerizo*.[50] Readers not accustomed to the Spanish pastoral tradition may thus be struck by my occasional translation of *pastor* as "shepherd" in cases in which texts represent herdsmen guarding other livestock than sheep. When I do this, it is to convey the ideologically motivated selection of *pastor*, which would be lost with the more neutral term "herdsman."

Since the objective of this book is both to show the preeminence of the figure of the shepherd to inform ethnocentric discourses in early modern Spain, and to analyse how it is manifested or contested in pastoral romances, this study is divided into two parts, each of them reflecting a different emphasis and approach. In order to show the impact that sheep herding had in early modern racial thinking and the imaginary of difference, the first part (chapters 1, 2, and 3) explores the three registers in which ethnocentrism operates: the racial, the cultural, and the historiographical. As already mentioned, it is not possible to delineate a clear distinction between them as such, but we can identify which tropes and genres privilege some of these aspects over the others.

Chapter 1 shows how sheep herding acquired an ethnocentric overtone in religious and political discourses as well as in visual artefacts dealing with the Christian community as a flock of sheep. This chapter is therefore very different in nature from the rest of the book, which focuses on the figure of the shepherd, but its inclusion is necessary because it foregrounds the importance that sheep herding had for the formation of early modern Spanish discourse on race, genealogy, and breeding. I begin with an exploration of the economics and logic of sheep herding in late medieval and early modern Spain, analysing how its vocabulary permeated racial discourse. Special attention is paid to how pastoral biblical images become emblems for a new concept of social engineering to advance the exclusion of *conversos* and the expulsion of Moriscos. The use of the figure of the Good Shepherd is thus reconfigured; while in John 10:16 it was meant to emphasize the inclusion into the Christian community, it was appropriated as a trope of exclusion by insisting on the need to separate good sheep from bad and to convey the idea that the "Spanish people" should be subjected to the same process of breeding selection that was applied to sheep. This appropriation affected as well the representation of other biblical figures associated with sheep herding, most notably that of Jacob, whose transformation of Laban's flock becomes the site in which competing pictorial and literary projects exploit its potential exemplarity for social policing in very different ways.

As mentioned above, the doctrine of blood purity in early modern Spain oscillates between the genealogical and the cultural as much as any racism does. Therefore, while chapter 1 analyses how concepts of heredity and genealogy are rooted in metaphorical discussions about sheep and flocks, chapter 2 explores the sartorial and linguistic elements that were used to build up the imaginary distance between the figure of the Moor and that of the "shepherd," so that the latter becomes a synonym for Old Christian.

By projecting genealogical ethno-religious difference into the cultural realm, pastoral performances served to naturalize the notion of an idealized community of Old Christians, and provided the symbolic means for excluding Moriscos (as well as Jews and Gypsies) from the representation of Spanish identity. The aestheticization of the figure of the shepherd is not incompatible with its ethnocentric instrumentalization; rather, it is its very possibility, and, as Balibar points out, the aestheticization of national human models is inherent to racism (57–8). In order to show the extent of this currency, chapter 2 includes in its analysis accounts of royal and local festivals, devotional texts on Marian apparitions to shepherds, and the representation of the Adoration of the Shepherds in literary and visual documents. I will discuss as well the applicability of Eric Hobsbawm's concept of "invention of the tradition," which he locates in the Enlightenment, along with the emergences of modern nationalism, but which can be applied as well to how traditions are invented in order to forge a historical continuity of the "Spanish people" in the early modern period.

Chapter 3 charts the continuities and discontinuities in the construction of Spain as a pastoral community in historiographical works and pastoral romances. It analyses how the emphasis on Iberian sheep herding across different historical times helped overcome anxieties about cultural continuity between ancient and early modern Spain and legitimize the alleged restoration of an anachronistic and imaginary Spain. While pastoral romances emphasized class difference and a model of civility in their embrace of classicism, they borrowed as well the construction of ethnic difference already prevalent in popular culture. Thus pastoral romances are contextualized by juxtaposing them with ethnic conflicts in Spain and the historiographical discourses that project an eternal Spanish identity since time immemorial. Classicism, I suggest, aims to supersede the legacy of Islamic Spain, mainly through the use of humanist cartography and historiography. Chapter 3 ends with an analysis of how early modern historiography tries to bridge the gap between the pagan religiosity of the pastoral world and the Christian identity of post-Tridentine Spain by crafting an etymology of "España" that derives it from the god Pan, who not by chance is also the main god presiding over pastoral fictions.

The second part of this book (chapters 4 and 5) focuses on the literary and ideological analysis of those pastoral romances that contest or at least complicate the exclusivist ideology of the Spanish Arcadia, unveiling the paradoxes imbedded in the process of cultural cleansing. Chapter 4 begins with an analysis of the place of the Moorish tale *El Abencerraje* in Montemayor's *La Diana* (c. 1559), which I read as a conscious intervention in debates over national self-representation. I argue, moreover, that this debate is not limited to the early modern period, but instead continues to haunt the editorial history of both texts up to the present day. I then move to Gaspar Mercader's *El prado de Valencia* (1600), which represents a different mode of inscription of Moorishness in Arcadia. *El prado de Valencia* presents itself as a pastoral romance even though its characters show a striking tendency to abandon the pastoral in favour of Moorish cultural practices, more in tune with the performance of aristocratic identity in Valencia. Finally, I argue that Cervantes mocks, through Don Quixote's pastoral project, the very possibility of cultural cleansing that the ethnocentric Arcadia represented. Since Cervantes's literary works, and especially the two parts of *Don Quixote* (published in 1605 and 1615) extend to the period that both predates and follows the expulsion of the Moriscos, his parodic attitude signals the transition towards the final rewriting of the genre.

Chapter 5 studies how the expulsion of the Moriscos between 1609 and 1614 affected the writing of pastoral. Lope de Vega's pastoral romance *Los pastores de Belén* (1612) combines the conventions of the genre with the biblical tradition of the Adoration of the Shepherds, thus emphasizing the Jewish origin of these religious pastoral motifs. Simultaneously, Lope de Vega conflates the Egyptians (who in his version welcome the Holy Family) with the Gypsies, precisely at a time when several voices were asking for the expulsion of the Gypsies as well as Moriscos from Spain. I then turn to late pastoral romances, such as Espinel Adorno's *El premio de la constancia* (1620) and Barrionuevo y Moya's *Primera parte de la soledad entretenida* (1638) to show how they contest the use of pastoral in the epic of the expulsion, most notably Gaspar Aguilar's *Expulsión de los moros de España* (1610), and initiate a period of nostalgia in which the Moor is included in the pastoral fiction.

Even though the allure that pastoral romances exerted in the Spanish imaginary largely collapsed or was redefined after the expulsion of the Moriscos, it is undeniable that the figure of the shepherd has lingered in the collective unconscious and is reenacted in times of anxiety over collective identity, as Unamuno's *En torno al casticismo* shows. Therefore, the conclusion explores how pastoral representation has evolved in conjunction with the idea of race from the seventeenth century through contemporary Spain, shifting from ethno-religious exclusion to modern phenotypic racism. During the Spanish Civil War (1936–9) and the subsequent dictatorship of Francisco Franco, the pastoral tradition was appropriated by both sides of the conflict in order to convey notions of national, political, and cultural identity, as is best exemplified in the literary and propagandistic debate between Ernesto Giménez Caballero and Miguel Hernández. By contrast, since the transition to democracy in 1975, the use of pastoral tradition has primarily reflected anxieties about African immigration to Spain, as I argue through an analysis of contemporary film, literature, rock music, and journalism.

I chose to begin this chapter with Unamuno, even though he is not the object of this book, because he encapsulates all the aspects that I want to address in this book, with one exception: Unamuno, like many of the texts with which the present book deals, talks extensively about shepherds, but seems to forget the reason why shepherds exist – the sheep – arguably because he takes their presence for granted. Yet sheep are not solely the economic background that explains the prevalence of shepherds in the imaginary of identity. Rather, sheep are the very reason why shepherds are taken as an emblem of cultural and genealogical purity in early modern Spain, as the next chapter explores.

PART ONE

Sheep Herding and Ethnocentrism in Early Modern Spain

1 Sheep Herding and Discourses on Race

¿Sería tan poético el sueño si en vez de ovejas merinas, de tan
fina calidad, las de la cabaña de Salicio hubieran sido de una sangre
menos pura, como las de cualquier gañán o rabadán que corre por
estos mundos?

[Would the dream be so poetic if, instead of Merino sheep, of such fine quality,
those in Salicio's herd had been of less pure stock, like those of any
rustic or shepherd that wanders around?]

Ramón Batlle Matabosch

Probably there are no pure breeds of sheep any more than pure races of men.

Robert Sabatino Lopez

What is the abject object of sheepishness?

Sarah Franklin, *Dolly Mixtures*

Sheep herding was the main economic activity in late medieval and early modern Spain. Long before the colonization of the New World, and well into the nineteenth century, the exportation of raw wool to textile factories in Flanders and Italy was the main source of foreign currency for Castile (Phillips and Phillips xi–xviii). While the importance of sheep herding and the pervasiveness of shepherds in early modern Spain have been amply studied by historians, literary scholars have paid only marginal attention to how sheep herding is reflected in the Spanish imaginary, nor have they pursued the economic and historical reasons for the success of pastoral romances. Even after Werner Krauss suggested in 1967 that the success of the pastoral romances was rooted in the interest of livestock owners in defending their legal and economic privileges ("Localización" 364–6), no scholar has seemed interested in pursuing this line of inquiry.[1] Without losing sight of the economic aspect brought forward by Krauss, this chapter argues that Spaniards also found in the materiality of sheep herding a rhetorical repertoire for expressing notions of breed and heredity even before the creation of a corpus of scientific knowledge of genetics and biological evolution.[2]

This chapter shows in the first place the interrelation of racial and sheep herding terminologies, and how the logic of sheep herding is transposed to the debates about the statutes of blood purity and the repression and expulsion of the Moriscos. It also explores how this conceptual convergence contaminates the interpretation of biblical images on which early modern Christian identity is based. Both the literary and pictorial representations of the Good Shepherd (John 10:16) and the time Jacob spent as a shepherd for his uncle Laban (Genesis 29–31) are thus impregnated with ethno-religious connotations that condition the production and reception of a new concept of power based on the management and improvement of population, at the same time that they help to advance and also to contest it.

Writing the History of Spanish Sheep

The extraordinary success that sheep herding enjoyed in Spain in the Middle Ages until the nineteenth century was largely due to a variety of factors, such as climate, low population density, and the long frontier wars with Muslims, which created ample uninhabited spaces and promoted the raising of livestock instead of agriculture. Undoubtedly, one of the factors that contributed to the success of Castilian wools in competing in Italian and Flemish markets was the introduction of a new race of sheep capable of producing fine white wool: the Merino sheep, which was the result of a careful zootechnical revolution.

As much as the success of the Merino sheep has been a source of national pride, its origin has sparked a heated controversy closely related to the debate from the sixteenth century onward about the legacy of the Islamic period. As I will explore further in chapter 3, early modern historians tended to emphasize an essential continuity between sheep herding in ancient Iberia and in early modern Spain, discarding any possible influence of al-Andalus in this economic activity.

Among modern historians of early modern Spain, this tale of continuity between ancient and early modern sheep herding was challenged by Julius Klein's *The Mesta: A Study in Spanish Economic History* (1920), which suggested the African origins of the Merino sheep. Klein stated that the term *merino* derived from the name of the North African tribe Banu-merin, who came to the Iberian Peninsula in 1146, arguing that this breed of sheep was unknown to Iberians before that date (4–7).[3] In 1953, Robert Sabatino Lopez reformulated the African hypothesis by suggesting that Genoese merchants first introduced the sheep from Tunisia to Spain, during the crisis of English wools in the mid-thirteenth century, in order to ensure resources for their own textile markets.[4] Such hypotheses of the African origins of the Merino have provoked heated reactions among Spanish historians. In a volume celebrating the Spanish economy after the victory of Francisco Franco in the Spanish Civil War, published in 1942, Ramón Batlle Matabosch defines the Merino sheep as being "de pura estirpe española, digan lo que quieran los genealogistas, extranjeros en su mayoría, que han querido buscarle unos antecedentes menos honrosos" (42) [of pure Spanish stock, no matter what genealogists say (foreigners for the most part), who have tried to ascribe to it less honourable heritage]. Similar patriotic underlying opinions can also be found in more circumspect academic works. In 1979 Antonio Sánchez Belda and María Sánchez Trujillano rejected the African hypotheses of both Julius Klein and Robert Sabatino Lopez, and revived the old thesis according to which the name of the Merino sheep comes from the Latin *maiorinus/merino*, the name for certain judicial offices in medieval Castile (294). Sánchez Belda and Sánchez Trujillano rarely provide evidence for their assumption that the Merino sheep is a "purely Spanish race" (293) whose origins date back to prehistoric Iberian peoples such as Turdetans and Tartessians, thus rendering their work highly speculative and situating it within an essentialist historiographical trend that looks for continuities in national identity from time immemorial.[5] In a reappraisal of both theories, Pedro García Martín suggests that the continuities between ancient Iberian and modern Spanish sheep herding are not found in the sheep breed itself, but in the willingness to improve the breed through an uninterrupted and millenarian process of interbreeding and genetic selection (109–20 and 134–44). This historian proposes that the most plausible explanation is that the Merino race is a cross-breed between ancient Iberian sheep, whose wool had a reddish colour, and North African rams.

Certainly, the aim of this chapter is not to commit to an opinion on the reality or genealogy of the Merino sheep, but to point out how historiography about sheep herding is not immune to the ambivalent attitudes with which Spaniards have faced the legacy of the Islamic period.[6] Even though Batlle Matabosch attributed the African hypothesis to foreign scholars aiming to discredit Spain, it had already been suggested as early as the eighteenth century by Spaniards themselves. In 1765 Fray Martín Sarmiento in a letter to the Duke of Medina-Sidonia complained about the importance of herding in Spain, considering it a trace of Moorish influence: "Es vergüenza que en España se haya introducido el modo de vivir de los sarracenos que, sin cultivar la tierra, andan vagabundos con sus ganados por los despoblados de la Lybia y la Arabia" (221) [It is unfortunate that the lifestyle of the Saracens, who do not cultivate the land, but instead wander with their herds about the deserts of Libya and Arabia, has been introduced in Spain]. Here, the African origins that had been systematically repressed in the early modern period were suddenly restored in order to discredit an economic model based on sheep herding, during a revolution in economic politics in favour of agriculture against herding. In a typical rhetorical strategy of Spanish identity discourse, everything can be conceived of as Moorish once it loses its social prestige. Clearly, whatever the real origins of Merino sheep, their treatment in Spanish historiography varies according to (and is highly indicative of) underlying discourses on national identity.

Sheep Herding and Racial Terminology

Arguably, the history of Merino sheep is permeated by concepts of national identity, not only because of its economic relevance but also because of the way in which the herd has traditionally served as a recurrent symbol for the composition of a community. Sánchez Belda and Sánchez Trujillano's book *Razas ovinas españolas* [*Spanish Ovine Breeds*] shows how the terminology of social difference is easily applied to sheep, as they utilize the descriptive vocabulary of race, such as the terms "phenotype," "race," "caste," and "ethnicity" in order to categorize and describe ovine breeds. The first two epigraphs selected for this chapter show how conceptions of the social fabric condition our understanding of the biological world, even if they express opposing viewpoints. For Batlle Matabosch, the Merino sheep are "de pura estirpe española" (42) [of pure Spanish stock], a statement that, uttered in 1942 Francoist Spain, might be read as a backdrop for the ascendant biopolitics of fascist states. Robert Sabatino Lopez instead reveals that dispelling myths on animal husbandry and purity leads to a similar conclusion in the human realm: "Probably there are no pure breeds of sheep any more than pure races of men, and crossbreeding is the best way to obtain fine specimens" (162). Published in 1953, Lopez's article on the history of the remote origins of a Spanish sheep breed turns out to be an intervention in ongoing racial debates: because it was published in the series Jewish Social Studies, he immediately evokes the recent European outburst of racial purity, but also, within the United States' academic context of the 1950s, he takes a position within the debates over the segregation laws of the Jim Crow era.

The preceding examples show how ethnocentric and zootechnical discourses constantly exchange their vocabulary and their underlying logic within an ideological framework that is anything but explicit. The slippage between zootechnical and racial terminologies, as we shall see, can be found as well in early modern Spain. Critical editions of early modern pastoral literary texts always devote part of their erudite annotations to explain unfamiliar sheep-herding terms to the average modern reader. While they provide lexical clarifications that treat each term as having a self-contained meaning, they generally do not explore the ideological connotations that sheep herding vocabulary might have had for early modern Spaniards. Indeed, because of its economic importance, its social ubiquity, and its ability to provide a visual model for issues of selected breeding and segregation, sheep herding furnished early modern racial thought with the terminology and logic needed to convey ethnocentric conceptions of social policy, mainly by borrowing its terminology from the vocabulary of marking ownership and evaluating wool quality.

Due to the growing numbers of flocks in late medieval and early modern Iberia, and the practice of allowing them to roam across the peninsula in search of winter pastures (*trashumancia*), a highly codified system of marks had to be implemented in order to ensure ownership. Along with branding, Spaniards employed a parallel system by marking each flock with a distinctive pattern of *almagre*, an ochre paste mixed with water, so they could be distinguished from a distance (Phillips and Phillips 112). The sheep-herding practice of *almagrar*, giving each different herd a distinctive pattern, was soon borrowed by the vocabulary of ethno-religious exclusion, as in Sebastián de Covarrubias's *Tesoro de la lengua castellana o española* (1611) [*Treasure of the Castilian or Spanish Language*], which states in the entry for "Jew":

> En tiempo del rey don Enrique, cerca de los años de mil trezientos y setenta ... se mandó que los judíos que habitavan en el reyno, mezclados con los christianos, truxessen cierta señal con que fuessen conocidos y diferenciados de los demás. Estos se llamaron judíos de señal ... Y de aquí entiendo les vino

> el llamarlos los enalmagrados, porque parecía señal de almagre, qual se pone al ganado para distinguir un hato de otro, y dende a tres años mandaron traer a los moros otra señal de paño açul. (719–20)
>
> [During the time of King Henry II of Castile, around 1370 ... it was ordered that Jews who inhabited the kingdom and mingled with Christians should wear a certain sign in order to be recognized and differentiated from the others. They were called "marked Jews" ... From this I understand that they came to be called "enalmagrados," because it recalled the mark made of *almagre* (red ochre), used for livestock to distinguish one flock from the other, and three years later they ordered that the Moors should wear another sign, of blue cloth.]

Sartorial laws requiring Muslims and Jews to wear distinctive marks and clothing had their origin in the 1216 Lateran Council, and were implemented repeatedly from the thirteenth through the fifteenth century, by Alfonso X in the *Siete partidas*, by Henry II of Castile in the 1371 Toro parliaments, and by Henry III of Castile in the 1405 Valladolid parliaments (Robert 58–72).

It turns out that *almagre* as metaphor for any collective distinguishing mark is not only applicable to sartorial impositions, but also to any cultural practices coming from both Jews and Muslims that set them apart. Thus Bernardo Pérez de Chinchón's *Antialcorano* (1532) states that God instructed biblical Jews to circumcise so they could be differentiated: "assí como el que tiene mucho ganado, señala sus ovejas con almagra, para conoscerlas entre otras ... assí dios señaló su pueblo, para que fuesse conoscido entre los gentiles" (184) [as whoever has a large number of livestock marks his sheep with *almagra* in order to differentiate them from other sheep ... so God marked his people, so they could be recognized among the Gentiles]. In the anonymous *Viaje de Turquía* (c. 1558), Pedro de Urdemalas, while describing the customs of Turkish women, establishes an analogy between the henna they use as a cosmetic and the marking of livestock: "se untan con la haleña el dedo pulgar de la mano derecha, y la media mano que lleban de fuera, que paresçen rabaño de ovejas almagradas" (407) [they paint their thumbs and half of their right hands with henna, so they looked like a herd of sheep painted with *almagre*]. The use of henna was hardly exotic in Spain, though; it was included among the Moorish cultural practices that Charles V aimed to ban in 1526 (Gallego Burín and Gámir Sandoval 203). Alonso Gutiérrez, in his 1588 report on the Moriscos, suggests enslaving them and branding their faces, and justifies such extreme measures precisely in an alleged balance between the desire to differentiate them from Old Christians and their own signs of cultural identity, conceiving *henna* as a form of *almagre* because after all, the Moriscas "por zelo se suelen labrar los rostros, braços y manos" (qtd. in Boronat y Barrachina 1:637) [they paint their faces, arms, and hands out of devotion]. The borrowing of the sheep-herding practice of *almagrar* as an analogy of the distinctive marks that Muslims and Jews had to wear in order to be distinguished from Christians during the late Middle Ages proves effective only so far as it refers to external, observable signs for identifying either sheep or ethno-religious communities. The mass conversion of Jews and Muslims during the late Middle Ages and the early modern period, along with the homogenization of sartorial practices, needed a new vocabulary in order to articulate and exteriorize an invisible, genealogical difference, which it would find in discourses of breeding the Merino sheep.

As we have seen, the Merino sheep was not a "natural," "pure" race (if there has ever been one), but the product of cross-breeding and selection orchestrated by humans, even without institutionalized biological knowledge. When ancient Iberian wools were mentioned in classical sources, they were praised for their quality and abundance, but declared unsuitable for dying because of the reddish or black colour, significantly reducing their market value. Hence, the plausible cross-breeding with African rams in order to whiten the breed. The creation of a breed that produced fine, abundant wool that was at the same time white was probably not reached until the fourteenth and fifteenth centuries, when European markets began to import Castilian wools in massive quantities (Phillips and Phillips 41). The collective success of animal husbandry was only relative, because the Merino breed still presented a high genetic instability that made the original dark pigmentation reappear at random. In this endeavour, the shepherds' task consisted primarily in watching over a never-ending process of breeding selection, eliminating any newborn lamb that possessed a regressive pigmentation (Phillips and Phillips 116–17; Wood 230–40).

Although this proto-scientific knowledge was a largely oral tradition (Laguna Sanz 125), it appears scattered in several early modern treatises devoted to agriculture. Gabriel Alonso de Herrera's *Obra de agricultura* (1513) instructs livestock owners to "siempre procurar que los carneros y ovejas sean blancos, porque la lana blanca es la demás prescio que la parda o prieta, porque sobre ella se puede dar cualquier color que quisieren, y sobre la prieta o parda ninguno" (306) [always seek to have rams and ewes that are white, because white wool is more appreciated than brown or black wool, since white wool can be dyed any colour, but the brown or black wool cannot]. Similarly, Miguel Agustín's *Libro de los secretos de agricultura, casa de campo y pastoril* (first published in Catalan in 1617; translated into Spanish in 1625) recommends buying "las ovejas que ... no tengan la lana grossera, o mezclada de diversas colores, por la insertitud del color" (444) [sheep ... that do not have coarse wool, or whose wool is mixed with several colours, because of the uncertainty of the colour].

When the entire economic fabric, linked to a global wool trade, depends on the constant vigilance over the genetic instability of sheep, it is easy to understand why the logic of sheep herding filters through every level of social discourse, which then allows for segregation to become accepted as common-sensical. It helps to explain the striking evidence that most of the Castilian vocabulary of racism comes from the semantic field of sheep herding. Even the overcharged term *raza* [race], seems to have been used first in sheep herding and in the wool trade, and only later was it adopted for the discourse of ethnic difference.[7] The first documented uses of the word *raza* in medieval Castilian suggest that its proper meaning was to point to the "defect" or the "stain" in woollen cloth, and that it was applied only metaphorically to the lineage, most frequently as a derogatory remark. In Juan Ruiz's fourteenth-century *Libro de buen amor*, the term appears first in the proverb "non ay paño sin raça" (94c) [there is no cloth without defect], and second as a complaint about how money is used to cover people's defects: "con el dinero cunplen sus menguas e sus raças" (504c) [people compensate for their deficiencies and defects with money]. By 1516 Antonio de Nebrija's *Vocabulario de romance en latín* still defines *raça* only as a textile term meaning a defect in the cloth: "raça de paño, panni raritas" (164). Even in the early seventeenth century, Covarrubias still gives the textile meaning as the primary one, only to later acknowledge that it was applied as well to the doctrine of blood purity: "Raza en el paño, la hilaza que se diferencia de los demás hilos de la trama ... Raza, en los linages se toma en mala parte, como tener alguna raza de moro o judío" (896–7) [*Raza* (stain) in the cloth, the thread that differs from the other threads in the weft ... Race, when speaking about lineages, is taken as a bad term, as having some *raza* (trace) of Moor or Jew]. As Covarrubias shows, the Spanish early modern racial system was not taxonomical: *raça* was conceived to be a negative mark that Old Christians lacked.

While *raça* was the most widely used term, there were other synonyms borrowed as well from the semantic fields of sheep herding and wool trade, such as *ralea* or *hilaza*. By the beginning of the seventeenth century, Francisco del Rosal's *Alfabeto primero* gives *ralea* as a synonym for *raça*: "Raça, falta en el paño, es *Ráritas*, y así la llamaron *Ralea* que era rareza de pelo, y después pasó a significar la falta de linage, que así también decimos hablando del linage: en el mejor paño cae la mancha" (528–9) [*Raça*, a defect in the cloth; it is *raritas* (scarcity), and hence they said *ralea* (bloodline), that was thinness or scarcity of fur/fleece, and later it came to mean the lack of lineage. And so we say, speaking about lineages: in the best cloth there is a defect]. Similar to *raça*, the term *hilaza* [thread] applies to both the imperfections in the woollen cloth and the stain in the lineage, as in the idiom "descubrir la hilaza," which Covarrubias equates to the proverbial phrase "En el mejor paño cae la raza" (851). The use of the term *hilaza* to refer to lineage is amply documented. Agustín de Salucio's *Discursos sobre los estatutos de limpieza de sangre* (c. 1600) complains about the excesses of the doctrine of blood purity that questioned any lineage that had converted to Christianity: "¿que familia avra que siendo infiel, y biviendo entre gente tan fiel, no descubra la hilaza en padres, ni en hijos, ni en nietos, ni bisnietos, por espacio de 100 años?" (37r–v) [what family could there be of non-believers, and living among believers, that in one hundred years will not discover the *hilaza* (thread) in fathers, sons, grandsons, or great grandsons?].

The recurrent racialization of sheep herding terms such as *almagrar*, *raça*, *hilaza*, or *ralea* reveals that this was the semantic field Spaniards first engaged with when attempting to conceptualize issues of genealogical difference by means of an already codified vocabulary of breeding selection and marking. The selection that Spaniards made of sheep herding terminology is also symptomatic of anxieties over blood purity. It is not by chance that their selected vocabulary referred to visible traits in sheep herding and the wool trade, ready for use in a racial theory that did not rely on physical differences – blood purity was a quasi-mystical concept based on the religion of ancestors. The borrowing of metaphors referring to observable traits in sheep and fabrics had the advantage of conferring ontological connotations to racial terms that would be otherwise hard to grasp, and of exteriorizing inner qualities that are otherwise invisible.[8]

The Good Shepherd and *limpieza de sangre*

The religio-political trope of the Good Shepherd is a widespread European rhetorical image used to exhort the monarch to care for the welfare of his subjects. It is already formulated by classical philosophers, such as in Plato's *Republic* (1:343a–b: 21) and Xenophon's *Memorabilia* (3:2; 73), and is ubiquitous in early modern Spanish political discourse (Márquez Villanueva, "Giovan Giorgio Trissino" 366n34; Milhou, "Propaganda" 59;

Magnier 367–93). It appears as well in the doctrinal discourse of ecclesiastical reform, reminding prelates of their obligation to provide pastoral care to their flock of believers (Tellechea Idígoras; Heliodoro García). The lack of a clear separation between church and state certainly allowed for the application of the same trope to both realms, as Antonio de Torquemada's *Coloquios satíricos* (1553) [*Satyrical Dialogues*] explicitly states: "no solamente los obispos y los otros perlados y pontífices son pastores ... pero que desa manera también se pueden llamar pastores los emperadores, reyes y príncipes, y los otros grandes señores, y todos aquéllos que tienen vasallos y súbditos con cargo de governarlos" (1:306) [not only bishops and other ecclesiastics are shepherds ... but also emperors, kings, and princes and other great lords could be called thus, and all those who have vassals and subjects whom they must govern]. On the one hand, the Catholic Church was a temporal power with its own economic system of vassals; on the other hand, the monarch often assumed the task of persecuting heresy and controlling religious orthodoxy.

The religious zeal of Spain's Catholic monarchy allowed for the political appropriation of scriptural prophecies in which the metaphor of the Good Shepherd was based. The imperial project brought to fruition with Charles V was especially receptive to the biblical prophecy of a time of completion, when the whole world would be ruled by only one shepherd and would have only one religious faith, as John 10:16 applies to the figure of Christ: "Tambien tengo otras ouejas que no son de este corral: aquellas tambien me conuiene traer; y oyran mi boz: y hazerse ha un corral, y un pastor" [I also have other sheep that are not from this flock, and I must bring them as well; and they will hear my voice, and there will be only one flock and one shepherd].[9] The prophecy was believed to match the universalizing aspiration of the Habsburg monarchy, fueled by the colonial enterprise in the New World and the massive evangelization of heathens. Its most famous expression is found in Hernando de Acuña's sonnet, which is usually read as a skilful appropriation of John's prophecy to praise the Spanish empire: "Ya se acerca, señor, o ya es llegada / la edad gloriosa en que promete el cielo / una grey y un pastor solo en el suelo, / por suerte a vuestros tiempos reservada" (328) [The glorious time when heaven promises that there will be only one herd and one shepherd in the world, and that is luckily reserved for our own times, is approaching, or has already come].[10] The colonial interpretation should not make us forget that the prophecy had also its own internal currency within Spain, in which religious homogeneity was repeatedly praised as a political goal with the expulsion of the non-converted Jews in 1492 and the forced conversion of the Moriscos during the first half of the sixteenth century. Read in this light, the temporal vacillation of Acuña's sonnet reproduces the uncertainty of whether the alleged religious unity has already been realized or whether it is still a forthcoming project. It is the quick switch from future reference ("Ya se acerca") to the uncertain past ("o ya es llegada") that allows the poem to not specify the methods through which such a unity had been or should be realized, and to elide whether its outcome is strictly religious or also cultural and genealogical.

Whereas Acuña's sonnet avoids facing this thorny question, there is a whole range of texts that openly explore the connotations of the biblical pastoral prophecy, spelling it out in very concrete terms. Indeed, the image of the Good Shepherd is used in the debates over blood purity as early as the mid-fifteenth century, as a reaction to Pero Sarmiento's attempt to create a statute of blood purity for the Cathedral of Toledo in 1449 in order to exclude the *conversos* from holding ecclesiastical positions. The writings in defence of *conversos* make ample use of the biblical metaphors to defend the full integration of New Christians into the ecclesiastical hierarchy. Alonso de Cartagena repeatedly evokes John 10:16 in his *Defensorium Unitatis Christiane* (c. 1449–50):

> El mismo Cristo ... quiso ser adorado por una y otra gente, tanto de los pastores de Israel como de los gentiles. Y finalmente en su predicación, después de haberse manifestado como enviado a las ovejas perdidas de la casa de Israel, añadió: "tengo otras ovejas que no son de este aprisco y es preciso que yo las traiga y se hará un solo redil y un solo pastor." (201)
>
> [Christ himself ... wished to be worshipped by every people, both by the shepherds of Israel and those of the Gentiles. And finally, in his sermon, after having revealed himself as the messenger to the lost sheep of the house of Israel, he said: "I have other sheep that are not from this herd, and I must bring them and there will be only one herd and one shepherd."]

Alonso de Oropesa's *Lumen ad revelationem gentium* (c. 1465) applies very similar terms in chapter 23 (457–61) to defend the integration of all the believers within the same herd and to criticize those who aim to guard the exclusivity of the church's "pastures" (460), in a clear allusion to the defenders of the statutes of blood purity.[11] Both Cartagena and Oropesa defend the Pauline interpretation of the Gospels, which claims that the inclusion in the Christian community erases all previous differences among their members.

While these late medieval treatises applied the biblical language of sheep herding to the ethno-religious conflicts of Spain to support integration and tolerance, the proponents of discriminatory practices co-opted this same pastoral discourse. The text that initiated this trend is a negative portrait of a bad shepherd, or, as Angus MacKay labelled it, "a tale of pastoral anarchy" (4). The anonymous *Coplas de Mingo Revulgo*, composed towards the end of the reign of Henry IV of Castile (1454–74), is a pastoral allegory that corresponded to the propaganda orchestrated by Isabel I of Castile against her half-brother, refiguring him as the bad shepherd Candaulo. One of the charges against him was the negligence in enforcing a distinction between the different religious herds:

> Modorrado con el sueño
> non las cura de almagrar
> por que no entiende de dar
> cuenta dellas a ningun dueño
> quanto yo no amoldaria
> lo de Christoval Mexia
> nin del otro tartamudo
> nin del meco moro agudo
> todo va por una via. (*Coplas* 154)

> [Drowsy with sleep, he takes no care in marking the sheep, since he does not have to account for them to any owner; because I could not differentiate Christoval Mexía's flock from that of the stutterer or from the cunning Moor from Mecca, all are jumbled together.]

The best-known and most widely disseminated gloss was that of Fernando del Pulgar (1485), who gave a clear political interpretation:

> Algunos acostumbran en los pueblos dar cargo a un pastor que guarde sus ouejas y cada uno señala las suyas con almagra de su señal que tiene conosçida y este señalar llaman los pastores amoldar. Quiere agora aqui dezir que tanta turbaçion ay en el hato, conuiene saber en el pueblo, que no se conoçerian las ouejas de Christobal Mexia. Estos son los christianos de Christo Mexia nuestro redentor. Ny menos se conoçerian ... los judios ... Ny menos se conosçerian ... los moros ... Y esta confusion dize que prouiene del sueño del pastor, y porque toca aqui en la poca diferençia que auia de los vnos a los otros, no plega a Dios que ... todos andouiesen rrebueltos y no se conoçiese ... quales eran christianos ny quales judios o moros ... esta copla quiere dezir que en los

abitos que deuen traer los judios y los moros señalados y apartados de los christianos no avya la diferençia que deue auer y que todos traen un avito. (*Coplas* 154–5)

[Many in the towns hire a shepherd to guard their sheep, and every one of them marks his own sheep with a distinct pattern of *almagre* (a red ochre); and they call this marking *amoldar*. (The poet) means that there is so much confusion in the flock, that is, the people, that the sheep of Christobal Mexia (Christ the Messiah) could not be recognized. These are the Christians of our Redeemer. The Jews ... would be even less recognized ... and so would the Moors ... And it says that this confusion comes about from the shepherd's distraction, and because it deals with the little difference that there was among them, it does not please God ... that they are all mixed up and that no one could distinguish ... who were Christians, who were Jews, and who were Moors. Therefore, this stanza means that there was no difference in the cloth that Jews and Moors should wear, different from Christians, as it used to be, and that all of them wear the same dress.]

While Fernando del Pulgar's gloss was notably the most influential interpretation during the sixteenth century and has become canonical (MacKay 27n72), there are other responses to the *Coplas de Mingo Revulgo*. Another fifteenth-century gloss, this one anonymous (manuscript BNM), interprets *almagrar* as a metaphor of the need for pastoral care, divorcing it from religious heterogeneity, which is not conceived as a threat to the survival of the kingdom (*Coplas* 155). Similarly, another anonymous gloss (manuscript R-M) also reads the *Coplas* in such a way that it can be understood as an exculpation of HenryIV and even a defence of religious freedom, contrary to Fernando del Pulgar's interpretation: "el Rey no es tenudo / a mas de dar leys / e fueros e dar / donde cada uno vjua bien saluo / si el pueblo dexa la fe por se re /boluer con generacion defen / dida en trato njn fabla e fa / zer comer de sus vjandas / en esto no tiene culpa / como cada vno ha de ser guardador de su alma" (*Coplas* 158) [the king is only supposed to make laws and ensure that each person lives safely; if the people abandon their faith to mix in conversation and speech and to offer their food to the prohibited people, they are not to be blamed, since each one should only look after his own soul].[12] Thus the same passage that Fernando del Pulgar interprets as a critique of the lack of ethno-religious segregation is appropriated in the anonymous manuscript R-M as an admonition to the monarch not to intervene in religious matters, and in the BNM manuscript, as a call for pastoral care for all.

At the beginning of the sixteenth century, in parallel with the massive catechization targeting both the descendants of the converted Jews of 1492 and the Moriscos forced to convert to Christianity in Castile (1502) and Aragon (1525), it is the inclusive Pauline vision of the Good Shepherd that predominates. Between 1535 and 1540, López Villalobos wrote in support of accepting *conversos* into the Order of San Francisco: "si son infieles, ¿por qué habéis querido ir contra el mandamiento del buen pastor, que mandó en su testamento que todas sus ovejas no tuviesen más de un corral, y todas debajo de un pastor, y manda que no haya acepción de personas?" (qtd. in Domínguez Ortiz, *La clase social* 250) [if they are infidels, why did you want to oppose the mandate of the good shepherd, who ordered in his testament that all sheep have no more than one corral, all of them under only one shepherd, and he orders that there should be no selection among people?]. In 1532, Bernardo Pérez de Chinchón, who devoted great effort towards the conversion of the Moriscos during the first half of the sixteenth century, addresses his sermons against the Koran to preachers who will use them to indoctrinate the Moriscos, insisting in his theological challenge that pastoral care has to be provided for the whole Christian herd: "vuestras son las ovejas conosceldas: porque el buen pastor: conosce sus ovejas, y pone por ellas su ánima" (78) [they are your sheep and you should get to know them, because the good shepherd knows his sheep, and risks his soul for them].

These debates on the figure of the Good Shepherd draw extensively on both the biblical passages and the knowledge of sheep herding, each side emphasizing a different aspect. While the defenders of the full integration of *conversos* relied on the Pauline interpretation of the Good Shepherd as one who takes care of his entire flock, defenders of the statutes of blood purity rewrote the trope by looking at the quotidian activity of sheep herding. If one of the tasks of the actual shepherd was selecting the animals that produce white and abundant wool, segregating and culling those considered defective (Phillips and Phillips 116–17), the defenders of the statutes follow this criterion of productivity, arguing that the same logic should be applied to the management of human population, eliminating from the beginning any "tainted" lamb that could threaten to "contaminate" the flock's purity.

The terms of the debate would remain unchanged throughout the sixteenth century, but the polemic was again brought to the fore with the reinvigoration of the statutes of blood purity in the middle of the sixteenth century. In 1547 Archbishop Juan Martínez de Silíceo managed to establish a new statute of blood purity in the Cathedral of Toledo, with the pope's approval in 1555 and Philip II's in 1556.[13] During these years, which also coincide with his appointment as cardinal and the royal succession from Charles V to Philip II, the polemics about the statutes frequently recurred to the pastoral metaphor of the Good Shepherd. Martínez de Silíceo was proclaimed cardinal on 20 December 1555, but the public celebration did not take place until three months later, on 26 March 1556. The three-month delay between his appointment as cardinal and the celebration could have been intended to draw it closer to the ceremony in the same city of Toledo in which Charles V abdicated in favour of his son Philip II on 19 April 1556. The festivities for the imposition of the cardinal's robe, as they are recorded in a contemporary account, included a pastoral play in which Martínez de Silíceo was represented as a shepherd taking care of his sheep.[14] From the scattered lines transcribed in the account, it is hard to find any direct reference to the statutes of blood purity in the performance. However, the polemic was certainly in the air, and it seems plausible to consider that this dramatic representation of Silíceo as a good shepherd would have been interpreted very differently depending on the audience (Surtz, "Cardinal" 230–2).

Less than a month later, during the coronation of Philip II, Fray Cipriano de la Huerga delivered a sermon on the Good Shepherd at the University of Alcalá, which was to a great extent a critique of Martínez de Silíceo's advancement of the statutes of blood purity and probably of his recent public role as the good shepherd himself.[15] Huerga's sermon emphasizes that the main virtue of the shepherd is his love for all his sheep (277–8) and that the application of the Pauline evangelical message must take precedence over any reason of state (280). He goes beyond the conventional biblical metaphors in order to make his point: "era el buen pastor Jesuchristo que, para curar la roña de sus ovejas, para dalles salud y libertad, sacó la miera de su sangre" (265) [Jesus Christ was the Good Shepherd, who, in order to cure the mange (roña) of his sheep and give them health and freedom, took the ointment (miera) from his own blood].[16] The Bible does not mention the *miera*, a juniper oil used to heal livestock, and the sermon goes on to list all kinds of maladies that actual sheep might suffer (274).

Following the tradition initiated by *Coplas de Mingo Revulgo*, Fray Cipriano de la Huerga mixes all kinds of pastoral metaphors taken from the Bible with the specialized terminology of sheep herding. While Huerga shows a specialized knowledge of sheep herding, the absences are even more striking. In his efforts to emphasize his opposition to the segregation of blood purity, Huerga is careful to select only the sheep-herding terminology that is in agreement with his defence of the Pauline evangelical message. He thus avoids addressing the need to select sheep for their wool productivity and the need to mark [*almagrar*] different herds. Huerga explains that, when he speaks about the sickness of livestock, he is referring to heresy conceived as a contagious malady (275–6), so that genealogical difference and ethno-religious origins are divorced from the pastoral metaphor. Since his sermon underlines that the monarch should always look for a cure for his flock, he does not mention the possibility of selecting and culling the sick sheep in order to preclude the whole herd from becoming infected.

In 1557 Luis Hurtado de Toledo published in Toledo the play *Cortes de la muerte* [*The Parliament of Death*], in which the Morisco Xarique encourages other fellow Muslims to convert to Christianity by quoting John's prophecy about the day in which "las leyex an de xer / todax vna sin auer / mas de vn paxtor y vn corral" (57v) [the religions should be one, without having more than one shepherd and one herd].[17] After a Portuguese *hidalgo* boasts of his blood purity, it is now the turn of the Jew Don Farón to reply to him, by quoting John 10:16, that religious homogenization erases previous ethnic differences: "tiempo ha de venir / quando se an de reduzir / las leyes todas a una / y todo será un pastor / y un rebaño y una grey" (59r) [the time will come when all laws shall be reduced to one and there will be but one shepherd, one herd, and one fold]. The Portuguese *hidalgo* reiterates that "judeos e un ruy ganado" (59r) [Jews are a despicable flock], so now it is the turn of Saint Augustine, who participated in the conversion of the Jews, to intercede against him, expanding the pastoral metaphor from John:

> Mas el es tan buen pastor
> que sus ouejas conosce
> y ellas a el con amor ...
> y que otras ouejas tiene
> que no son deste rebaño
> las quales traer conuiene
> y oyran su boz quando suene
> de que no les verna daño. (59v)

[However, he is such a good shepherd, who knows his sheep, and they know him with love ... and he has other sheep that are not from this herd, and he must bring them, and they will hear his voice when it sounds, and no harm will come to them.]

Thus, although the play acknowledges the initial existence of a diversity of "herds," it promotes the view that the final objective is precisely to merge them into a single one to make their initial differences disappear. Published in Toledo in 1557, when the discussion about the application of the statutes of blood purity was still strong, the *Cortes de la muerte* participated in the ideological debate by promoting the full integration of the *conversos* into the national body. The inclusion of the Morisco Xarique in the play also marks a shift of focus in Old Christian ethnocentrism: while the first half of the sixteenth century was characterized by the Inquisitorial persecution of the converts from Jewish origin, by the 1550s cultural and religious pressure was applied to the converts from Islam.

Yet, at least at the initial stage, the conversion of Moriscos was viewed optimistically, with a renewed effort at evangelization and their inclusion in the Christian fold. Mármol Carvajal, in his *Historia del Rebelión y castigo de los moriscos del Reino de Granada* (1600) refers to the first evangelization project by Hernando de Talavera in Granada by the end of the fifteenth century: "Bueno fue por cierto el consejo que tomaron los Católicos Reyes ... en encomendar aquel nuevo ganado cerril, no usado al yugo suave de Dios, a pastor [Hernando de Talavera] tan antiguo y tan ejercitado en su ley, para que por medio suyo viniessen a juntarse a su rebaño" (152) [The Catholic Kings took good counsel ... when they assigned this wild herd, not accustomed to the gentle yoke of God, to a shepherd so old and exercised in their faith, so he could manage to bring them into the same flock]. In Diego Sánchez de Badajoz's *Farsa de la Yglesia* (published posthumously in 1554), the shepherd who baptizes the Moor announces his entrance into the Church by comparing him to a sheep: "Pastor: ¿Reñegas casa de Meca?/Moro: Xí./Pastor: Tú quedas hecho oveja" (468) [Shepherd: "Do you renounce the house of Mecca?" Moor: "Yes." Shepherd: "You then become a sheep"]. It is in the context of Morisco evangelization that, according to Cyril Mérique, Juan Timoneda was commissioned to write the *Aucto de la oveja perdida*, staged in Valencia in 1557, although there is no explicit mention of Moriscos in the text (69–71). Ecclesiastic authors repeatedly lament the abandonment of Moriscos by comparing them with sheep, such as the bishop of Segorbe, who says in 1587 that "viven como ovejas sin pastor ques grande lástima" (qtd. in Boronat y Barrachina 1:622) [they live like sheep without a shepherd, which is a great pity], or Pedro de Valencia, who complains in his *Tratado acerca de los Moriscos de España* (1606) that priests "los olvidan como si no fuesen del rebaño" (127) [forget them as if they were not from this herd].

However, the initial optimism dissipated as soon as ecclesiastical authorities realized that Moriscos formed compact isolated communities, in many cases protected by Christian feudal lords. In this context, traditional pastoral texts such as the *Coplas de Mingo Revulgo* were still being glossed as if the status of the Moriscos, now formally Christian, had not changed. When Juan Martínez de Barros writes his own gloss on the *Coplas de Mingo Revulgo* in 1564, he includes in the tenth stanza a long erudite note which is obviously not an anachronistic critique of the reign of Henry IV of Castile, but an intervention in the debate over the statutes of blood purity and the situation of the Moriscos (*Coplas* 156–8). He begins his commentary by appropriating the line of interpretation suggested in the fifteenth century by Fernando del Pulgar, for whom the activity of *almagrar* to the shepherd's livestock refers to the need of separating and marking the different religious communities, recalling medieval laws promulgated on this issue. Nevertheless, Martínez de Barros seems to suddenly realize that times have changed and, at least in theory, there is no such religious difference by the time he writes his gloss: "Aunque esta diferencia, gracias a Dios en estos Reynos esta quitada; porque ... Don Fernando y Doña Ysabel ... mandaron salir a los Yudios, y Moros de estos Reynos, salvo aquellos que se tornasen Christianos" (*Coplas* 157) [Although this difference, thanks be to God, has disappeared from these kingdoms, because ... Ferdinand and Isabel ... ordered that all Jews and Moors should leave these kingdoms, with the exception of those who converted to Christianity]. However, this brief clarification does not preclude Martínez de Barros from listing medieval laws on the matter and pointing out where to find them, as if, truly speaking, mass conversions to Christianity had not really happened, or at least as if they had not changed the institutional need to "mark" ethnic difference (157–8).[18]

Fray Luis de León, a disciple of Fray Cipriano de la Huerga (whose defence of the Pauline message we have already seen), makes the most explicit critique of the statutes of blood purity in conjunction with the figure of the Good Shepherd in *De los nombres de Cristo* (1583) [*On Christ's Names*], under the name "Rey de Dios" [King of God].[19] Immediately after reiterating the Pauline evangelical message that baptism and conversion erase previous genealogical differences (375), the text, set in the form of a dialogue, directly confronts royal support for the statutes of blood purity. Sabino, one of the interlocutors, raises the question of how those kings who spread infamy about their subjects are to be praised, to which Juliano responds that these kinds of rulers: "ninguna cosa son menos que reyes ... Porque si son cabeças, ¿qué honra es ser cabeça de un cuerpo disforme y vil? Y si son pastores, ¿qué les vale un ganado roñoso?" (376) [are anything but kings ... Because, if they are heads [of state], what honour is there in being head of a deformed and corrupted body? If they are shepherds, what is the use of mangy livestock?].[20] Fray Luis here inverts the use of sheep herding metaphors, so the *roña* is not a manifestation of the *raça*, but rather of the infamy imposed on the descendants of the *conversos* in the Christian flock. The *roña* is thus conceived of not as a congenital sickness of a certain strand of sheep, but as the sickness introduced into the flock by the shepherd himself.[21] Thus Fray Luis promotes a vision of the Good Shepherd, in accordance with Huerga's use of the metaphor as a Pauline symbol for integration.

Fray Agustín Salucio is probably the first to acknowledge Fray Luis's attack against the statutes of blood purity in his own proposal for reform, *Discurso sobre los estatutos de limpieza de sangre* (c. 1600) [*Discourse about the Statutes of Blood Purity*]. Even though Salucio cautiously frames his proposal as an acclamation of the statutes and the work of the Inquisition, he quotes the aforementioned passage by Fray Luis de León approvingly, criticizing the monarch who allows the spread of infamy among his subjects (26r–7r).

Fray Luis's attack against the statutes was first criticized by Álvaro Piçario de Palacios in a letter addressed to the Inquisition in 1609 (Sicroff 308–9; Perea Siller 1134–5). It does not seem a coincidence that this denunciation occurred the very same year in which Philip III decrees the expulsion of the Moriscos. Unlike Piçario de Palacios, Marcos de Guadalajara y Xavier's *Memorable expulsión y justísimo destierro de los moriscos de España* (1613) [*The Memorable Expulsion and Very Justified Exile of the Moriscos from Spain*], instead of refuting Fray Luis's authority, appropriates it by reinterpreting *De los nombres de Cristo* as a praise of the segregation policy that had led to the expulsion of the Moriscos, and therefore as a justification of Philip III's decree:

> tomemos [por aviso] el que nos da el Padre fray Luys de León en los *Nombres de Cristo*: "Que ser Rey propriamente, es, no tener vassallos viles y afrentados ... porque si son cabeças, qué honra es, ser cabeça de un cuerpo disforme y vil? Y si son pastores, qué les vale un ganado roñoso?" (74v)
>
> [We should take Fray Luis de León's advice in *Nombres de Cristo*: "Properly speaking, to be king is not to have wicked and infamous subjects. Because, if they are heads of state, what honour is there in being head of a deformed and corrupted body? If they are shepherds, what is the use of mangy livestock?"]

Fray Luis's pastoral strategy was thus co-opted by the same interpretation of the Good Shepherd against which he was writing, transforming it into the logic of segregation. The goal of the Good Shepherd is no longer to cure the "ganado roñoso" (the Moriscos), but rather to eliminate it "para la conservación y seguridad del ganado Cathólico" (Guadalajara y Xavier 12r) [for the conservation and security of the Catholic herd].[22] Guadalajara y Xavier is arguably the apologist of the expulsion who best exemplifies this new vision of the trope of the Good Shepherd as a figure who needs to protect the flock by identifying and excising undesirable elements. Instead of protecting the Christian flock from an external threat, the Good Shepherd is now supposed to protect the flock from itself, and therefore vassals are conceived of as a population that has to be constantly regulated for its own good.[23]

Guadalajara y Xavier also laments the dangerous consequences that cultural assimilation had for properly differentiating between Old Christians and Moriscos in Castile:

> estos viven entre Christianos, y los más dellos hablan bien nuestro vulgar, y visten como nosotros; siendo tan Moros en la ley como los del Reyno de Valencia ... de que resulta no poco escrúpulo para sus Obispos y Curas, pues no conocen el rostro de su ganado, y no lo pueden conocer; quiero decir, la ley y costumbres en que viven. (83v)

> [they live among Christians, and most of them can speak our language, and they dress like us, being as much Moors in their faith as those in the kingdom of Valencia ... and from this comes a lot of suspicion of their bishops and priests, since they do not know the face of their herd, and they cannot know it; that is, they cannot know what faith and customs they practise.]

In spite of Guadalajara y Xavier's lament, the disappearance of ethnic markers after Philip II's prohibition of Morisco customs in 1566 was soon replaced by institutional recording of Morisco lineages. On the one hand, the methods through which genealogical infamy was used to discriminate against those accused of Judaism after 1492 were applied as well to the Moriscos: the *sambenitos*, the defamatory clothing of those punished by the Inquisition, were exhibited on the walls of the churches in order to maintain the memory of those lineages accused of Judaizing, Islamicizing, or engaging in other heresies. At the same time, there is a radical difference in the way descendants of Jewish origin and those of Muslim origin were categorized by both religious and civil authorities.

In the case of *conversos* from Jewish origin, the memory of their lineages was maintained, after the fifteenth century, if at all, in public opinion, and written record was made only when an individual, or even an entire family, was prosecuted by the Inquisition for relapsing into Judaism. By contrast, in the case of the Moriscos, by the 1570s, not only were records kept of those who were accused of Islamicizing, but there was also an institutional effort to elaborate a census of all of them. Parish archives registered the ethnic origins of the children who were baptized (Méchoulan 196), and Philip II mandated the creation of several censuses on the Morisco population between the 1570s and 1580s.[24] As James Tueller points out, after the 1580s, "the suggestion to expel was changed to an order to count, name and register" (101). To a great extent, the expulsion of the Moriscos in 1609, more than forty years after the eradication of Morisco cultural practices, was largely possible because of Philip II's compilation of such massive information about Spain's population. This bureaucratization of ethnic difference implied a radical transformation of power. Until the late sixteenth century, the monarchy had simply allowed for the establishment of private and individual initiatives to impose statutes of blood purity, without actually involving the state in their imposition. By ordering censuses based on genealogy, the state assumed a completely new role in which detailed and substantive information about its subjects was used to manage the population.

This bureaucratic approach to dealing with Moriscos, considered in conjunction with the ethnocentric interpretation of the concept of the Good Shepherd, informs a new understanding of power that can be seen in royal policies and in art commissioned by the monarchy.[25] Philip III actively embraced this new conception of power, initiated by his father Philip II, in the way he nurtured his royal image and his political actions as a Good Shepherd. It is probable that the king made the decision to expel the Moriscos in January 1602, although the expulsion was delayed for practical reasons until a truce with the Low Countries was established in 1609 (Harvey 303–4). In 1603, soon after the decision was made, the queen commissioned the court painter Pantoja de la Cruz to create several paintings for the royal chapel, including one Adoration of the Shepherds in which the queen herself was cast as the Virgin Mary, and Philip III as one of the shepherds (Kusche 221–2) (figure 1). Pantoja de la Cruz's painting entailed a radical change in royal representation that affected not so much the image of the monarch himself, but the notion of how monarchical power should be exercised. Even though his painting was originally designed for private use and the monarch's self-contemplation, the representation

Figure 1 Juan Pantoja de la Cruz. *The Birth of Christ with the Royal Family* (1603). Courtesy Museo del Prado, Madrid.

of Philip III as a shepherd is consistent with his role as the Good Shepherd, revealing his eagerness to exploit pastoral iconography as a viable representational trope of royal power. Read both within the vogue of early modern courtly pastoral literature and as a micro-representation of the national space that was to be perfected, the court becomes a realization of pastoral Arcadia – without losing its religious iconographic undertones. Philip III disguised as a shepherd is an imposture predicating a symbolic identification between the king and "the people," conceived as the constituency that legitimates power, as I will show in chapter 2.

By reconfiguring himself as a "real" shepherd, Philip III adopted the zootechnical logic of sheep herding inherent in the discourse of blood purity, culling "bad" sheep (those with *raça*) and protecting "good" ones. Maria Kusche argues that the picture was commissioned to induce the queen to conceive a male heir after having delivered two girls (109–12). Yet the painting's promotion of royal fertility is not incompatible with the king's representation of the shepherd as an agent of death as well; if Philip III is represented as an agent of life, it is because he is the administrator of both birth and death, inhabiting a new concept of the state that requires the management of population, exercising control over segregation and reproduction.

Although both Charles V and Philip II had been cast as Good Shepherds by their subjects, similar invocations addressed to Philip III exceed the traditional trope and clearly articulate a new concept of power. An anonymous ballad published immediately after the expulsion of the Moriscos praises the measure to expel dangerous elements taken by Philip III and executed by the Marquis of San Germán, "que como buenos pastores / tan bien guardan su ganado, / apartando del que es bueno / el que es insolente y malo" (Ruiz Lagos 224) [they guard their flock well, as good shepherds do, separating the good (sheep) from the insolent and bad]. Juan Márquez's political treatise *El governador christiano* (1615) [*The Christian Governor*] does not state so explicitly that the mission of the Good Shepherd is to separate the good sheep from the bad. However, he lays out the conventional vision of the monarch as a Good Shepherd (48–9) after defending the recent expulsion of the Moriscos and promoting the expulsion of the Gypsies (16–17), which indicates that, at least for him, the figure of the Good Shepherd is compatible with the removal of entire communities from Spanish territory.

Similarly, the *arbitrista* Pedro Salazar de Mendoza proposes in 1618 to continue with the expulsion of the Gypsies, arguing that "entre Moriscos y Gitanos, como ruin ganado, ay poco que escoger" (qtd. in Sánchez Ortega 87) [between Moriscos and Gypsies, as vile livestock, there is little from which to choose]. Francisco López de Zárate praises Philip III in 1619 for behaving "[c]omo pastor astuto, que separa / del ganado luzido el sospechoso" (41r) [as a clever shepherd, who separates the healthy from the suspect flock]. He continues with a eulogy to both the king and the Duke of Lerma: "El uno, y consejero poderoso / el otro, de catolicos rebaños; / desterraron el daño, al Moro fiero" (41r) [The king, and the powerful counsellor, (herders) of Catholic flocks, expelled the harm, and the ferocious Moor].[26] The existence of so many proposals addressed to the king to expel the Moriscos and the Gypsies reveals that there is a consensus by which the monarchy is considered the site for such appeals. There is therefore a general perception that population management, including ethnic cleansing, is a duty of government.

This early modern responsibility of the monarch bears a close resemblance to Michel Foucault's "governmentality," outlined in his lectures at the Collège de France, a conception of power which he distinguishes from what he called "pastoral power" (*Security* 115–253). Foucault identified a radical shift in early modern political philosophy, from the old protective pastorate based on the figure of the Good Shepherd to a different system of theories, practices, and technologies of government that constitute a new phase in which racism stands as

> a way of fragmenting the field of the biological that power controls ... of establishing a biological-type caesura within a population that appears to be a biological domain. This will allow power to treat that population as a mixture of races, or to be more accurate, to treat the species, to subdivide the species it controls, into the subspecies known, precisely, as races. That is the function of racism: to fragment, to create caesuras within the biological continuum addressed by biopower. (*Society* 255)

While I find Foucault's conception of biopower very suggestive for structuring my own analysis, I agree with David Nirenberg, who critiques Foucault for locating the idea of "racism" and the emergence of biopower in the nineteenth century. Rather, as Nirenberg suggests, we can trace how early modern racial terminology, even if using a different set of criteria, was "already embedded in identifiably biological ideas about breeding and reproduction" ("Race" 79).[27] As my analysis of early modern texts dealing with the figure of the Good Shepherd demonstrates, even though they borrow their language from the traditional biblical and political repertoire, they actually look for the practices of the proto-scientific achievements of sheep breeding techniques, most notably the need to select and cull the unsuitable newborn lambs with *raça* in order to keep at bay the genetic instability of the breed. They also combine this technique with the macroeconomic discourses that the international wool trade entails, enlisting the development of statistics to measure and control populations through ethnic censuses. This is precisely the kind of racism that, in Foucault's definition, "justifies the death-function in the economy of biopower by appealing to the principle that the death of others makes one biologically stronger insofar as one is a member of a race or a population, insofar as one is an element in a unitary living plurality" (*Society* 258).[28] Thus, pace Foucault, biopower emerges in early modern Spain within the very traditional figure of the Good Shepherd, instead of in opposition to it, rooting itself in biblical authority, both in the Old Testament and the New.

Representing Jacob's Livestock

Since scripture is not immune to this biopolitical interpretation, it is not surprising to find that other biblical passages associated with sheep herding were read within this paradigm. During the seventeenth century, there was a growing interest in the figure of the Old Testament shepherd Jacob, who was construed as the prefiguration of Christ and the Christian Church (Mulcahy 76), and thus became as well a prefiguration of the Good Shepherd (Taggard 87–8). However, the extraordinary interest attending the figure of Jacob in seventeenth-century Spain seems to go beyond theological explanations, providing another biblical prefiguration of biopower in the miraculous tale of how to intervene in the breeding of the herd. As with the figure of the Good Shepherd, the way early modern authors represent the scriptural figure, emphasizing or suppressing different aspects within the biblical source, reveals their positions within the ethno-religious debate.

As told in Genesis 29–31, Jacob, after several years working as a shepherd for his father-in-law, Laban, decides to return to his own land and asks for the pay he is due. He proposes the following deal: he will keep only the spotted and speckled sheep and goats, while Laban will keep the white livestock: "yo pasaré hoy por todas tu ovejas para quitar de ahí toda oveja pintada o manchada y todo carnero bermejo en los carneros; y lo pintado y manchado en las cabras, y esto será mi salario" (Genesis 30:32) [I will pass through all your flock today, removing all the speckled and spotted sheep, and all the reddish rams among the rams, and the spotted and speckled among the goats, and that will be my salary]. Laban is quick to accept the deal, presumably because the spotted sheep have less market value than the sheep that produce white wool. He is, of course, unaware that Jacob will transform this apparently self-prejudicial deal into a trick to seize all his livestock. Inspired by God, Jacob then invents a trick to intervene in the reproduction of the flock when they were mating: "tomóse Jacob varas de álamo verdes y de almendro y de castaño y descortezó en ellas unas mondaduras blancas, descubriendo la blancura de las varas" (Genesis 30:37) [Jacob took him rods of green poplar, and of the hazel and chestnut tree; and pilled white streaks in them, and made the white appear which was in the rods]. By placing the rods in the watering troughs, Jacob manages to make the ewes conceive only spotted and speckled sheep: "y calentábanse las ovejas delante de las varas y parían las ovejas cinchados, pintados y manchados" (Genesis 30:39) [And the flocks conceived before the rods, and the ewes delivered striped, speckled, and spotted lambs]. By manipulating the breeding process he finally takes possession of Laban's entire herd, in which there are hardly any white sheep left.

The motif of Laban's livestock was used as a metaphor for miscegenation in early modern Europe, and was commonly associated with the belief that the imagination can leave an impression on women during conception, as the biblical passage was read in terms of phenotypical difference.[29] In Francisco del Vallés's *De Sacra Philosophia* (1587) [*On Sacred Philosophy*], the episode of the transformation of Laban's livestock is used to exemplify how imagination can exert its influence over biological nature, making it clear that what happens with animals can happen as well to humans (125–33). Following a similar line of thought, Cosme Gómez Tejada's *El filósofo* (1650) [*The Philosopher*] puts the two questions together: "Por que de padres negros como Etiopes, suelen nacer hijos blancos, hijos negros? Y por que mirando las ouejas, y carneros varas descortezadas al tiempo de la generacion (industria de Iacob) nacian los hijos manchados, siendo los padres negros, o blancos" (116r) [Why is it that from parents who are black like Ethiopians, there are often born white and black sons? And why, when the ewes and the rams looked at the peeled rods of the poplar tree at the moment of conception (Jacob's ruse), their offspring was born spotted, whether the parents were black or white]. Both Vallés and Gómez Tejada seem to be in agreement with their European counterparts in reading this biblical episode to explain perceived anomalies in reproduction, and how imagination and sight can influence generation.[30]

All the renditions of the biblical story of Jacob and Laban in seventeenth-century texts and paintings show an underlying anxiety about how to represent Jacob's livestock.[31] The ambiguity of *raça /mancha* in early modern Spain, which could refer either to the physical stain in the wool or to human ancestry, may serve to shed some light on the striking uneasiness with which early modern writers and painters deal with the story of Jacob and Laban. In Luis Vélez de Guevara's biblical play *La hermosura de Raquel* (published in 1615) [*The Beauty of Rachel*], Jacob tells his brother Esau about his life in Laban's land, and that he agreed to remain in his service with one condition:

Que fue, Esau, deste modo,
que de sus blancas manadas
fuessen mios los corderos
que le naciessen con manchas.
Lauan, acepto el concierto,
y de una industria estraña
use, para acrecentar
la hazienda que faltaua.
Tomando de los almendros,
y de otros arboles, varas
descorteçadas a trechos,
y a las presas de las aguas,
adonde yuan a beuer
los ganados enramana
con las varas sus orillas,
en cuya vista las mansas
Ouejas, luego se hazian,
como acontece, preñadas,
porque quando a beuer llegan
conciben de mejor gana.
Y como estauan mirando
las varas llenas de manchas,
parian de aquella vez
todas las crias manchadas.
Quitaualas otras vezes,
por dar a Lauan ganancia,
y desta suerte, parece
que ... [sic] ¿duermes? O viua tabla
de la muerte, el se ha dormido. (160r)

[It happened, Esau, in this way: that from his white herds, the spotted lambs would be mine. Laban accepted the deal, and I used a strange ruse in order to increase the wealth that I lacked. Taking rods from the almond tree and others,

> I peeled them in several parts and put them around the watering troughs, where the livestock went together to drink. In sight of them, the docile ewes became pregnant, because they conceive better when they drink at will. And as they were looking at the spotted rods, they delivered from that time only spotted lambs. Sometimes I removed the rods, so Laban could gain something, and this way ... Are you sleeping? Oh, vivid image of death, he has fallen asleep.]

Not only is the story of how Jacob managed to take possession of Laban's livestock irrelevant to the main plot of the play, but it is told only to remain unfinished because his brother Esau falls asleep, which is a rather whimsical plot trick to prevent Jacob from making explicit the deal's outcome.

At first sight, it seems to be a case of authorial self-censure: Vélez de Guevara may have felt that there was something scandalous in the story for an early modern Spanish audience and he withdrew it before having to confront its implications. If this is so, then we should wonder why he mentions such an irrelevant anecdote in the first place. Even if Vélez de Guevara refrains from offering a moral conclusion, the play raises the issue that Jacob, who incarnates the prefiguration of the Christian Church, proudly possessed livestock that was entirely comprised of spotted sheep. In the linguistic economies of sheep herding and racialism, in which *mancha* and *raça* are synonyms used indiscriminately, the composition of Jacob's herd contradicts the rhetorical efforts to present ethnic cleansing as a necessary sacrifice to preserve the purity of the Catholic flock.

Yet, even though Jacob's sheep were "impure" by sixteenth-century standards and therefore the story does not seem adequate to convey the ideology of blood purity, it was continuously appropriated throughout the seventeenth century, conveniently cleansed of its irksome elements. Vélez de Guevara's account of the biblical story of Jacob and Laban, though unfinished, is still respectful of its source; by contrast, Cristóbal de Monroy y Silva's play *El pastor más perseguido* [*The Most Persecuted Shepherd*] offers a rendering of Jacob's ruse (told in this case in Leah's words) that opts for manipulating the biblical story:

> Concerto con nuestro padre,
> que el ganado que naciesse
> manchado, havia de ser suyo,
> y fue la cria de suerte,
> que los corderos vestidos
> de evano sobre la nieve

de diferentes colores,
fueron vivos ramilletes.
Mudó el concierto Laban,
y el tuvo la misma suerte,
pues de una color nacieron. (20)

[He agreed with our father that the lambs born spotted would be his (of Jacob). And the conception happened in such a way that the lambs dressed of ebony over snow were of several different colours. When Laban changed the agreement, he had the same fate, since the lambs were all born of one colour.]

What is significant in this rendering of the Bible is that, although the play acknowledges that Jacob managed to make Laban's sheep conceive only spotted lambs at first, the text nonetheless suggests that at some point the deal was reversed by Laban and that Jacob then made the sheep conceive lambs of one colour; although this does not necessarily entail that the sheep were white, it dispels the threat of the spotted sheep.

In order to dispel the threatening composition of Jacob's livestock, Monroy y Silva overemphasizes a potential reversal of the deal that in scripture is only suggested in passing: "Si él decía así: Los pintados serán tu salario; entonces todas las ovejas parían pintados; y si decía así: Los cinchados serán tu salario, entonces todas las ovejas parían cinchados" (Genesis 31:8) [If he said: "The spotted lambs will be your salary," then the ewes conceived only spotted; and if he said "the striped will be your salary," then the ewes conceived only striped]. Furthermore, the playwright manipulates the alternatives raised in the biblical verse, which is between "striped" [cinchados] or "spotted" [pintados], and not between "spotted" [manchados] and "all of one colour" [de una color] as Monroy y Silva puts it. He also omits the fact that the final composition of the herd will be the one prophesied by the angel in Jacob's dream: "Alza ahora tus ojos y verás todos los machos que suben sobre las ovejas cinchados, pintados y petriscados, porque yo he visto todo lo que Labán te ha hecho" (31:12) [Raise your eyes and you will see that all the rams mounting the ewes are striped, spotted, and mixed, because I have seen what Laban has done to you]. As in the case of Vélez de Guevara's play *La hermosura de Raquel*, in Monroy y Silva's *El pastor más perseguido* the plot does not require telling the story of Jacob's livestock, which seems to be a mere digression. Monroy y Silva is apparently raising an unpleasant issue, which he could easily have avoided, with the sole purpose of offering his own take. Rather than merely a bad translation or misinterpretation, *El pastor*

Figure 2 Pedro Orrente. *Jacob Watering Laban's Sheep before Peeled Branches* (c. 1612–22). Courtesy North Carolina Museum of Art, Raleigh. Purchased with funds from the State of North Carolina.

más perseguido reflects a calculated selection and manipulation of the biblical source in order to make a statement about the interpretation of a biblical passage that was problematic and at the same time necessary.

If these written documents and literary texts had their own rhetorical ways of circumventing the thorny question of Jacob's livestock, contemporary pictorial renderings of the same biblical passage resorted in most cases to blatant denial of scriptural textuality. The motif of Jacob's life was quite successful in early modern European painting. However, as Mindy Nancarrow Taggard remarks, in contrast to their European counterparts, seventeenth-century Spanish artists show an unusually strong interest in representing the episode of the peeling of the rods and in emphasizing the

Figure 3 Bartolomé Esteban Murillo. *Jacob Laying Peeled Rods before the Flocks of Laban* (c. 1665). Oil on canvas. Courtesy Meadows Museum, SMU, Dallas. Algur H. Meadows Collection, MM.67.27.

role of Jacob's livestock (88). Pedro Orrente (1580–1645) was probably the initiator of this pictorial trend. He painted several canvases on the life of Jacob that became extremely popular throughout the seventeenth and eighteenth centuries, being widely copied and imitated throughout Spain.[32] In *Jacob Watering Laban's Sheep* (1612–14) (figure 2), which depicts Jacob in the act of peeling the rods before Laban's flocks, Orrente inaugurated a recurrent pictorial theme of seventeenth-century Spanish painting. Later in the mid-seventeenth century, Bartolomé Esteban Murillo (1617–82) was inspired by Pedro Orrente for his own canvas series on the life of Jacob, which includes *Jacob Laying Peeled Rods before the Flocks of Laban* (1660–5) (figure 3).[33] As Orrente and his followers did, Murillo underlined the importance of Jacob's ruse in manipulating the phenotype of Laban's herd. According to Taggard, the presence of two copulating sheep in the painting is the proof that Murillo was following the biblical passage closely (74 and 87). By including this "earthy detail," as Taggard labels it,

Murillo underlines the reproductive message of the tale. In this way, both paintings become emblems of how human manipulation can alter natural animal reproduction, sanctioned, if not promoted by the divinity. Taking into account how flocks of sheep are an immediate metaphor for the Christian community, these images also imply that monarchs can manipulate their subjects' genealogy and composition at will. These paintings are in themselves a eulogy of a new form of practice of power, and of the set of relations and obligations that monarchs exercise over the populations they rule.

At the same time that both Orrente and Murillo emphasize the control of reproduction and the importance of sheep in their compositions, they also share the same denial. While their paintings take an obvious jouissance in including livestock as an allegory of the Christian flock by representing the sheep as all white, they fail to address the episode's most important trait, which is that Jacob will end up owning only spotted and speckled sheep. The depiction of white, "pure" sheep might certainly be justified in the representations of the peeling of the rods, in which Jacob is still trying to manipulate the sheep's phenotype for his own benefit. The problem lies, though, in that the visual representations of biblical scenes that take place *after* Jacob's ruse are still careful to depict perfectly white sheep. By painting white sheep in a later scene in which Laban pursues Jacob in the return to his land, Orrente actually contradicts the scriptural story of Jacob's appropriation of his father-in-law's livestock (figures 4 and 5). In the representation of the same scene, Murillo suddenly relegates the livestock to the background (figures 6 and 7), in a distant fog that complicates the visualization of the sheep and which might indicate a desire to avoid addressing this polemical issue.[34]

The thorny outcome of the biblical story is counterproductive in a society in which the language of genealogical purity ensures the reproduction of only white offspring from Merino sheep, repressing the genetic instability of colour. Therefore, even when the biblical scriptures require the depiction of spotted sheep ("ovejas con raça"), early modern artists avoid doing so by any means. This is probably because, if Jacob is to be read as the prefiguration of the Good Shepherd, and his livestock as the prefiguration of the Christian flock, the story mandates an ethnic cleansing opposite to the one promoted in early modern Spain. By manipulating the representation of sheep, these paintings avoid an inherently dangerous interpretation: that Philip III (and later Philip IV) should be identified not with Jacob, but rather with Laban, the greedy livestock owner who wants to get rid of the speckled sheep and keep the white ones. Since the preference for white sheep eventually brings Laban's ruin, the literalness of the story's exemplarity would turn on its head customary praises of the monarch, such as when Lope de Vega says about Philip III that "el mayoral Helipe / su blanco ganado guarda" (qtd. in Perceval, "Lope de Vega" 74) [the shepherd Philip guards his white flock].

Although when conveniently manipulated biblical imagery serves to legitimate the biopolitical discourse, by the same token it may also be the site of contestation. While Murillo followed Orrente in his own composition of the subject and in denying the colour of the sheep, Jusepe Ribera (1591–1652) stands alone in his own depiction of Jacob's livestock (figures 8 and 9). According to Taggard "Ribera's two paintings describe a placid Jacob patiently awaiting the fruits of his labors" (88). However, the presence of clearly spotted sheep reveals a different temporality, for those two scenes necessarily occur *after* Jacob has performed his ruse and has transformed Laban's livestock into a spotted herd. Ribera also stands apart in the closeness with which Jacob embraces his new livestock, especially the newborn lamb with the indelible *raça* marking its head, in sharp contrast to the detached shepherds of Orrente and Murillo, who manipulate nature without even looking at their herd.

Thus Ribera is contesting previous treatments of the subject with his depiction of spotted lamb and sheep – although his reasons are uncertain. It might be the case that his long stay in Naples prevented him from being aware of the connotations that "spotted sheep" had in Spain, although this is an implausible explanation, since Naples belonged to the Spanish empire and his works were commissioned and exhibited in Spain and the court.[35] It could be argued that these paintings, made in the 1630s, served to support the reform of the statutes of blood purity promoted during the government of Count-Duke of Olivares (1621–43), who, according to his opponents, even tinkered with the idea of allowing the return of Sephardic Jews.[36] Regardless of Ribera's agency (or that of his patrons), it is evident that, within the context of early modern Spanish imagery of sheep herding, his radically "scriptural" spotted sheep reveal what is missing in the other painters' depictions of Jacob's livestock.

As these examples demonstrate, the tradition of sheep herding is the privileged symbolic site from which to launch an apprehensible and enduring image of the Old Christian community as a community of pure white sheep. Sheep herding thus serves as a rhetorical and iconographic commonplace to convey notions of "purification" and ethnocide. The trope transforms the traditional "pastoral power" into a new concept of biopower, that is, in Foucault's terms, into the regimentation of the life and

Figure 4 Pedro Orrente. *Laban Catches up with Jacob* (c. 1620–5). Museo del Prado, Madrid. Courtesy Art Resource, NY.

Figure 5 Pedro Orrente. *Laban Catches up with Jacob* (c. 1620–5) (detail). Museo del Prado, Madrid. Courtesy Art Resource, NY.

death of entire populations. The economic message does not override the ethnocentric subtext, but rather reinforces it. From the moment that pastoral representation draws on a system that emphasizes breeding as a metaphor for both economic prosperity and ethnic cleansing, the two concepts are tightly interrelated, as the foregoing examples show. The wool trade was an integral part of the world trade system in which the Spanish empire was actively engaged. By placing ethnic politics under the rhetorical umbrella of sheep herding, ethno-religious diversity can be conceptualized in terms of statistics, profits, and risks, attributing to the monarchy the role of managing such accounting. Thus the discourse of economics takes over traditional ethnocentric policies of exclusion, transforming the monarchy into the site of a new concept of biopower.

Though they clearly overlap, there are certainly significant differences between the concept of biopower proposed by Foucault for the nineteenth century and that incarnated by Philip III. One primary difference is that early modern Spanish biopower was still highly dependent on traditional political iconography, which it modified but felt forced to embrace. Be it

Figure 6 Bartolomé Esteban Murillo. *Laban Searching for His Stolen Household Gods* (c. 1665–70). Oil on canvas, 243 x 362 cm. © The Cleveland Museum of Art, Gift of the John Huntington Art and Polytechnic Trust 1965.469. Courtesy The Cleveland Museum of Art.

the political image of the Good Shepherd or the life of Jacob, its iconographic logic of power proves somewhat hesitant, still in search of a justifying theological precedent and therefore insecure about imposing its own logic on its own terms.[37] The defensive strategy of the emergence of biopower in early modern Spain is understandable because its logic has been borrowed from a proto-scientific knowledge of animal husbandry that, even though it had amply proven its practical efficacy in the international trade system through a body of oral knowledge among shepherds, had not produced an institutionalized body of knowledge that could serve as symbolic capital by itself. But both discourses complement each other, so that when biblical imagery is insufficient to sublimate ethnocide,

Figure 7 Bartolomé Esteban Murillo. *Laban Searching for His Stolen Household Gods* (c. 1665–70) (detail). Oil on canvas, 243 x 362 cm. © The Cleveland Museum of Art, Gift of the John Huntington Art and Polytechnic Trust 1965.469. Courtesy The Cleveland Museum of Art.

the quotidian ecocide that mandates the culling of every Merino lamb with *raça* reinforces the logic that expelling undesirable elements from the Christian herd is the common-sensical solution.

What is the relation between the discourse of blood purity, sheep herding, and pastoral literary representations? Critics seem to agree that sheep are ubiquitous in pastoral romances and poetry, that the very existence of the genre is constituted by their presence, and at the same time, sheep are blatantly ignored, mentioned only in passing as part of the setting. Julio Baena notes that the artificiality of early modern Spanish pastoral romances, whose ignorance of sheep herding contrasts with the ubiquity of real sheep herding and wool trade at every level of the Spanish economy, should be read as an imposture, as a pretence of not knowing what sheep herding is really about ("De Belisa" 79). At this point, we should add that there is a second imposture in pastoral literature in regard to its dependence on the presence of sheep: the pretence of not knowing that discourses about sheep and shepherds were inextricably linked to discourses of race and genealogy. It is precisely in this dual denial that the effectiveness of the naturalization of both pastoral biopower and pastoral literature resides. It is only in the twentieth century that Batlle Matabosch can complain in his fervent defence of the "Spanishness" of the Merino sheep: "¿sería tan poético el

Figure 8 Jusepe Ribera. *Jacob among the Sheep* (1632). Museo de El Escorial. Courtesy Art Resource, NY.

sueño si en vez de ovejas merinas, de tan fina calidad, las de la cabaña de Salicio hubieran sido de una sangre menos pura, como las de cualquier gañán o rabadán que corre por estos mundos?" (41) [Would the dream be so poetic if, instead of Merino sheep, of such fine quality, those in the herd of Salicio had been of a less pure stock, like those of any rustic or shepherd that wanders around?]. We can only wonder how early modern Spaniards would have answered this question – more important, we can assume that they would have never put the question in such explicit terms. It is only centuries later, when the discourse of sheep herding has been replaced by other discourses as the privileged trope to convey the image of race and human difference, that its latent logic can be exposed.

For Batlle Matabosch, the courtly shepherd and the rustic one are differentiated by the quality of the sheep they guard. Early modern Spaniards

Figure 9 Jusepe Ribera. *Jacob with the Herd of Laban* (c. 1638). Courtesy Museo Cerralbo, Madrid.

would have never put the difference in those terms, arguably because they knew that rustic shepherds guarded Merino sheep while courtly ones simply did not exist outside of literary fictions, and therefore, it would not be the quality, even less the purity of the sheep that distinguished them. But, in spite of their differences, both early modern and modern idealizations of sheep herding recur to this contiguity between the sheep and the shepherd, although they articulate it in different ways and for different symbolic purposes. As effective as metaphors related to sheep were at conveying the need to mark, separate, and even dispose of entire populations, they were of limited use when ethnocentrism focused on the imaginary of cultural identity. For this, a constructed figure of the shepherd (no less constructed than the imaginary of white sheep) provides an essentialized model of collective identity, flexible enough to account for the difference between courtly "shepherds" and rustic ones, but sufficiently rigid so that what they share in common differentiates them unmistakably from other people marked as non-shepherds, as the next chapter shows.

2 Rustic Culture and the Invention of the Spanish People

porque los poetas aún no habían advertido
que entre los moros hay pastores.

[because poets have not yet realized that among
Moors there are shepherds as well.]

Mateo Luján de Sayavedra, *Segunda parte de la vida del pícaro Guzmán de Alfarache* (1602)

As the previous chapter has shown, the traditional trope that conceived of a people as a community of sheep was used in early modern Spain to discuss notions of racial purity and social engineering. In these metaphors, the shepherd is usually an isolated figure that stands for the monarch, described only in relation to the herd that he guards. There is, however, another array of pastoral symbols that focus on shepherds as forming a community. When dealing with either sheep or shepherds, these are taken as rhetorical constructs that serve as symbols for a collective identity. Rephrasing Benedict Anderson's well-known definition of the nation as an imagined community (15), we can say that early modern Spaniards conceived of Spain either as an imagined community of sheep or, alternatively, as an imagined community of shepherds, and the present chapter explores the ideological significance of the latter. There are basic differences between these two concepts: while conceiving of populations as sheep allows for expressing genealogical and reproductive issues, the community of the shepherd is geared towards representing cultural difference and ethnicity.

As racial thought bases its effectiveness precisely in its conflation of nature and culture, moving easily between these two levels of human expression, so pastoral imaginary naturalizes and essentializes cultural difference at the same time that it strives to provide visual, perceptible evidence for abstract and invisible conceptions of genealogical difference. Therefore, it is not possible to establish a neat differentiation between the community of the shepherd and the community of the sheep, thanks to the cultural premise that sheep and shepherds are tightly related to each other by the mandates of "purity." Each imagined community is supposed to safeguard that evasive concept of "purity" in different realms. If Spain was a country full of sheep, it follows logically that it was full of shepherds as well, but shepherds themselves (just like sheep) have little or nothing to do with how others appropriate them as an image for their society. The purpose of this chapter is to analyse how in the sixteenth and seventeenth centuries the reality of being a country of shepherds is elaborated, refigured, and appropriated to construe a new concept of cultural difference and collective identity. Theatrical, literary, and pictorial representations of rustic shepherds usually present themselves as being based on genuine folkloric and pseudo-ethnographic observations – and they are often interpreted as such. However, I argue that these representations of rustic people are not innocent, but rather imbued with ideological motivations that aim at inventing a perfected image of the Spanish people in opposition to Moriscos and *conversos*.

By tracing the ever-changing imaginary boundaries between the shepherd and the Moor, it is possible to reveal how they are historically and culturally situated within the evolution of Spanish racial discourses. As Étienne Balibar (56) points out, the construction of a model for national identity and for a "true national" is achieved through a complex system of exclusions, denegations, and projections. In the case of early modern Spain, the belief in an idealized and homogenous community of Old Christians relied on parallel stereotypes of idealized and homogeneous communities of Jews and Muslims (and their descendants). In this way, basing identity on the denial of a previously constructed Otherness transforms the search for the "national race" into an unattainable objective, since, as Balibar states for a different case, "the fact that the 'false' is too visible will never guarantee that the 'true' is visible enough" (60). By projecting itself into the figure of the shepherd as the not-Morisco and not-*converso*, Old Christian identity manages to conceal its purely oppositional, negative nature, while it remains in need of a specific, positive content that guarantees its visibility as a reproducible model of collective identity.

If the symbolic opposition between Moors and shepherds is so dominant in the Spanish imaginary, if "shepherd" becomes a synonym for "Christian" and therefore the opposite of "Moor" (and also of "Jew"), it is precisely because it remains what Pierre Bourdieu defines as a *habitus*:

> systems of durable, transposable dispositions, structured structures predisposed to function as structuring structures, that is, as principles which generate and organize practices and representations that can be objectively adapted to their outcomes without presupposing a conscious aiming at ends or an express mastery of the operations necessary in order to attain them. (53)

Following Bourdieu, what we can call "pastoral habitus" is constituted by a repertoire of assumptions, visible in cultural artefacts, that, without being made explicit by those who practise them, link sheep herding with Christian (or even Old Christian) identity.

Precisely because the effectiviness of these assumptions lay in remaining unarticulated, it is very rare to find self-reflective moments in which the imaginary opposition is exposed. Probably such a moment is found in Mateo Luján de Sayavedra's *Segunda Parte de la vida del pícaro Guzmán de Alfarache* (1602), situated at the apex of pastoral's popularity. In this continuation of the popular first part of Mateo Alemán's *Guzmán de Alfarache* (1599), the rogue Guzmán meets an eccentric playwright who is preparing "una jornada pastoril a la morisca de allá de África, que es una maravilla; porque los poetas aún no habían advertido que entre los moros hay pastores" (422) [a pastoral play in the Moorish fashion from Africa, which is a rarity, because poets have not yet realized that among Moors there are shepherds as well]. Although the remark that North African Muslims can also be shepherds may seem blatantly obvious to us, it is nonetheless perceived by the other characters as indecorous, since in early modern Spanish imaginary Muslims and shepherds were mutually exclusive identities.[1]

This conceptual opposition is of course insufficient by itself, and needs to be materialized through the codification of visible signs. Therefore, the issue of clothing immediately arises as the main objection for blending "Moors" and "shepherds" on stage when Guzmán and his companions question the feasibility of such a literary experiment: "¿Cómo se habrán de vestir esos pastores, que los pellicos que usamos en España no les podrán convenir?" (422) [What will those shepherds wear, for the *pellicos* (shearling jackets) we wear in Spain won't suit them?]. The objection reveals the underlying assumption that, since pastoral garments are central to the exhibition of a Spanish "Christian" identity, they are therefore inappropriate for representing North African Muslims. Struck by the objection, the playwright hesitates for a while and then answers that "bien nos podríamos informar en Valencia, de muchos que han estado cautivos en Argel, qué vestido usan por allá la gente serrana y pastoril" (422) [We could gather information in Valencia from those who have been captives in Algiers, about what kind of dress mountain and pastoral people wear over there]. Thus, while the playwright initially proposed a play in which the use of pastoral clothing suggested sameness, he is finally convinced to reinstate difference by inventing pastoral clothing that unmistakably marks Muslim shepherds as Muslims. As we will see by the end of this chapter, the *pellicos* [shearling jackets] are not necessarily a specific garment, but constitute to a great extent an empty signifier without exact referential content and whose main meaning seems to be that of Christian clothing as opposed to an equally idealized and essentialized Moorish dress.

This passage in Luján de Sayavedra's novel perfectly illustrates the inherent ideological opposition between the Moorish and the pastoral that this book aims to describe. Even if the episode ends with the playwright scorned by Guzmán and his fellows, portrayed as eccentric, this marginal character serves to articulate and therefore to expose the ideological dichotomy whose effectiveness lay in remaining implicit. This denaturalization of the "pastoral habitus" reveals the ideological tenets underlying the figure of the shepherd, an opposition that was constructed and manifested in early modern Spanish society, and that we will now explore.

The Adoration of the Shepherds

One of the biblical subtexts that grounds the importance of the shepherd in early modern Spain is the Adoration of the Shepherds, an instantiation of which portrays Philip III and his court, as we have seen in chapter 1 (figure 1). It reflects a passage taken from Luke 2:8–20 in which an angel announces Christ's birth to the shepherds, who are then the first to worship the newborn. The basic elements of the motif are already contained in Luke's Gospel, but the episode was amplified in patristic literature and Renaissance theological and catechistical texts (Valiñas López 129–48). The Adoration was traditionally used to convey the message that Christian evangelization was universal and that it reached all social classes, starting with the most humble. Yet, as I have already shown in the cases of the figure of the Good Shepherd and the story of Jacob, scriptural passages related to sheep herding tend to be interpreted in an ethnocentric way in early modern Spain.

The growing relevance of rustic figures in Spanish society in relation to the ideology of blood purity is revealed in the evolution of pictorial representations of the Nativity spanning from the fifteenth through the seventeenth centuries.[2] As Francisco Javier Sánchez Cantón (47–67) shows, when the tradition of painting the Adoration of the Shepherds is borrowed from northern Europe, early in the fifteenth century and through the first half of the sixteenth century, the Annunciation to the Shepherds appears only in the background, while the Nativity takes the central role, as exemplified in a painting by Juan de Borgoña from the first half of the sixteenth century (figure 10). From the mid-sixteenth century on, however, the shepherds begin to take prominence in the composition, sharing the space with the Holy Family, as evinced in the Adoration by Luis de Morales (figure 11). The Adoration gathers momentum when it is promoted by the Counter-Reformation as the ideal imagery for fostering the catechization in the Catholic faith of the most humble socio-economic classes.[3] By the end of the sixteenth century and during the whole seventeenth century, shepherds move to the foreground, actually invading the space of the painting and, in some instances, even displacing the Holy Family, as clearly shown in Bartolomé Murillo's Adoration (figure 12). While this iconographic evolution does reflect pan-European artistic trends in the painting of the Nativity,[4] in Spain, these three stages, which clearly tend towards granting a higher prominence to the shepherds, parallel the dynamic of the debate over blood purity along with their symbolic identification with Old Christians. Art historians face the indeterminacy of how to interpret the representation of these shepherds vis-à-vis their social context: while Julián Gállego suggests that the rising of the pictorial Adoration of the Shepherds may be linked to the doctrine of blood purity (257), Richard Mann argues that El Greco's *Adoration* was intended to counteract the statutes of blood purity in Toledo (34). Lacking contextual information about patrons and artists, as well as more grounded and detailed knowledge of iconographic conventions, all we can do is speculate about the relation that paintings of the Adoration establish with social concerns about blood purity.

This connection seems more obvious in the similar evolution of rustic characters in the first theatrical manifestations, which also derive from the Adoration. As compared with other parts of Europe, Castile lacked a theatrical tradition during the Middle Ages, and the few surviving texts seem to derive from the *officium pastorum*, representations of the Nativity that took place in a liturgical context (McKendrick 6–10). The emergence of Castilian drama can be located at a very specific moment, immediately after 1492, with the fall of Granada, the expulsion of the Jews, and the discovery of the New World. It is in 1492 that Juan del Encina, the founder of Castilian theatre, began staging his plays in the palace of Alba de Tormes for the court of Fadrique Álvarez de Toledo, second Duke of Alba; he was soon followed by playwrights such as Lucas Fernández, Diego Sánchez de Badajoz, Gil Vicente, and Bartolomé Torres Naharro.

Figure 10 Juan de Borgoña. *Nativity* (first half of 16th century). Palacio Arzobispal de Madrid. Album. Courtesy Art Resource, NY.

Figure 11 Luis de Morales. *Adoration of the Shepherds* (c. 1560–3). Iglesia de Nuestra Señora de la Asunción, Arroyo de la Luz, Cáceres. Album. Courtesy Art Resource, NY.

Figure 12 Bartolomé Esteban Murillo. *The Adoration of the Shepherds* (c. 1645–50). Museo del Prado, Madrid. Alfredo Dagli Orti. Courtesy Art Resource, NY.

In these early dramas shepherds and other rustic characters took prominence on stage and, in spite of the tendency to deal with profane motifs, they were still related to the cycle of the Nativity. This rustic dramatic character is very often an ignorant commoner who speaks a debased dialect, dresses poorly, devours pork and wine, and boasts of possessing blood purity that makes him superior to *conversos*. Such dramatic shepherds exhibit their rustic genealogies, reciting long lists of relatives and ancestors with ridiculous names and including their places of residence and professions.[5]

Yet, as happens with painting, there is no agreement on the motivation that lies behind the representation of rustics. Since Américo Castro's suggestion that most of these playwrights might be of *converso* origin ("Lo hispánico" 57–63), many scholars have followed the same path of inquiry.[6] Among these scholars, it is largely understood that the representation of shepherds and rustic characters on stage is an attack directed towards the mob of Old Christians who boasted of their humble origins and "purity of blood" to promote their own social advancement. More nuanced interpretation, however, suggests that anti-Semitic passages uttered by rustic characters, such as those found in Lucas Fernández, may not represent the author's scorn of the character, but a strategy of self-defence that casts rustics in a positive light (Gitlitz, "Lucas Fernández" 510; Yarbro-Bejarano 16–17). This indeterminacy towards blood purity in any given play and the ambiguity of the figure of the shepherd remain, and may well have been experienced by early modern audiences. Thus Ronald Surtz suggests that the 1556 pastoral performance celebrating the appointment of Archbishop Martínez Silíceo as cardinal, and in which he was represented as a shepherd, may have been interpreted in very different ways by the audience: while those supporting the blood purity statutes would have identified with the character of the shepherd, those opposing them would have seen the confirmation that their enemies were just a rustic mob ("Cardinal" 230–2). The important point to be made here is that, regardless of the original purpose of each play, either to mock the pretensions of social mobility of Old Christians or to locate them as the "true national," both positions agreed in identifying them with the figure of the shepherd. This hermeneutical uncertainty seems to have been decided by the end of the sixteenth century in favour of Old Christians. As Francisco Márquez Villanueva points out, while the theatrical drama of the sixteenth century tends to cast shepherds and rustic characters in a negative light, they are dignified by the end of Philip II's reign, in parallel with the general imposition of the statutes of blood purity (*Fuentes* 69–83).

This evolution of rustic characters in theatre parallels the increasing frequency with which patrons commissioned portraits as shepherds, most notably in the seventeenth century, which also confirms the consolidation of the shepherd as a positive figure. Certainly, it is not always possible to identify who commissioned a painting and why, and therefore the identification of the shepherd with the patron is often largely a matter of speculation. But in some instances the patron's choice to be depicted as a shepherd is documented: the Aragonese bishop Juan de Palafox explicitly commissioned his portrait as a shepherd in Pedro García Ferrer's work in the cathedral of Puebla de los Ángeles (Mexico), between 1646 and 1649 (Gali Boadella 473–4).[7]

The identification of patrons with biblical figures in "retratos a lo divino" [divine-style portraits] is a broader phenomenon that goes beyond paintings of the Adoration of the Shepherds (Orozco Díaz 29–35), but arguably the selection of this motif had its own specific implications. Because of the connotations that the figure of the shepherd had attained, patrons achieved a twofold objective by being projected into the Adoration: on the one hand, they called attention to their blood purity, while on the other they presented a humble image of themselves that reinforced an imaginary bond with their community. This identification of the ruling classes with the people, represented in the figure of the shepherd, reaches almost every level, as we have seen in the 1603 *Adoration* of Pantoja de la Cruz (figure 1), in which Philip III and other members of his court appear as shepherds. Presenting Philip III as a lowly shepherd creates the illusion that there is a cultural bond between him and his vassals, and that therefore his power is legitimized by them.[8]

The identification is also attested in theatrical performances, like the aforementioned 1556 play in which a rustic character represents the archbishop of Toledo Martínez de Siliceo in his presence, celebrating his promotion to cardinal, which seems to indicate that he approved if not requested such a link. And the same goes for the pastoral romances, which have been usually read as roman à clef in which the fictional shepherds stand for actual courtiers, poets, and patrons (Finello, *Evolution* 51–9). This search for the real name behind shepherds and shepherdesses was customary with early modern audiences: in 1603, when returning from a trip to Old Castile, Philip III and Margarita of Austria stayed in Valencia de León, where the Marquis of Las Navas announced to them that their host was the person behind Diana in Montemayor's *La Diana* (Menéndez Pelayo 249).

In spite of its increasing association with Old Christians, the scriptural source does not allow full resolution of the indeterminacy of the figure of the shepherd. The dignification of commoners often passed through their association with the shepherds of the Adoration, drafting an imaginary lineage linking them with the first people who came to worship Christ. Thus Gaspar Gutiérrez de los Ríos, in his *Noticia general de la estimación de las artes* (1600) [*General Report on the Reputation of Sciences*], refers in his defence of peasants to the argument of a religious piety that is supposed to be immemorial: "Quien hay que sea mas perseuerante en la religion que ellos? Los primeros que conocieron a Dios fueron pastores" (236) [Who is more persistent in their religiosity than they? The first to know Christ were shepherds]. However, any collective identification with the characters of the Adoration is troubled by the ethnicity of those primeval shepherds, since they were actually Jews.[9] The symbolism of the Adoration of the Shepherds as a metaphor for the Old Christian community is based on the denial of this obvious fact, which seems to be erased in most of the theological and catechetical texts that broach the issue.[10] This erasure, though, exists in counterpoint with other texts that early on exploit the Jewishness of the first shepherds to defend the integration of *conversos*, as we will see in the case of Lope de Vega's *Pastores de Belén* (chapter 5).

So much contextual information is missing about patrons, intended destinations, and audiences that in many cases it is almost impossible to determine whether the figure of the shepherd is used to defend the statutes of blood purity or to criticize them, whether the imagined community of shepherds evoked in these images is to be understood as an inclusive or exclusive collective identity. Sheep herding becomes again and again the trope in which blood purity can be promoted but also denied, and therefore the shared symbolic space in which competing versions of the Spanish people and the Christian community collide. What seems clear, though, is that the cultivation of the Adoration of the Shepherds contributed to the creation of a specific concept of the cultural and religious identity of the Spanish people.

Apparitions to Shepherds

To these semi-liturgical performances and iconographic cycles that derived in one way or another from the motif of the Adoration we should add the tales of Marian apparitions that prompted the erection of shrines scattered throughout Spain and the resulting celebration of *romerías* [devotional processions], in which shepherds often had a prominent role. Many, if not most, of these stories follow the same pattern: an image of the Virgin, buried during the Islamic period to protect it from the infidel, appears after the territory has been conquered by Christians. In Castile, most, if not all, the apparitions happened either to shepherds or to people who worked in professions related to wool trading, very often in places where sheep herding had an important economic role (Christian 10). Such was the case of the Virgin of Guadalupe, in the version found in *Libro de la invención de esta Santa Imagen de Guadalupe*, written by Fray Diego de Écija by the early sixteenth century. In this tale, a shepherd allegedly discovered in 1337 an image of the Virgin in Guadalupe (Cáceres) that was buried during the Islamic period, and for this he was rewarded by Alfonso XI with aristocratic status (49). A similar case is that of Nuestra Señora de la Cabeza (Our Lady of Cabeza) in Andújar. Of the many such apparitions, I mention these two – the Virgin of Guadalupe and the Virgin of Cabeza – because they were the most successful as devotional images, spreading far beyond their local origins, most notably in the New World, and their popularity was reflected in many seventeenth-century literary texts.[11]

While these legends of apparitions to shepherds were a pan-European phenomenon, Spain's particularity is that these stories were almost always related to the ideology of the *reconquest* and the role that sheep herding had in the frontier socio-economic system. Indeed, in spite of its symbolic instrumentalism, there is some reality supporting the medieval encounters between Muslim raiders and shepherds, which might have contributed to the creation of an ideological divide between them. In order to repopulate frontier lands, livestock owners were granted privileges and titles of nobility in exchange for taking the obligation to defend their territories from Muslim raids. They were called *caballeros villanos* [commoner knights or nobles], an apparent oxymoron that underscores that the combination of sheep herding and the fight against the Muslim was one of the instances in which the expansion of Christian kingdoms to the south allowed for upward social mobility.

This newly promoted class became the origin of an influential aristocracy of livestock owners; they soon found that the most profitable economic activity was the raising of livestock, a moving commodity that could be safeguarded in case of attack.[12] Even though raids occurred on both sides of the frontier and Christian "shepherds" were not particularly defenceless, they were usually portrayed in collective memory as the primary victims of Muslims. Such is, for example, the case of the medieval ballad known as "La merienda del moro" [The Snack of the Moor], which recounts a Muslim raid in the frontier territory of Jaén: "Por los campos de Jaén van los moros peleando; / tierra por donde ellos iban todo quedaba arrasado: / no dejan cabra ni oveja, ni pastor con su rebaño" (Correa 1:452–53) [Through the fields of Jaén the Moors go fighting; in the lands they go through everything is devastated; they leave neither goats nor sheep, nor shepherds with their herds].[13] These *caballeros villanos* had therefore very good reasons to present shepherds as the symbols for the Christianization of the new conquered territory, using the stories of divine apparitions as foundational myths. Indeed, as Peter Linehan notes, it is not by chance that they adopted the Virgin of Guadalupe as their patroness, since this legend encapsulates a convenient combination of herding, fight against the Muslim, and social mobility (301).

Figure 13 Bernardo Asturiano. *Pilgrimage to the Virgin of Cabeza* (17th century). Courtesy Museo de Nuestra Señora de la Cabeza. Andújar (Spain).

Even though many of these legends are supposed to have circulated in oral form as part of popular religion since the Middle Ages, they were compiled, expanded, and in many cases reinvented in local hagiographies written during the early modern period (Christian 7–15).[14] They usually borrowed the iconography of the Nativity, and their repetitive narrative shows that they imitated each other (Christian 34, 209–10). Some of these legends even produced their own pictorial cycles, influenced by the Adoration of the Shepherds. This is especially true in the case of Our Lady of Cabeza, whose devotion caused its own pictorial cycle in which the shepherd is always present (Domínguez Cubero). In one of these paintings, made in the late seventeenth century by Bernardo Asturiano (figure 13), we can see two different historical layers overlapping: to the left there is the Marian apparition to the shepherd that allegedly took place in the thirteenth century; the rest of the canvas depicts the pilgrimage to the shrine of Our Lady of Cabeza as it was held in the seventeenth century.[15]

In all these legends the ethno-religious confrontation is latent and is subject to similar denials: while in the Adoration of the Shepherds they figure as the first Christians who engage in the supersession of Judaism (as far as their Jewish origin is obviated), in these medieval apparitions the shepherds serve as the embodiment of a militant quasi-biological confrontation against Islam.[16] However, since the apparitions took place in territories newly conquered by Christians, there was always the possibility that these shepherds were *Mudéjares* (Muslims living under Christian rule), or even Muslims who converted and were integrated into the Christian fold. This is why early modern expansions of these legends tried to dispel the anxiety about the genealogical origins of these medieval shepherds. The first written account of the apparition of Our Lady of Cabeza, the 1633 manuscript by Juan de Ledesma, only says succinctly that, shortly after Fernando III conquered the city at the beginning of the thirteenth century, the image was found by "un pastor buscando una vaca" (qtd. in Rodríguez Becerra 67) [a herdsman looking for a cow]. While Ledesma is obviously replicating the story of the Virgin of Guadalupe, Manuel de Salcedo Olid's *Panegírico historial de Nuestra Señora de la Cabeza* (1677) expands the legend making this figure a shepherd (not a cowherd) and a former captive of Muslims to further emphasize his antagonism to them: "entre los Christianos que apacentavan ovejas en aquellos montes, avia un pastor natural de Colomera (que entonces era de Moros) hijo de cautivos Christianos de aquella villa" (208) [among the Christians that tended sheep in those mountains there was a shepherd born in Colomera (which belonged to Moors in that time), son of Christian captives from that town]. In this way, the ethnicity of the medieval shepherd was rewritten in the seventeenth century (or at least established beyond any doubt with little or no evidence) to emphasize his belonging to the Old Christian community and to reinforce local identities.[17]

A very telling exception to this identification of the shepherd of the apparitions as Old Christian is the case of Our Lady of Lluc (Majorca). As the established version holds, right after the conquest of the island by Jaume I in 1229, the image appeared to a shepherd named Lluc, who had recently converted from Islam to Christianity (Camps i Moranta and Seguí i Trobat 32–3). Yet, even though the cult to Our Lady of Lluc already existed in the Middle Ages, the exact moment in which the tradition of the legend was established is not clear. During an investigation held in 1642 to determine the miracles of this image, several individuals recounted their versions of the oral legend; while some of them mentioned the participation in the discovery of a young shepherd from Muslim origin, other simply stated that he was a servant or a slave, or do not even account for the shepherd.[18]

The first written source to unmistakably state that the shepherd was a convert from Islam is Rafel Busquets's *Llibre de la invenció y miracles de Nostre Señora de Lluc* (1684) who describes him as: "un miñó de poca edat, y de los que quedaren esclaus en la Conquista que de esta isla … después de haver rebut, ab les aygües del Baptisma, la Fee, y ab ella el nom de Lluch, fent offici de pastor, y apacentant per aquella campaña les ovelles" (qtd. in Camps i Moranta and Seguí i Trobat 65–6) [a boy of young age, one of those who remained as slaves after the conquest of this island … after having received faith and the name of Lluch with the waters of baptism, working as a shepherd, and tending that field with the sheep]. The context in which Busquets published his account of the apparition is very significant. The end of the seventeenth century was a period of intense persecution of the *chuetas* – as the *conversos* from Jewish origin were known in Majorca – reinvigorated in 1679 by a series of auto-da-fés against those accused of Judaizing (Porcel 48–51; Stallaert, *Ni una gota* 147–9). Busquets's selection of all the oral variants to present the foundational shepherd as a convert from Islam might be an intervention on the social debates of the island, either to defend *conversos* or, more probably (since the book was published by a printer associated with the Inquisition), to emphasize the need of true conversion to Christianity and justify the Inquisition's activity.

Thus early modern authors rewrote the medieval legends of Marian apparitions so that the ethnicity of the foundational shepherd could be significant for the community of believers, linking his ethnicity to both the iconography of the Nativity and the myths of the *reconquest.* These local devotions, disseminated throughout Spain and its colonial territories, contributed critically to spreading the popularity of the figure of the shepherd and its militant ethno-religious symbolism across the social spectrum, coexisting and often interacting with other pastoral manifestations.

Performing Rustic Culture

Along with the prevalence of rustic characters in pilgrimage, theatrical performances, and visual arts related to religious themes, the figure of the shepherd also plays a prominent role in a wide array of both secular and religious festivities all over Spain. Pastoral performances are recorded for the celebrations of the Dukes of Sessa in Madrid in 1541 (*Relaciones de los reinados* 2:158), in the festivities held in Bayona in 1565 to celebrate the presence of Isabel de Valois and Philip II (*Relaciones históricas* 103), as well as in the festivities of the Holy Sacrament held in Valladolid in 1583 (Fernández Martín 16) and in Seville in 1594 (Alenda y Mira 1:105). For the occasion of the wedding between Philip III and Marguerite of Austria in 1599, pastoral performances were organized when they visited Valencia (Gauna 2:499 and 2:559) and Zaragoza (Gauna 2:931–2). In the 1550s in Toledo, peasants' ballets seem to have been a common element in public celebrations (Salomon 377–8), precisely at the same time and place in which the archbishop, Martínez de Siliceo, was represented by the figure of a rustic shepherd in 1556.

In addition to their inclusion in theatre and public celebrations, pastoral performances were also staged in local courts. Dominick Finello, conflating literary representations with accounts of public festivals, concludes that "[e]vidence of pastoral disguise from novels and eclogues, coupled with that of the historical record, proves that the bucolic mask is more familiar to public celebrations than once was thought" (*Pastoral* 164).[19] The scattered documentation for pastoral festivals (or rather, for pastoral elements in public celebrations) is not impressive, at least when compared to the massive documentation that exists on other courtly festivals like the game of canes.[20] This is why Maxime Chevalier, finding that courtly festivals in which aristocrats dress up as shepherds are much less numerous than equestrian spectacles in which they dress up as Moors, argues that nobles found the latter a more suitable vehicle for expressing their class status (45). We may still wonder, along with Finello, whether pastoral performances were much more common than it has seemed, whether they were not recorded either because pastoral was precisely the most common element and the least spectacular in the festivities or because in most of the cases they were included in celebrations in which the nobility was not involved and therefore no poet or chronicler was there to record them.

What is the relevance of these public expressions of pastoral to the world of real shepherds and peasants? For Finello there is some sort of reality behind the shepherd disguise, so that "[t]hrough the poetic Arcadia the reality of a geography crossed by sheepwalks was visible, and the sheep traffic itself allowed writers to acquaint themselves with a reality in which the song of the authentic shepherd could be heard" (*Pastoral* 165–6). It is questionable, however, whether the aim of these urban and courtly performances would be to hear "the song of the authentic shepherd." Even if Spain was a land in which sheep herding was specially cultivated, as indeed it was, public displays of sheep-herding identity, rather than aiming at reconstructing the "authentic" pastoral culture, imposed a layer of fictionality upon reality. Even though each context raises particular class interests, both kinds of performances refer to a shared metaphor of sheep herding because both kinds of performers are in need of a common model of collective self-representation. Herein are echoed the implicit continuities between popular and courtly expressions of pastoral stated by Finello.

By questioning the divide between the rustic, "authentic" shepherd, and the courtly, "fictional" one, it becomes clear that they share more than the elements that separate them, since neither has anything to do with the culture of "real shepherds" – but both refer to such shepherds as the source of cultural legitimacy. To be sure, this divide is exploited by early modern Spaniards, and it is therefore easy to be misguided by their accounts, such as the one made of the festivities held on the occasion of the royal wedding between Philip II and Anne of Austria in 1570. Before reaching Segovia, the queen first passed through the village of Valverde, where people welcomed her "con danças y cantares, y con aquellos atauíos y adereços que la senzilla rusticidad tenía ... vino toda aquella buena gente con gran roydo de gaytas y tamborinos y panderos, instrumentos de la música aldeana" (qtd. in López Estrada, "La poesía pastoril" 312) [with dances and songs, and with those attires and decorations that the simple rustics had ... these good people came playing bagpipes, drums, and tamborines, all of them peasant instruments].

The author of the account makes sure to underline that the festivities of Valverde were performed by real rustics, and against this backdrop he contrasts another pastoral performance made during the entrance of the monarchs in Segovia. When both Anne of Austria and Philip II enter the cathedral of Segovia, they are welcomed by a quite different pastoral performance related to the *officium pastorum*:

> salieron nueue pastorcicos escogidos entre los moços de coro y de la yglesia, muy apuestos, con sayos de diferentes colores y sedas y sus melenas ruuias cogidas en cintas de oro y plata ... llegaron en presencia de Su Magestad ... El villancico era compuesto muy prima y graciosamente en canto de órgano y eran sus palabras en términos pastoriles mostrando regocijo por la venida de la reyna nuestra señora. (qtd. in López Estrada, "La poesía pastoril" 313)

> [nine little shepherds showed up, chosen among the church chorus boys, very gallant, with tunics of different colours and kinds of silk, and their blond hair tied up with gold and silver ribbons. They arrived in the presence of His Majesty, and their *villancico* (carol) was exquisitely composed in organ tune, and they spoke in pastoral terms, rejoicing the arrival of our queen.]

As López Estrada points out, the interest of this account lies in the juxtaposition of two pastoral performances of very different kinds. The first rustic performance in the village of Valverde contrasts vividly with the ecclesiastical, urban pastoral performance in Segovia, although both of them resort to the same iconographic motifs in a region that specializes in sheep herding and cloth making. When comparing both performances, López Estrada states that "[e]l marco de las fiestas celebradas por las bodas de los reyes permitía este acercamiento entre verdaderos aldeanos y fingidos pastores" ("La poesía pastoril" 317) [the context of the festivity celebrated to honour the monarchs would allow this rapprochement between the true villagers and the fake shepherds]. While I agree with López Estrada's assertion stressing that pastoral is a shared iconographic language whose symbolism crosses class lines, it is however highly problematic to conceive this difference by pitting the "true villagers" against the "fake shepherds," labelling the former as "manifestación del folklore auténtico" (López Estrada, "La poesía pastoril" 317) [an expression of authentic folklore].

Even if the actors involved in the first performance of Valverde were actual peasants rather than comfortable if not wealthy townspeople, it does not follow that there is anything "authentic" in their performance. As Luis Díaz Viana warns, although folklorism presents itself as a conservative act of preserving tradition, it is paradoxically an agent of change, since the observer "authenticates" what is truly traditional and distinctive in a popular performance, imposing upon participants a codification of and awareness of identity in cultural practices (14). As José María Perceval notes, regarding urban royal spectacles involving rustic figures, the "invention of the people" is imposed from above upon this very same people that it is supposed to represent, at the same time that power legitimates itself by identifying with this fabricated image of the popular ("Lope" 72n27). It is through this identification that, as William Childers states "this penetration of a centralized authority into all aspects of everyday life does not take the external form of a tyrannical imposition, but rather appears as a preference of the people themselves" (*Transnational Cervantes* 183). The importance of the figure of the shepherd serves thus to reinforce the illusion of "the people" as the legitimizing source of power. The currency of this fictitious political bond helps explain the apparent interest of the Spanish monarchy and the ruling classes in popular culture.[21] At the same time, the presence of the observer (both the author of the account and the queen in the case of Valverde) makes the performers aware of the value of their act when they rehearse the homogenizing form of cultural identity they are expected to inhabit, thus opening a space for their agency.[22]

One characteristic that seems to bring together both public and theatrical performances is that, in their schematic typology of characters, the shepherd is usually pitted agains the Jew and the Muslim. While the early Castilian drama reflected the tensions between Old Christians and *conversos* from Jewish origin, the figure of the shepherd is soon integrated into an eschatological confrontation between Christianity and Islam. Thus in Hernán López de Yanguas's *Farsa turquesana* (c. 1529) [*Turkish Farce*] the Turkish ambassador Mahometo, on his way to deliver a message from the Ottoman emperor to Pope Clemente VI, meets two shepherds who ridicule him. The stage directions stress the equivalence between Turks and Moriscos, since the Turkish emperor goes "vestido a la morisca" [dressed in the Morisco style] while the attire of the shepherds is described tautologically: "los pastores como pastores" (1) [the shepherds as shepherds]. Bernal Díaz del Castillo's *Historia verdadera de la conquista de la Nueva España* (c. 1568) [*True History of the Conquest of New Spain*] describes a fight between Turks and shepherds for the festivities held in Mexico in 1538 to celebrate the truce between France and Spain (653). In Sánchez de Badajoz's *Farsa de la Yglesia* (published in 1554) [*Farce of the Church*], the only non-allegorical characters are a shepherd and a Muslim, who is referred to in the stage directions as a Moor even after his baptism and conversion. Similarly, the confrontation of Moors and rustic characters is found in other plays from the second half of the sixteenth century, such as the anonymous *Farsa del Sacramento llamada de los lenguajes* [*Farse of the Sacrament, also Called Farse of Languages*] (Rouanet 3:331–5) and Luis Hurtado de Toledo's *Cortes de la muerte* [*The Parliament of Death*] (1557). The opposition between rustic people and Moriscos is clearly instrumental in the festivities organized by the Duke of Lerma in 1617, at which Antonio Mira de Amescua's now lost *Máscara de la expulsión de los moriscos* [*Masque of the Expulsion of the Moriscos*] opposed "rustic people" and Moriscos to celebrate the recent expulsion dictated by Philip III and his favourite. Arguably, the Duke of Lerma commissioned this piece of propaganda to present his decision as a popular initiative and to defend himself from the voices that complained against the measure (Marchante-Aragón 100). This constant pitting of shepherds against Muslims thus serves to widen the imaginary divide between shepherd as the representative identity of the Spanish people and its Others.[23]

Beyond the pastoral content of early modern plays, both performance and historical accounts of the origins of the genre signal the confrontational value of the shepherd. In Cervantes's play *Los baños de Argel* [*The Bagnios of Algiers*], written during the early seventeenth century, Christian captives in Algiers stage a Nativity pastoral dialogue by Lope de Rueda. Here, the choice of the genre is not arbitrary, since pastoral serves the Christian characters as an affirmation of identity whose confrontational nature passes unrecognized to their Muslim captors (Irigoyen-García, "La música" 52–9). Early modern Spaniards also articulated a narrative that made sense of the causal relationship between the world-changing events of 1492 and the beginnings of their own theatrical tradition. Agustín de Rojas de Villandrando's *Viaje entretenido* [*The Entertaining Journey*] (1603), offers a beginning in which the rise of Castilian drama is anchored in the fall of the Nasrid kingdom of Granada and the repression of Jewishness after 1492: "en la dichosa era / que ... don Fernando e Isabel ... / de echar de España acababan / todos los moriscos, que eran / de aquel reino de Granada, / y entonces se daba en ella / principio a la Inquisición, / se le dio a nuestra comedia" (148) [In the happy age when ... Ferdinand and Isabel ... had just expelled from Spain all of the Moriscos from the kingdom of Granada and began the Inquisition, our theatre started]. By emphasizing the simultaneity of the creation of a dramatic tradition and the events around 1492, Rojas de Villandrando conceives of Spanish theatre as part of a teleological project of ethnic cleansing in which every stage has its unequivocal position in the chain of events.

On all these levels, then, the figure of the shepherd articulates the opposition of an Old Christian identity and its ethno-religious Other. As we have already noted, conceptions of race based on genealogy rather than phenotype confront the problem of rendering their imagined difference visible. Since blood purity is an inner trait that cannot be seen, it has to be projected into a repertoire of highly ritualized objects and cultural practices. The codification of the figure of the shepherd allows for a rehearsing of an Old Christian identity that audiences, beholders, and readers can easily identify as signifying that inner invisible trait. It is precisely through these signs that beliefs regarding fictions of lineage can be articulated and convey (or rather impose) a certain effect of reality. In what follows, we will see how clothing and speech served as the main elements through which fictions of ancestry were materialized and projected into the realm of cultural representation.[24]

Dressing the Shepherd

The preoccupation with pastoral clothing as a marker for the Christian community is reiterated in theological treatises, in which early modern Spaniards again interpreted the scriptures according to their specific cultural anxieties. Strikingly, one of the sources used for predicating pastoral clothing as the proper dress for Christians is again the prophecy of the Good Shepherd of John 10:14–16, in which there is no reference to clothing: "Yo soy el buen Pastor: y conozco mis ouejas, las mias me conocen ... Y oyran mi boz, y hazerse ha un corral y un pastor" [I am the good shepherd and I know my sheep, and they know me. They will listen to my voice and there will be only one flock and one shepherd]. Juan de Ávila (1499–1569) rephrases the scriptural text in one of his sermons: "soy buen pastor, que me he vestido del vestido de mis ovejas y conocen mi voz" (2:261) [I am a good shepherd, since I have dressed in the same attire as my sheep and they know my voice], later expanding the circular sartorial bond between the shepherd and the herd: "Yo las conozco a ellas. Ellas andan vestidas de mi vestido, y yo, como buen pastor, vestido del suyo" (2:264) [I know them. They wear my attire and I, as a good shepherd, wear theirs]. His interpolation reveals how, due to the mandates of public exhibition of identity, religious belief is not enough to become a "Christian shepherd," who needs to be clearly identifiable, not by such a fragile sign as the voice, but by the material visibility of sartorial uniformity.[25]

Similarly, and around the same time that Juan de Ávila was preaching in Andalusia, the 1535 treatise *Espejo de religiosos* also echoes John 10:14–16 in the allegorical dialogue between Desseosso and a shepherd, emphasizing the shepherds' garments as a sign of religious recognizability:

> – Y essa çamarra ¿para que es?
> – Esta – dixo el – es la vestidura y habito de nosotros, ca si no vestiesse çamarra, no seria conoscido por pastor.
> – ¿Y de que la hazeys essa çamarra? –dixo Desseoso.
> – De pellejos de ouejas ... pues me veen vestido de su pellejo mesmo, amanme y vanse tras mi. (qtd. in López Estrada, *Notas* 34b)
>
> [– What is this *çamarra* (sheepskin jacket) for?
> – This – he said – is the attire that we wear, because if I were not wearing a *çamarra* I would not be recognized as a shepherd.
> – And what is the *çamarra* made of? – said Desseoso.
> – From sheepskins, and since they see me dressed with their same skin, they love me and follow me.]

Unmistakable shepherd clothing includes the *zamarra* and the *pellico*. In theory, both are made of sheepskin and have in common that they retain the fleece, but early modern texts show a high degree of variation, and reaching a fixed definition does not seem possible.

According to Carmen Bernis, "La zamarra, más rústica, dejaba la piel al descubierto; el pellico utilizaba la piel como forro" (*El traje* 408) [the *zamarra*, more rustic, left the skin on the outside, while the *pellico* uses the skin as lining], following the lexicographer Sebastián de Covarrubias, who states that the *pellico* is "el çamarro del pastor hecho de pieles" (860) [the shepherd's *çamarro* made of skins]. Other documents seem to contradict Covarrubias, such as Antonio de Torquemada's *Coloquios satíricos* (1553), in which, according to one of the characters, both garments were reversible, depending on the time of the year when they were used: "traemos encima las çamarras y pellicos en el invierno con el pelo adentro, que nos pone mucho calor, y en verano afuera por que la lana nos defienda del sol, y el pellejo es para nosotros templado" (1:310) [during the winter we bring the *çamarras* and the *pellicos* with the fleece to the inside because it warms us; during the summer we put the fleece to the outside so the wool defends us from the sun, and because the skin is cooler].

Early modern categorizations of the *pellico* and the *zamarras* were also complicated because they were appropriated by the upper classes. As Diego de Guadix states "ya hazen estas çamarras o çamarros tan curiosos que no son ábito de pobre sino regalo y abrigo de muy ricos" (481–2) [Now they are making these *zamarras* or *zamarros* so luxurious that they are no longer attire for the poor people, but delight and garments for the very rich].[26] Similarly, Covarrubias distinguishes between *zamarros* and *zamarras*, based on the social class that wears them: "Çamarro. Vestidura de pieles de corderunas o abortos, que son delgadas y tienen el pelo blando y corto. Éstos son los çamarros de gente regalada, a otros llaman çamarras, propio hábito de pastores, de mayores pieles, aunque atufadas" (391) [*Çamarro*. Garment made out of the sheepskin of newborn lambs, which is thin and has soft and short fleece. These are the *çamarros* of rich people; the other are called *çamarras*, which are a common shepherd garment, made out of larger skins but coarser].

Therefore, even though *pellicos* and *zamarras* were clearly conceptualized as having a rustic origin, aristocratic versions were included under those terms and were considered to derive from them. The idealized shepherds of pastoral romances use lavish *pellicos* and *zamarros*, while the shepherds in the earlier Castilian stage were characterized by the use of coarse *zamarras*. Even though the finish of the garment serves to maintain social differences between both kinds of shepherds, the shared shearling underscored the common attachment to sheep, since the shepherds wore their sheepskin in some way, as mandated by Juan de Ávila.[27] Rusticity was thus conceived as being in compliance with the exhibition of Christian identity.

In the early sixteenth-century, courtly shepherds still pretend to be wearing rustic garments, as in *Libro en que se qüentan los amores de Viraldo y Florindo* [*Book that Tells the Loves of Viraldo and Florindo*] (c. 1541), in which the shepherd Viraldo wears *pellico* and *zamarra* (105), and the manuscript's illustrations recreate how his clothing is made out of raw sheep skin (figure 14). But by the second half of the sixteenth century references to the *pellico* in the pastoral romances seem to refer to a completely different garment. In Bernardo de la Vega's *El pastor de Iberia* (1591), the noble Filardo, wanting to take on a new life, engages in the pastoral convention of leaving the courtly dress behind to take up the humble *pellico* instead: "Ocupad el lugar simple Pellico / del fanfarron y recamado sayo" (8r) [Oh humble *pellico*, take the place of the pretentious and lined tunic]. However, when he later reencounters his beloved Marfisa after having swapped his dress with another ruder shepherd, she wonders "[c]omo agora la tosca xerga viste, / El que traxo pellico de brocado?" (96v) [how is it that he is wearing coarse cloth, when he was wearing a brocade *pellico*?]. Marfisa thus reveals that, when Filardo initially adopted the so-called *pellico*, he did not intend by any means to cross class boundaries, since the *brocado*, as opposed to the *xerga*, was a lavish material that had nothing to do with the raw wool that rustic shepherds wore.[28]

The distance between rustic and courtly pastoral garments is clearly revealed in Luis Gálvez de Montalvo's *El pastor de Fílida* (1582), in which the characters – all of them from noble origins – are described wearing *pellicos* of different shapes and materials (499). To offer only one example, it is clear that the description of Filena's dress is not referring to rustic garments made out of raw wool: "la hallaron vestida de una grana fina, con pellico azul de palmilla, pespuntado de pardo y lazadas verdes; camisa labrada de blanco y negro" (524) [they found her dressed up in a fine red dress, with a blue *pellico* made of a rich interlaced cloth, with brown and green ribbons, and a shirt embroidered in black and white]. These courtly shepherds are surprised by the arrival of one "pastor serrano" [wild shepherd] whose clothing vividly contrasts with that of the inhabitants of this fake Arcadia: "llegó un pastor robusto con un cayado; dejó un sayo tosco, sin pliegues, hasta los pies, y en el brazo izquierdo un zurrón de lana, cinto ancho de piel de cabra y caperuza baja de buriel. Serrano era el traje y el color del rostro más" (577) [a robust shepherd arrived with a crook; he took off his coarse tunic, with no pleat, down to his feet, and with a woollen pouch on his left arm, a wide belt made of goat skin and a tawny low hood. His clothing was of the mountains, and the colour of his face even more so].

Figure 14 *Libro en que se qüentan los amores de Viraldo y Florindo* (c. 1541). Photograph by Martín García Pérez. Courtesy Enrique Martín.

This issue of shepherds' clothing is illuminated by the corpus of self-reflective theoretical commentary provoked by the importance of the Adoration of the Shepherds in painting. In his *Diálogos de la pintura* (1633) Vicente Carducho reflects on the proper way to depict shepherds, criticizing what he considers two equally undesirable extremes:

> no digo que se pinte al pastor con pellico de blanco armiño, ni la cayada del sacro laurel, o cedro, ni a la pastora con rayos del Sol por cabellos, y dos luceros por ojos, ni por calzados coturnos ricos ... y otras cosas semejantes de que usan los Poetas, con sonoras voces a los oídos: pero tampoco han de bajarse tanto, que al pastor le pinten con los pies desnudos asquerosamente, como algunos lo han usado. (qtd. in Calvo Serraller 295)
>
> [I am not saying that the shepherd should be depicted with a shearling of white ermine, nor with a crook of sacred laurel, nor the shepherdess with sun rays in her hair, and two stars instead of eyes, or wearing luxurious buskins ... and so many similar things that the poets use, with melodious voices for the ears: but neither should they descend so low that the shepherd is depicted barefoot and filthy, as some have done.]

Between these two extremes in representation, Carducho proposes a "proper" model that must not to be confused with realism. What he suggests is not driven by the observation of real shepherds, but constitutes a conventional model of manners that tries to emphasize both humility and some degree of social standing at the same time.[29]

These two extremes are well represented in two of the Adorations selected for this book. Murillo's (figure 12) reflects in the foreground a shepherd with a humble woollen shearling and filthy feet; by contrast, Pantoja de la Cruz's *Adoration* (figure 1) portrays Philip III and his court as shepherds, but their lavish *pellicos* clearly distinguish them from commoners. Yet, in spite of their obvious differences, both garments are comprised under the same name *pellicos*, since both are associated with a form of sheep herding and Christian identity. We might be tempted at first to see in this contrast an opposition between real and fake shepherds. However, such a divide is in itself integral to the ideologies of representation.

Even if real shepherds wore the kind of woollen shearling of Murillo's *Adoration*, the painting does not intend to be an ethnographical reflection of shepherds' life. Indeed, the woollen shearling is perceived through the myriad of discourses that construe it as the opposite of both urban affectation and cultural contagion. Murillo's shepherd may be barefoot and wear rags to underscore the humility of the figure, but the raw wool that shows through is strikingly white and without *raça* – when, in fact, if shepherds dressed in raw wool, they would only be allowed by livestock owners to take that of lambs rejected because of their pigmentation (not the white ones used for export). Since the Nativity is supposed to take place in the winter, it is improbable that the fleece would be worn to the outside, which is more proper for summer, as Antonio de Torquemada pointed out. Besides, shepherds would certainly not be barefoot to wander with their herds over the Castilian plains, a detail that can only be included to emphasize penance. These inconsistencies in the representations of rustic shepherds are crucial and reveal how they are ruled by a set of conventions that are ideologically motivated, even if they do not and cannot refer us to some kind of "real shepherd."

This conflation of class difference through shepherds' clothing also erases cultural differences between the different kingdoms of the Iberian Peninsula. Noël Salomon warns against considering theatrical dress as a reliable ethnographic document, stating that temptation should be avoided to read at face value expressions like "a la usanza de esta tierra" (432) [in the customs of the land]. As Salomon further observes, the theatre ignored the enormous sartorial variation of early modern Spain, imposing a sense of uniformity while constructing the convention of the "rustic" dress on the basis of the garments used by peasants in the region of Madrid and Toledo (428).[30] The same could be said of pastoral romances, and, one may guess, of the pastoral performances held throughout Spain, from Galicia to Seville and from Segovia to Valencia. The identification with the figure of the shepherd works, thus, as a device of cultural homogenization within Spain at the same time that it privileges a very specific region as the naturalized reference for the construction of Spanish culture.

Furthermore, rustic clothing provided a certain feeling of antiquity, and therefore a sense of cultural continuity with the past. In his analysis of rustic clothing used in urban performances, Salomon suggests that "por sus aspectos más coloridos y aparatosos el traje aldeano podía atraer a los ciudadanos y a los nobles que manifestaban algún gusto por lo arcaico. En efecto, el traje rústico de fiesta no era a menudo más que un antiguo traje noble despreciado" (405) [because of their colourful and flamboyant look, rustic attires may attract citizens and aristocrats who showed some taste for archaic elements. In fact, the festival rustic attire was often just an old disparaged aristocratic garment]. Therefore, what audiences may recognize in rustic clothing is its archaic form, its condition as a relic, and the possibility of using it to perform a connection with a constructed view of the past.

How far back in time this past was located depended on every particular author's concept of ethnogenesis, the time frame chosen as the alleged origin of the community. For some, it can be as far away as in Alonso de Villegas's devotional treatise *Quinta parte del Flos Sanctorum* (1594), which recommends sartorial moderation by suggesting that the *zamarras* and *pellicos* were the garments that God originally gave Eve and Adam when they were expelled from the earthly Paradise: "Y en tal sazon dioles Dios dos vestidos de pieles de animales, al talle de çamarros, o pellicos pastoriles: y a este vestido dado de Dios sucedio todo lo que de presente se usa" (56v) [In that time God gave them two attires made of animal skins, similar to the shepherd *çamarros* or *pellicos*, and from this garment bequeathed by God comes all the clothing that we use today].[31] Positing a somewhat less remote origin, Baltasar de Echave's *Discursos de la antigüedad de la lengua cantabra bascongada* (1607) forges a Basque etymology for *sayo* [rustic tunic] that dates it back to the mythical foundations of Spain by Noah's grandson Tubal (13–14). Similarly, Andrés de Poza, in his *Antigua Lengua de las Españas* (1587), claims that the clothing of the Basque peasants was the same as those worn in biblical times (Juaristi 88–99). Thus rustic clothing became an icon of primitivism and cultural purity, providing the possibility to reenact the customs of the ancient Spaniards long prior to 711 and the Muslim period, and to convey the feeling of belonging to a community that had not been altered since time immemorial, as they were assumed not to have been contaminated by "foreign fashions."

As we saw with Juan de Ávila's rephrasing of John 10:14–16, shepherds' garments – mainly the *pellico* and the *zamarra* – are conceived as signs of both cultural and religious recognition, as the sartorial bond between the shepherd and his sheep. If the shepherd has to look like his flock, and the flock needs to look like its shepherd, the circular logic of the materiality of the pastoral community excludes any other "herd" that distinguishes itself by different sartorial practices. Thus the theatrical shepherds always perceive the change of dress as a threatening influence coming from the ethno-religious Other. In Juan del Encina's *Égloga de Mingo, Gil y Pascuala*, the rustic shepherd Mingo resists abandoning his pastoral garment in order to wear courtly dress, complaining about any piece of clothing that he tries on, until he finally utters his utmost fear: "¡Ha, pareceré judío!" (*Teatro completo* 184–5) [Ah, I will look like a Jew!]. In Gil Vicente's *Auto da fé*, Benito is not even able to recognize the appearance of Christian Faith, wondering about the identity of what he thinks to be a woman "quellotrada a la morisca" (*Compilaçam* 73) [dressed in the Morisco style]. In the *Coplas de Mingo Revulgo*, in which shepherds' garments and instruments serve as symbols of royal authority (MacKay 29), Henry IV of Castile is stripped of both his pastoral and royal attributes as the initial step in refiguring him as a Moorish and illegitimate monarch: "Uno le quiebra el cayado / otro le toma el çurron / otro le quita el çamarron" (122) [one of them breaks his crook; the other takes his pouch; and the other steals from him his *çamarrón* (sheepskin jacket)].

The sartorial opposition between rustic people and Moriscos becomes clear in the aforementioned account of the festivities organized by the Duke of Lerma in 1617 to celebrate the expulsion. Even though Mira de Amescua's *Máscara de la expulsión de los moriscos* is now lost, its content can be retrieved by the extant accounts commissioned by the Duke of Lerma. One account of the festivities by Francisco Hernández Caso reveals the staging of two dances: "una de moriscos y moriscas, y otras de villanos, que con bailes y danças maravillosas, adornados de ricos y proprios trajes, nos representaron una guerra en que los moriscos fueron vencidos, presos y ... desterrados" (qtd. in Ferrer Valls, *Nobleza* 261) [one from Moriscos and Moriscas, and another from rustic people, who, with magnificent dances, dressed with luxurious and proper garments, staged a combat in which the Moriscos were defeated, imprisoned ... and exiled]. This short account of Mira de Amescua's play already underlines its essential trait: it reenacted the symbolic opposition between "rustic people" and Moriscos in order to craft an exclusionist vision of Spanish identity.

Another account by Pedro de Herrera shows in detail how Mira de Amescua's play carefully constructed difference. The peasants' performance is described first:

> Salieron bailando con castañuelas y bueltas aldeanas de mucho regozijo ... Traían los villanos sayos y caperuzas de terciopelo verde, guarnecidos de passamanos de oro ... salió una labradora, vestida de sayuelo y vasquiña de tela azul y encarnada, con mucha guarnición de molinillos y franjas ... Dezía como los moriscos maquinavan su levantamiento para renovar la perdición que huvo en tiempo de la Cava, y pues aora no reinava Rodrigo sino rey tan pío y valeroso, no diesse lugar a segunda destruición, antes, ya que los moriscos generalmente eran culpados, saliessen todos de su señorío.
>
> (qtd. in Ferrer Valls, *Nobleza* 272)

> [They appeared dancing with castanets and very delightful peasant-like movements ... The rustics wore tunics and hoods of green velvet adorned with golden stripes ... A peasant woman came, dressed in a smock and skirt of blue and red fabric, well adorned with edgings and stripes ... She said that the Moriscos were plotting to renew the destruction that happened in the times of the Cava, and, since Rodrigo was no longer reigning, but such a pious and brave king, he should not allow this second destruction; rather, since

the Moriscos were all of them guilty, all of them should leave his kingdom.]

As the mention of lavish fabrics reveals, the peasant-style dress has little to do with the actual garments that rustic people wore in the countryside, but with the conventions of early modern courtly representations of rustic people on stage. By emphasizing the role played by stylized rustics, the performance conveyed the message that the decision of Philip III and the Duke of Lerma to expel the Moriscos did indeed reflect the desire of the Spanish people, here represented as rustics by their dress – not because the opinion of a real Castilian peasant counted, for that matter, but because powerful interest groups identified with rustic people and shepherds as their imaginary ancestors. These notions of genealogical purity and national culture, figured by the shepherd, dominated the political language that different groups used to negotiate power – this is what it means to say that the concept of Old Christian identity was part of the hegemonic discourse of early modern Spain.[32]

By contrast, Moriscos in Mira de Amescua's play are equally stereotyped in their dress and their speech, with musicians "vestidos de moros, turbantes, marlotas y capellares" (qtd. in Ferrer Valls, *Nobleza* 273) [dressed as Moors, with turbans and Moorish tunics and hoods] and dancers lavishly dressed "a lo africano" [in the African fashion], playing and dancing the *zambra*, the typical Morisco dance. Another Morisco joins the performance, announcing the imminent expulsion and characterized as "un moro ridículo (por el vestido ... y lenguaje aljamiado de su voz)" (qtd. in Ferrer Valls, *Nobleza* 273) [a ridiculous Moor, because of the way he dressed ... and the Morisco dialect he spoke]. Thus the rustics and the Moriscos are equally stereotyped as the most antithetical characters in early modern Spain through a careful construction of very distinctive ethnic marks, in both the sartorial and the linguistic registers. However, as Marchante-Aragón remarks, even though the play aims at construing Old Christian peasants and Moriscos as two perfectly separate identities, positioning peasants as an emblem of cultural and racial purity in Spain, "this construction of the 'Self' is hybrid in its birth, since it draws upon the presence of the newly fashioned 'Other' in order to establish its existence and the origins of its authenticity" (123).

Yet the persistent representation of shepherds wearing *pellicos* and *zamarras* and Moriscos wearing *marlotas* does not accurately reflect the sartorial reality of early modern Spain. On the one hand, Muslims are always depicted wearing *marlotas* – a kind of long Moorish tunic – turbans, and other "Moorish" attire, especially in ballads and literary texts such as *El*

Abencerraje (c. 1550–60) or Pérez de Hita's *Guerras civiles de Granada* (first part 1595, second part 1619).[33] Theatrical texts, when describing Moriscos, often depict them like literary Moors, even though most of the Moriscos were not differentiated from Old Christians in their sartorial practices (Carrasco 120–4). Even in Granada, where the identity value of Morisco dress was an issue prompting confrontation, Morisco men tended to dress in the Castilian manner by the mid sixteenth-century; clothing identified as Moorish was only being worn by certain women or was circumscribed to specific ritual festivities such as weddings (Núñez Muley 211–12).

On the other hand, and in parallel with the stereotyping of Morisco sartorial practices, the *pellicos* and the *zamarras* seem to be overrepresented in literary texts. Bernis holds that the *zamarras* should have been common garments, based on the abundant references to them found in pastoral literature (*Trajes* 174). However, as we have seen, we should be cautious when dealing with literary representations, especially with such an idealized genre like the pastoral romance and even in the no-less-idealized representations of rustic shepherds. It seems that, in reality, shepherds wore mostly *sayos*, long rustic tunics typical of peasants. The absence of the *sayo* in written references to shepherds' clothing may well be due to its lack of effectiveness in providing a radical sartorial differentiation between Moriscos and Old Christians, because of its supposed similarity with the *marlota*, also revealed in the reversibility of sartorial terminology in contemporary lexicons.

The lexicographer Sebastián de Covarrubias, who largely reflects the ideology of sartorial difference by defining *marlota* as "vestido de moros," nonetheless acknowledges its similarities with rustic attire, suggesting that it is a garment similar to "sayo vaquero" (790). Properly speaking, the *sayo vaquero* was a Moorish coat of Turkish origin, mostly used in the game of canes, but the signifier alludes to rustic culture, since it literally means "cowherd tunic." The lexicographer Francisco del Rosal, even when pointing to an Arabic etymology for *marlota*, states that its actual use in early modern Spain is independent of its cultural origin: "ropa aforrada, o piel con lana, habito de Pastores" (443) [lined cloth, or skin with wool, shepherds' attire]. The ease with which the *marlota* and the *sayo* (and even the *sayo vaquero*) can be identified as the same piece of clothing collapses the alleged sartorial distinction between Moriscos and Old Christians.[34] The reason for preferring the *pellicos* and *zamarras* over the *sayos* in the pastoral imaginary may well be directly related to displacements in representation that safeguarded the figure of the shepherd as exclusively identified with Old Christians.

Foreign visitors, ignorant of the highly codified yet unconscious "pastoral habitus," did not respect Spanish imaginary of difference, revealing its culturally constructed nature. Christoph Weiditz's *Trachtenbuch* (1529) depicts a Castilian shepherd wearing a *sayo*, probably unaware that a Spaniard would have preferred to depict him with a *zamarra* (figure 15). Ironically, the German traveller Johannes Lange, describing in 1526 the clothing of the Granadan Moriscos, concludes that they dress "como nuestros pastores aldeanos" (qtd. in Arié 128n1) [as our rustic shepherds]. The contrast between local and foreign perceptions of sartorial difference within Spain serves to shed light on the linguistic ideology that drives early modern terminology concerning clothing. While foreigners tended to see Spanish customs as essentially Moorish, bringing their own preconceptions about the Iberian Peninsula, Spaniards usually saw ethno-religious differences as depending on the genealogy of the bearers, and not on clothing itself. Feliciano Chaves criticizes the distortion of trying to impose a sharp distinction between "Morisco" and "Castilian/Christian" fashions, arguing that "[s]artorial items commonly called 'Morisco' ... oscillated between the realms of Iberian clothing (daily wear) and costume (theatrical, performance-oriented), depending on the wearers, the circumstances and the intended audience" (134). Feliciano Chaves's observation about early modern categorizations of clothing serves to illustrate the opposition between *pellicos* and *marlotas*. Although there was probably a wider difference between rustic and courtly *pellicos* than between rustic *sayos* and *marlotas*, Spaniards preferred to categorize clothing according to the expected ethno-religious identity of the bearer (rather than by the materiality of the clothing itself), either by combining different garments under the same term (such as the polyvalent "pellico") or, inversely, by giving two separate names when a very similar garment was used by different ethnic groups (like the difference between *sayo* and *marlota*).

At this point we should briefly return to the passage with which I opened this chapter, about the eccentric playwright who is writing a play with Moorish shepherds in Mateo Luján de Sayavedra's *Segunda parte de Guzmán de Alfarache*. When the other characters ask him how the Moorish shepherds should dress because "los pellicos que usamos en España no les podrán convenir" (422) [the *pellicos* we wear in Spain won't suit them], the playwright proposes to inquire about it among the North African captives in Valencia. After the playwright recites his play only to be ridiculed, Guzmán and his companions head to the city of Valencia and never encounter him again. Therefore the question of how should Moorish shepherds dress is apparently left unresolved.

Figure 15 Christoph Weiditz. "Das ist ain kastilianischer Hiertt" (This is a Castilian Shepherd). Plate 20v of *Trachtenbuch* (c. 1530–40). Hs 22474. Photograph by L. Streppelhoff. Courtesy © Germanisches Nationalmuseum.

Yet the narrative still gestures to the issue in an indirect way. As soon as Guzmán arrives in Valencia, he goes to watch the entrance of Margarite of Austria for her wedding to Philip III. It is telling that Guzmán begins the next chapter mentioning the exact spot where he will watch the pageantries: "Salíme al portal de los Serranos, por donde había de ser la felicísima entrada de la reina nuestra señora" (423) [I went out to the gate of Serranos, where the happy entrance of the queen our lady will take place]. This emblematic location in the city of Valencia was indeed the place where the royal entrance took place (Gauna 1:26–7). Yet, even if Guzmán's description is accurate according to historical facts, the reference to this concrete location happens in the text right after using the term "serrano" to talk about the rustic people in Africa.[35] While this might be a coincidence, it is the first sign that indicates that the question opened in the previous chapter about the sartorial propriety of rustic clothing is still lingering in the text.

Guzmán goes on describing at length the pageantries for the royal wedding, focusing, as many other accounts of the celebration did, on the lavish clothing worn by the aristocracy, among them the *marlotas* of the participants in the game of canes.[36] At some point, though, he describes them as "vestidos todos con marlotas o sayos vaqueros de grana y pasamentos de seda" (424) [all of them dressed with red *marlotas* or Turkish coats with silk braids]. This use of *marlota* and *sayo vaquero* as plain synonyms is made in passing, and it is therefore easily missed. However, when we see them in connection with the deconstruction of pastoral clothing made earlier in the text, it becomes clear that the author is trying to articulate facts and fiction in order to make a point about the sartorial ideologies of early modern Spain.

At first, the playwright who composes the play about Moorish shepherds is criticized for not considering that North African shepherds cannot be staged wearing the same clothing of Spanish shepherds. Those who object to his literary project implicitly claim that the *pellicos* are an essential feature of Spanishness that cannot be transferred to other cultures. Immediately after that, Guzmán goes to watch the royal entrance in Valencia in a spot whose name evokes the point of view of rustic people, only to describe Spanish aristocracy dressing as Moors but with garments whose signifier points to rustic culture. The question raised for the eccentric playwright is therefore solved in the most unexpected place, by implying that, when the Spanish aristocracy pretends to be dressing up as Moors, they are in fact appropriating a lavish version of rustic identity. Thus both passages reflect the inconsistent conceptualization of Moorish and rustic clothing and end up mirroring each other, illustrating how the opposition of categories such as "Moorish" and "rustic" is rather nominalistic.

Rustic Speech as Relic

Another characteristic of the rustic shepherd is the use of a very specific jargon usually called *sayagués*, purportedly a rural Castilian dialect of the area of Sayago (Salamanca), which was soon stereotyped as a literary jargon for any Spanish rustic character (Stern, "Sayago"). Even though it served to ridicule the characters who spoke it, it was as well a marker of Old Christian identity, since it contrasted with the conventional theatrical jargons used to depict Moriscos, Jews, and Blacks.

The evolution of the terms applied to this theatrical speech parallels the reassessment of the figure of the shepherd during the sixteenth century. While this jargon was labelled *rústico* or *pastoril* during the first half of the sixteenth century, those terms were replaced by *sayagués* at least from 1567 on, when it is first mentioned in Juan de Timoneda's *Coloquios pastoriles* (Stern, "Sayago" 232–3). The connection of this distorted version of Castilian to a specific geographical area and its dissociation from shepherds takes place during the period when *lenguaje pastoril* stopped being identified merely with rustic people. After the poetic projects of Garcilaso de la Vega and Jorge de Montemayor, "pastoral language" became a stylized and conventional courtly code that was used to convey the sentimental and civilized quality of pastoral lovers.[37] Thus the change of the terminology reflects the shift in the social dignification of the figure of the shepherd by the second half of the sixteenth century.

In the second half of the sixteenth century, precisely at the same time that the interest in antiquities grew in some erudite circles (Rallo Grus, *Los libras* ix–lxxvii; López Trujillo 51–74), the representation of rustic language is entangled with the conception that the primitivism of rural people grants a privileged way to study the cultural origins of the nation. While rustic jargon was originally a kind of speech invented to characterize shepherds and rustics in Castilian drama, by the second half of the sixteenth century it became a pseudohistorical tool. Juan de la Cueva published in 1587 his *Coro Febeo de romances historiales*, a collection of historical ballads composed entirely in *fabla antigua* [old speech], which was in fact a version of the conventional *sayagués* (Cebrián 101–14). The communication of temporality in these literary experiments and historiographical forgeries was possible because there was a mass of available medieval texts from which writers could borrow for their compositions and which was shared with their audiences, who would immediately identify them as such.[38] Medieval texts such as *Conde Lucanor*, the *Coplas de Mingo Revulgo*, and the *Siete partidas* were reedited several times during the sixteenth century, making

Spanish authors aware of the historical evolution of Castilian. Furthermore, although Juan de la Cueva's poetic project was regarded as an extravagant literary experiment by his own contemporaries, he was not alone in cultivating this language as a suitable literary dialect.

On the contrary, the *fabla antigua* was used in historical plays as a way to recreate the purported speech of "ancient Spaniards." The most famous example of the use of this "old speech" is Lope de Vega's *Las Batuecas del Duque de Alba*, which recounts the discovery of an isolated territory in Extremadura in which the inhabitants, the direct descendants of the Visigoths, managed to preserve their "ancient Castilian." As Leo Cabranes-Grant points out, Lope's aim to recreate "ancient Castilian" on stage is driven by the intuition that archaic language conveys a certain image of national history (152), which ultimately serves to reinforce the ethno-religious "purity" of Spain (158). The play dates the discovery not in Lope de Vega's time, but in 1492, thus linking it to the fall of Granada, the expulsion of the Jews, and the discovery of the New World, and so the rise of the empire, the process of ethnic cleansing, and the providential finding of the primitive "true national" are conceived as simultaneous and complementary events (Vega Ramos 177). Lope's invention of the discovery of the "true Spaniards" in the Batuecas was taken so seriously that it gave rise to a series of historiographical treatises on the subject (Vega Ramos 184–9). The influence of this play showcases how historiography, ethnography, sheep herding, and literature blended together with the aim of reconstructing the alleged origins of Spain, and how the artificial rustic jargon of the shepherds of the first half of the sixteenth century was appropriated by the turn of the century as archaeological proof.

As we have seen in this chapter, through a careful selection and codification of rustic culture, early modern playwrights, historians, performers, and painters participated in construing the figure of the shepherd as a symbol for the "Spanish people." The figure of the shepherd is therefore part of a larger "invention of tradition," a process that Eric Hobsbawm defines as "a set of practices, normally governed by overtly or tacitly accepted rules and of a ritual or symbolic nature, which seek to inculcate certain values and norms of behavior by repetition, which automatically implies continuity with the past" (1).[39] This invention of tradition in relation to the cultural trope of the shepherd worked at multiple levels. With the assumption that rustic culture had remained unaltered, it served as a document of the past, as well as the proof of continuity of Spanish culture from time immemorial.[40] By raising one conventional (but locally based) model of pan-Iberian culture, it made an abstraction of cultural heterogeneity among Christians. And third, by creating a model (with variants) that could be shared by both commoners and aristocracy alike (at least nominally, on the sartorial level), it helped to make abstraction of social differences for the sake of fostering a common Spanish identity. In this way, what we can call "the invention of the shepherd" served to homogenize at once space, time, and the social spectrum, by creating a model that was supposed to be unalterable, adopted equally throughout Spain, and engaged in by all social classes alike.

As Ton Otto and Poul Pedersen remark, Hobsbawm's categorization is based on a dated ideological divide between "traditional" and "modern" societies, in which the former are supposed to be "genuine" while the latter are conceived to be "invented" (28–9). In order to move away from the divide between "authenticity" and "fabrication" while still saving Hobsbawm's otherwise productive concept, Otto and Pedersen focus instead on the explicitness with which a tradition establishes a link with the past of the community and how "assumed continuity with the past is part of their validating principle" (29). For the purposes of this book, both high and low versions of the figure of the shepherd are considered to belong to the same process of invention of tradition that seeks to build a determined Spanish identity in the early modern period by appealing to a certain continuity with the past. The divide between "authenticity" and "fabrication" may still be valid, but only insofar as we apply it not to the reality of practices, but to perceptions and strategies. Seen in this light, Juan del Encina's or Mira de Amescua's rustic characters are no less artificial than those of Philip III's court as depicted by Juan Pantoja de la Cruz or those of pastoral romances. While "rustic" shepherds are supposed to convey the notion of primeval authenticity, embodying the alleged cultural continuity of the Spanish people, courtly shepherds exhibit the "artificiality" of civility and social manners expected in the classes in charge of ruling the empire.

"Popular culture" was appropriated, managed, and consumed by avid humanist scholars who invested their efforts in crafting an image of an idealized individual who could embody Christian and Spanish identity – on the way excluding Moriscos and *conversos*, but also Blacks and Gypsies, as I will explore in chapter 5 and in the conclusion. The next chapter explores the twofold dialectics of attachment to the primeval past and opposition to the legacy of the Islamic period, showing the fundamental affinity between both high and low versions of the figure of the shepherd, between "popular culture" and more elevated genres like historiography and pastoral romances. Even though all pastoral expressions are characterized by ethnocentrism, they adopt in each case specific strategies of differentiation. Thus while popular constructions of the figure of the shepherd focused on a pseudo-ethnographic study of Spanish (viz. Castilian) cultural practices, as this chapter has shown, idealized courtly versions of pastoral relied on the developments of humanistic historiographical and archaeological projects, grounding their construction of the pastoral Arcadia in particular conceptions of the relation between collective identity in the present and the past, as the next chapter suggests.

3 In the Land of Pan: Pastoral Classicism and Historiography

entre Salicio y Nemoroso se habían de hacer
mis diligencias, que allí tengo mi Montaña,
mi Galicia, mi Vizcaya y mis Asturias.

[between Salicio and Nemoroso diligent scrutiny
of my lineage had to be made, since there
I have my Montaña, my Galicia, my Biscay, and my Asturias.]

Luis Vélez de Guevara, *El Diablo Cojuelo*

If theatrical and pictorial displays of rustic shepherds promoted them as a symbol of "the Spanish people," they coexisted with an idealized and stylized version of a literary shepherd that extended its identification to a different social level. In contrast with rustic characters, a more idealized figure of the shepherd emerged in pastoral romances during the second half of the sixteenth century. Yet how do pastoral romances construct an ideal of Spain that denies its Islamic and Jewish legacies if their seemingly atemporal idyll exists outside history and therefore they do not explicitly address the issue?

Pastoral romances share some common traits with the conformation of rustic shepherds as representative of the nation, such as the pastoral clothing analysed in chapter 2. But there are certainly many differences between the idealized pastoral romances and the rustic theatrical shepherds. These differences lie not only in the class status they are supposed to represent, but also in the particular emphasis that each places on different aspects of the representation of collective identity. Whereas the rustic shepherd invokes a notion of the people and therefore embarks in a proto-anthropological search of "popular culture," pastoral romances concentrate on imagining a classicist and classist refiguring of both the landscape of Spain and its history, drawing from humanistic historiographical inquiry. Pastoral romances invest their ideological capital in refiguring time and space, which becomes the focus of their fictions, in opposition to the pastoral dramas in which the focus of the performance is on the codification of the cultural practices of rustic characters. Thus the emphasis on the individual lineage of theatrical shepherds takes a collective turn in the pastoral romances, which seek to represent the genealogical and cultural purity of their characters through a particular historical view of the Arcadia.

Upon first perusal, we find that history is precisely the element that is absent in pastoral Arcadias, which seem to exist in a timeless idealized limbo. Their apparent ahistoricity is actually a strategy for naturalizing a determined vision of Spanish history which aimed at negating the Islamic period and establishing an essential continuity between the immemorial inhabitants of the Iberian Peninsula (anachronistically called Spaniards) and early modern Spain. In spite of the seeming disconnect between both discourses, Spanish historiography did influence the pastoral romances. Historiography provided the frame and the signs of temporality through which pastoral romances constructed an idealized landscape of a primeval "Spain": it supplied a repertoire of ancient place names with which the Arabized Iberian landscape could be classicized, and the idea of a remote community of shepherds characterized as Gentiles and therefore as the remote ancestors of Old Christians. Pastoral romances and historiographical discourses must be analysed side by side because historiography also uses the overarching prestige of the figure of the shepherd as a model with which to overcome the discipline's limitation to recreate an image of Spanish culture in the past.

Ancient Place Names and Pastoral Cartographies

Within the process of cultural cleansing of everything Moorish, place names attained a very special identity value. As Alonso Hernández del Portillo states in his *Historia de la muy noble y más leal ciudad de Gibraltar* (c. 1610–22) [*History of the Very Noble and Loyal City of Gibraltar*], when dealing with the ancient names of Gibraltar "ha permanecido más el nombre árabe ... como lo son casi todos los nombres de las más ciudades de España" (56–7) [the Arabic name has prevailed ... as in most of the city names in Spain]. While Hernández del Portillo's remark that most of the city names in Spain are Arabic is largely an exaggeration, it reveals nonetheless the perception that early modern Spaniards had about the imprint of the Arabic language on Iberian toponymy. Right after the expulsion of the Moriscos between 1609 and 1614, Francisco Núñez de Velasco's *Diálogos de contención entre la milicia y la ciencia* (1614) [*Dialogues of Opposition between Arms and Letters*] presents a proposal to eradicate all Arabic place names in Spain arguing that "seria acertado, pues esta infiel y Barbara nacion se ha expelido de España raer y cancelar de todo punto su memoria excluyendo de entre nosotros todos los vocablos que de aquella nacion se han recibido" (346r–v) [since these heathen and barbaric people have been expelled from Spain, it would be appropriate to completely erase and cancel their memory by ridding ourselves of all of the terms that we have received from them]. As his proposal shows, the attack against Arabic place names occurs in a larger context of ethnic and cultural cleansing.[1] Núñez de Velasco does not himself clarify what place names should replace the Arabic ones, but Spanish historiography had been dealing with this issue since the Middle Ages.

Even though the recovery of ancient Iberian place names can already be located in the thirteenth-century historiographical project of Alfonso X of Castile and in the works of fifteenth-century humanists, the definitive push for more systematic research on toponymy takes place in the sixteenth century. After the fall of the last Muslim kingdom of Granada in 1492, Isabel of Castile and Ferdinand of Aragon were in need of a historiographical narrative to justify territorial expansion and to prove that they were the legitimate inheritors of a previously unified Iberian Peninsula.[2] This would demonstrate that their conquest of Muslim territory was also a cultural restoration, and that the "original" place names needed to be retrieved, and if possible, Arabic place names should be replaced. The internal colonization of the Iberian Peninsula during and after the so-called *reconquest* made use of different naming strategies that were at best sloppy: sometimes Christian conquerors replaced Arabic place names with religious or military names that served to commemorate their victory, or gave a Romance translation; but in many cases, Christian settlers simply adopted the existing Arabic place name for the territory, an attitude that according to early modern humanists contradicted the very concept of "restauración" [restoration].[3]

As most early modern humanists conceived it, in order to compensate for the void between the political and military incorporation of the territory and the alleged cultural reintegration, it was necessary to research the "original" place name that in theory legitimated its conquest.[4] Early modern historians complained that, after eight centuries of Islamic presence in the Iberian Peninsula, the "original" names were no longer remembered. Thus Gonzalo Argote de Molina, in his introductory eulogy to Ambrosio de Morales's *Las antigüedades de las ciudades de España* (1575) [*The Antiquities of Spain's Cities*], laments the lack of memory in Spain: "Tus pueblos destruydos / sin dellos quedar nombre, / y de otros el renombre, / apenas con los sitios conocidos" (n.p.) [Your cities destroyed without even their name remaining; and of others, their renown with their location scarcely known]. Similarly, Diego de Colmenares's *Historia de la insigne ciudad de Segovia* (1637) [*History of the Famous City of Segovia*] states that Islamic presence "assolo pueblos, y ciudades, sin dexar rastros de sus nonbres, ni sitios: causa de la confussion que en la antigua Topografia de España tienen los escritores naturales y estraños" (72) [wiped out towns and cities, leaving no trace of their names or location, which is the cause of the confusion that native and foreign authors have with the ancient topography of Spain].

Due to the indeterminacy in identifying ancient toponyms, some humanists rejected recovery of the "original" place names of Spain as a futile project. Pietro Martire d'Anghiera explicitly criticized the recourse to the alleged ancient place names. In a letter from 1510 to the Count of Tendilla, he explained why he did not use the ancient place names for the Spanish cities and towns, pointing to the uncertainty of such a recovery: "Vértigo sentirá, efectivamente, todo aquel que pretenda entender algo de las cosas granatenses, si llamo *Illiberis* a Granada, *Abdera* a Almería, tal como lo hicieron muchos, pues la duda está principalmente en si aquélla era en realidad Illíberis o ésta Abdera" (335) [anyone who attempts to understand something about Granada will be dizzy if I call Granada *Illiberis* and Almería *Abdera*, as many others have done, because the main doubt is whether the former was really Illiberis and the latter Abdera].

In spite of uncertainty over the possibility of identifying the location of ancient place names based on classical sources, which were at best imprecise – if not merely fantasized topographies about a distant and exotic Hispania – Spanish humanists devoted themselves to amassing great amounts of information that would ultimately lead to the supersession of the highly defective Ptolemaic map of Iberia, although they often based their revisions on such unreliable sources as Annio de Viterbo's pseudo-Beroso (Caro Baroja, *Las falsificaciones* 65–8). Almost every historian in early modern Spain crafted his own reconstruction of the old place names of Spain, which was transmitted from one historian to the next. The interest in ancient place names was revived in Elio Antonio de Nebrija's *Muestra de la istoria de las antigüedades de España* (1499), and in his geographical glossary *Dictionarum oppidorum*, included in the 1512 edition of his Latin-Spanish dictionary. In 1535 Miguel Servet published an edition of Ptolomeo's *Geography* including his own list of Iberian ancient place names.[5]

The task was later pursued by many humanists throughout the sixteenth century, mainly by royal historians such as Florián de Ocampo's *Corónica general de España* (1543), Pedro de Medina's *Libro de grandezas y cosas memorables de España* (1548), and Ambrosio de Morales's *Las antigüedades de las ciudades de España* (1575), who were soon followed by a myriad of local intellectuals.[6] By the end of the sixteenth century, Philip II commissioned the Flemish printer and cartographer Abraham Ortelius to create a map of ancient Iberia, *Hispaniae Veteris Descriptio*, for which he had the help of the Spanish humanist Benito Arias Montano. The map was issued in 1586 and was subsequently included in the *Theatrum orbis terrarum* after 1590, providing a chart for the easy identification of ancient place names.[7]

Between the end of the sixteenth century and the beginning of the seventeenth, several historiographical and lexicographic works compiled and expanded previous findings, such as Diego de Guadix's *Recopilación de algunos nombres arábigos que los árabes pusieron a algunas ciudades* (c. 1593), Bernardo de Aldrete's *Del origen y principio de la lengua castellana* (1606), and Sebastián de Covarrubias's *Tesoro de la lengua castellana* (1611). Juan de Mariana says in the prologue to his *Historia General de España* (1601) that "[a] las ciudades, montes, ríos y otros lugares señalamos los nombres que tuvieron antiguamente en tiempo de romanos" (1:li) [we point out the ancient Roman names for the cities, mountains, rivers, and other places], apparently without finding any methodological inconvenience for this project. Like Martire d'Anghiera, Gaspar Escolano is more cautious about the possibility of reconstructing the "original" place names in his *Segunda parte de la década primera de la historia ... de Valencia* (1611): "Si para hallar el rastro perdido destos pueblos y Rios, queremos seruirnos de los autores antiguos de perros de muestra, son menester otros tantos para sacarlos a ellos de rastro: tanta es la confusion y babylonia de sus escritos" (2) [If we want to use ancient authors as hunting dogs in order to find traces of these cities and rivers, we would need as many (hunting dogs) to take them back to the trace; such is the confusion and the Babylon of their writings]. However, he quickly concludes, quite optimistically, that the mistakes in the ancient sources can be remedied with the knowledge of the real topography of Valencia (4).

The "discovery of the original name" would be the exclusive domain of a reduced circle of scholars interested in history and archaeology but for the fact that the alleged ancient names were being disseminated to wider audiences through other genres such as pastoral romances, where place names play an essential role in the construction of a Spanish Arcadia. It becomes necessary to de-familiarize ourselves, as readers of early modern Spanish literature, with an act of naming that, by its ubiquity, has become naturalized. For example, the act of saying "Betis" instead of "Guadalquivir" is so frequent in early modern Spanish literature that it passes unnoticed by any modern reader. It is, however, an appropriation that erases the trace of the people that dwelled there before the act of naming took place, conveying an imaginary continuity with the territory's past in which Christian settlers are conceived as the descendants of its primeval inhabitants.[8]

The opening line of Jorge de Montemayor's *La Diana* (c. 1559), the originating pastoral romance in Spain, begins by indicating the location even before the name of the characters: "Bajaba de las montañas de León el olvidado Sireno" (109) [Sireno the forgotten was descending from the mountains of León].[9] This information is not a mere note on the geographical origins of the character. León has a particular symbolic value for the discourses of blood purity: the inhabitants of the mountains of León – along with those of Old Castile, Galicia, and Navarre – were supposed to have avoided any contact with Muslims and were therefore considered to have blood purity, and people requesting the status of *hidalguía* usually claimed that their ancestors came from those areas (Sáez 28). The opening lines of *La Diana* would therefore be interpreted by Spanish readers as a statement about Sireno's lineage.[10] While the reality of the geographical ethnic distribution in early modern Spain is of course much more complicated than the alleged division of north-south, it is nonetheless an operative fiction.[11]

The link between genealogy and pastoral literature is revealed in parodic form in Luis Vélez de Guevara's *El Diablo Cojuelo* (1641). During a debate over the obsession of blood purity and the genealogical falsifications that it causes, Don Cleofás concludes enigmatically that he keeps the proof of his lineage in the works of the initiators of Spanish pastoral poetry: "tengo mi ejecutoria en las obras sueltas de Boscán y Garcilaso" (98) [I have my probative document among the uncollected works of Boscán and Garcilaso]. As he quickly clarifies, there is no such document, or rather, the document is the book itself: "Si a mí me hicieran merced … entre Salicio y Nemoroso se habían de hacer mis diligencias, que no me habían de costar cien reales; que allí tengo mi Montaña, mi Galicia, mi Vizcaya y mis Asturias" (98) [If they were to grant me the favour … between Salicio and Nemoroso diligent scrutiny of my lineage had to be made, and this would be very cheap for me, since I have there my Montaña, my Galicia, my Biscay, and my Asturias]. Presenting the shepherds of Garcilaso's first eclogue as the metonymical guarantors of the blood purity of the owner of the book, equating them with the mythical geography of blood purity, is of course nonsensical. But the absurdity reveals the ideological assumption that underlines the production and circulation of the pastoral genre. As Don Cleofás fantasizes about defending his lineage by having a "book of shepherds" in his possession, so the authors of pastoral romances, by presenting themselves – along with their literary circles, patrons, and buyers – as shepherds, aim at symbolically proving that they belong to the imaginary community of Old Christians.[12]

Similar to Sireno, the shepherdess Belisa comes from Galicia, and none of the Galicians have their blood purity challenged in *La Diana*.[13] By contrast, characters like Felismena and Felis, coming from Andalusia, need to classicize their place names of origin before they can enter the pastoral community. Felismena introduces herself by saying that "mi naturaleza es de la gran Vandalia, provincia no muy remota de ésta donde estamos, nacida en una ciudad llamada Soldina" (193) [my birthright comes from the great Vandalia, a region not very far away from where we are now, and I was born in a city named Soldina]. "Vandalia" is clearly the Latin equivalent of Andalusia – although it is arguably a late Latin name, given to the eponymous Germanic tribe.[14] The case of "Soldina" is less obvious: since Narciso Alonso Cortés first suggested that this fancy name might refer to Seville (136), critics have reproduced this identification, even though it was merely a conjecture. Another candidate might be Écija, as found in Diego Dávalos y Figueroa's *Primera parte de la miscelánea austral* (1602): "ay tradicion que esta ciudad tuuo por nombre Soldina Augusta, y otros dizen gran Soldina" (173v) [there is a tradition that says that this city was named Soldina Augusta, and others say great Soldina].[15] The classicization of place names parallels in this way the mythical geography of blood purity. While the locations in the north of Spain appear with their common place names (Duero, Pisuerga, León, Galicia), those in the south, especially if they are or merely sound Arabic, are excluded from pastoral representation unless they adopt a classical name (or a name that sounds classical). The literary selectivity of "ancient place names" reveals that the primary goal of classicism is the substitution of Arabic names, and not a full "restoration" of ancient Spain.

This erasure of Arabic place names affects almost every pastoral romance, and not even Madrid or Alcalá de Henares, political and cultural centres of the empire, are free from suspicion. In Cervantes's *Galatea* (1585), when the shepherds Tirsi and Damón burst into the pastoral setting, the shepherdesses Galatea and Florinda are surprised [suspensas]. The shepherdess Teolinda, who does know Tirsi and Damón, introduces them in the best way she can to sooth the anxiety towards the outsiders:

> tenéis hoy en vuestras riberas a los dos nombrados y famosos pastores Tirsi y Damón, naturales de mi patria; a lo menos Tirsi, que en la famosa Cómpluto, villa fundada en las riberas de nuestro Henares, fue nacido; y Damón ... si no estoy mal informada, de las montañas de León trae su origen y en la nombrada Mantua Carpentanea fue criado. (251)
>
> [you have today on your shores the two well known shepherds Tirsi and Damón, who are from my own land; at least Tirsi, who was born in the famous Complutum, a city founded on the shores of the river Henares; and Damón ... if I am not badly informed, brings his lineage from the mountains of León, and was raised in the notorious Mantua Carpetana (Madrid).]

In both cases, the newcomers are integrated into the pastoral community only by concealing the Arabic place name of their respective places of origin. *Complutum* was the alleged Latin name of Alcalá de Henares, a place name that was easily identifiable as Arabic,[16] and this substitution is found in many other pastoral romances.[17]

On the other hand, *Mantua Carpetana* refers in this text to Madrid; although there is still some uncertainty of whether Madrid is an Arabic name or not, most early modern authors believed so.[18] *Mantua Carpetana* was actually used in medieval Castilian historiography for the territory comprising the surroundings of Toledo, whereas the ancient place name of Madrid was considered to be "Ursaria" [Land of Bears]. It is not until the middle of the sixteenth century, that is, by the time that Madrid was about to become the permanent capital of Spain, that *Mantua Carpetana* was conceived to be its "original name," with the aim of forging a prestigious foundational origin in competition with other European capitals (Delage 64–5).[19] In this we clearly see how the repertoire of "ancient place names" was carefully constructed and rearranged in early modern Spain.[20]

As mentioned above, the place name that is cited most often in pastoral literature and almost in any literary text is "Betis." It serves as a cover-up for "Guadalquivir" which is an easily recognizable Arabic name meaning "Great River."[21] In many cases, Betis also serves as a metonym for Seville or even for all of Andalusia, such as in Cervantes's *La Galatea* (1585), where the shepherd Lisandro says that he was born "[e]n las riberas del Betis" (*Galatea* 188–9) [on the shores of the Betis].[22]

Finally, the case of the river Turia – which replaces the Arabic name "Guadalaviar" – in Valencia is probably the clearest case study in which place-name substitution was effective, revealing that the actual removal of Arabic place names was far from being a mere poetic licence and implied an underlying transformative desire. As Vicenç Rosselló i Verger demonstrates (42–4), the "ancient," place name Turia was in reality an invention of Pere Antoni Beuter's *Primera parte de la història de València* (1538). The way Beuter tells it, it would seem that Guadalaviar was a past name that was no longer in use by the first half of the sixteenth century: "Los moros, en lo temps que ocuparen l'Espanya, lo nomenaren Güetalaviar, que vol dir riu blanc" (88) [During that time in which they occupied Spain, the Moors called it Güetalaviar, which means white river]. While I have not been able to document when "Turia" officially replaced "Guadalaviar," it becomes clear that early modern non-literary texts referred to the river by its Arabic name well into the seventeenth century, and that the official substitution did not happen probably until the nineteenth century.[23]

As Martire d'Anghiera had complained, Beuter's identification of the ancient place named Turia with Guadalaviar is based on a loose interpretation and selection of the otherwise vague and contradictory classical sources. Notwithstanding, while Beuter's "discovery" was a matter of controversy in geography and historiography, it was enthusiastically and immediately embraced in pastoral fiction, perhaps contributing to the consolidation of "Turia" as the "true" name of the river of Valencia. In Gaspar Gil Polo's *Diana enamorada* (1564) characters talk about the "famoso Turia, río principal de aquellos campos" (143) [the famous Turia, main river of those fields], although they still alternate it with the name Guadalaviar (126). Lope de Vega's *Arcadia* (1598) names it Turia, introducing an erudite note to clarify that it is a "río de Valencia, llamado de los moros Guadalaviar" (107n117) [river of Valencia, named by the Moors Guadalaviar]. Two years later, Gaspar Mercader begins *El prado de Valencia* (1600) "[e]n las riberas verdes y floridos linderos del famoso Turia" (13) [on the green shores full of flowers of the famous Turia], without ever mentioning the name Guadalaviar.

As we have seen, the substitution of classical place names for Arabic ones should be laid alongside the fact that place names not suspected of being of Arabic origin are not replaced in pastoral fictions – even when classical versions were also available for them. This selectivity of the classicization strategy reveals that its only functionality is the erasure of the legacy of the Islamic period, and there is not a real interest in fully recreating or "restoring" an ancient immemorial Spain just for the sake of doing so.[24] Furthermore, the use of ancient place names serves as a temporal sign: it conveys the image of an idealized community of shepherds and nymphs that inhabit a classical Iberian Arcadia where the Islamic period has never taken place, where there is a perfect continuity in the identity of ancient Iberians and early modern Spaniards.

Inhabiting the Past with Shepherds

In order to grasp the influence of certain conceptions of history in the construction of the pastoral Arcadia, it is necessary to consider pastoral romances within a wider intellectual context and, conversely, to analyse how the writing of history is equally influenced by the idealizing drive of the pastoral community – both in its rustic and courtly versions. A signal characteristic of early modern Spanish historiography is the search to recreate an image of Iberian prehistory in as remote a past as possible. Starting with the Old Testament account of how the three sons of Noah – Shem, Ham, and Japheth – spread their lineages throughout the world (Genesis 9), there are several histories that seek to prove that Tubal, son of Japheth, was the founder of Spain. The arrival of Tubal in Spain was taken from the *Antiquities of the Jews*, by the first-century historian Josephus, and the reference was amplified by medieval historiography through Saint Isidore, Jiménez de Rada, and Alfonso X (Lida de Malkiel 11–20). The definitive boost for the invention of a mythic ancient past was Annio de Viterbo, who, in his *Commentaria super opera auctorum diversorum de antiquitatibus loquentium* (1498), elaborated a list of the twenty-four kings of Spain who descended from Tubal (Estévez Sola 213–17). The historiographical fabulations of Viterbo had a tremendous impact in Spain, where they were used to demonstrating that the ancient Iberian Peninsula was already a civilization on its own, far before the Roman period.[25] These myths entered Spain through humanists such as Antonio de Nebrija's *Muestra de la istoria de las antigüedades de España* (1499) and Lucio Marineo's *De rebus Hispaniae memorabilibus* (1532), and they were reproduced with uneven enthusiasm by Spanish historians, although, as both Chantal Grell (239–41) and Pablo Fernández Albaladejo (125–37) point out, Annio de Viterbo only summarized and rearranged the myths already invented by medieval Castilian historiography.[26]

The account of military deeds and the compilation of royal genealogies, while providing prestigious precedents for early modern empire, were insufficient to convey an image of the ancient inhabitants of the Iberian Peninsula that would be recognizable for their contemporary audiences.

The material culture of ancient Iberians still needed to be reconstructed in order to demonstrate the cultural continuities between them and the early modern inhabitants of Spain. When confronted with these quandaries, early modern historians found in sheep herding the material for the construction of an image of ancient Iberia that helped convey a sense of uninterrupted cultural identity within an essential concept of the history of Spain.

This pastoralization of history, however, even though trying to impose a unifying vision of the Spanish people, is informed by competing political agendas and different conceptions of collective identity. The cultural prestige attained by the image of the shepherd provided a repertoire of images that historians could borrow for the reconstruction of the past in a meaningful conceptual frame. While the pastoral was widely recognized as a prestigious pan-European courtly trope, sheep herding served a more specific local role, because the production of wool and its export to the rest of Europe was the main economic activity in Castile since the late Middle Ages. The emphasis on the continuity of Iberian sheep herding across different historical periods helped historians overcome anxieties about cultural continuity between ancient and early modern Spain at the same time that it favoured one specific economic activity over others. Thus it contributed to a proto-national political discourse which obtained most of its legitimacy from the "restoration" of an earlier, largely imaginary "Spain" temporarily lost during the Islamic period.

Because of the influence that sheep herding had on many aspects of material culture, it permitted identification with a past that was presented as a quotidian, familiar experience. Such an effect of familiarity in the blending of both past and present sheep herding is evinced when the historian Ambrosio de Morales, in his *Las antigüedades de las ciudades de España* (1575), addresses the importance of ovine livestock in the Iberian Peninsula under the Roman empire. After quoting ancient historians like Strabo and Pliny praising Iberian wools (31v–2r), he suggests that the vibrant wool industry of the past has to be understood through the lens of the present: "Mas para bien comprehender la multitud y bondad de los ganados de España, es bien considerar las muchas lanas que sin jamas cessar se sacan de España para Italia y para Flandes ... Y todo es testimonio de la infinita multitud de los ganados de España" (32r) [But in order to fully comprehend the multitude of goodness of Spain's livestock, it is necessary to consider the large quantity of wool that is endlessly taken out of Spain for Italy and Flanders ... And all is evidence of the infinite multitude of livestock in Spain]. Here Morales deliberately conflates past and present times, not, as it was customary, to say that the prestige of the past serves to praise a present activity, but rather to argue that the familiar present is the visual evidence ["testimonio"] that ultimately corroborates his retroactive description of the past. This pastoralization of Iberian history establishes an equation between timeless sheep herding and Spanish identity which serves to impose a concept of cultural continuity in the social imaginary.[27] The association between sheep herding and historic identity was constructed by emphasizing the role of shepherds who were mentioned in previous historical sources.

Pastoralization both in time and space can be conceived as a tool of cultural and historical homogenization. Just as different Iberian territories were subjected to a process of cultural homogenization, so were different layers of the past made to conform to a single unifying cultural image. Probably because the eight centuries of Islamic presence were construed as interference to the natural development of Spanish history, early modern historians channelled most of their efforts towards stressing the continuities between ancient and medieval Iberia. This colonization of the past is an extension of the logic behind the *reconquest*, an attempt to restore the essential sameness which sixteenth-century audiences shared with their ancestors. As we have seen in the case of Ambrosio de Morales, the similarity is created retroactively by infusing their ancestors with characteristics that sixteenth-century Spaniards could identify as their own. This temporal colonization is achieved by stressing the role of existing shepherds in Iberian history, or by transforming a historical (or pseudohistorical) character into a shepherd, crowding the past with them and turning them into the bearers of Spanishness.

Significantly, historical sources tend to refer to the presence of shepherds in the moments most fraught with anxieties over Spanish identity. Already in late medieval historiography, the legends about the fall of Spain ("la pérdida de España"), which describe the defeat of the last Visigothic King Rodrigo against Muslims in 711, are amplified by the apocryphal episode of the king's penance. In the *Crónica sarracina* (c. 1430) [*Chronicle of the Saracens*], by Pedro del Corral, the fugitive king is finally welcomed by a *mayoral*, who prescribes for him the protocol for his penance (2:400–4). Although the character is clearly a priest and there is no mention of livestock in the passage, the term *mayoral*, like the equivalent term *pastor*, could refer either to a shepherd or to the clergy charged with congregational care.[28] The polysemy of Corral's *mayoral* is exploited in the anonymous ballad "Después que el rey don Rodrigo," which splits him into two different characters, a shepherd and a hermit, somehow implying that both herdsmen and clergy helped each other in preserving Christian identity under Islamic rule, once the Visigothic monarchy disappears with Rodrigo's death. The first character that King Rodrigo encounters in the ballad when he is fleeing the Muslims is an anonymous shepherd; he is a rustic man who shares his bread with the defeated king, and directs him to the hermit who will impose on him his deadly penance (Menéndez Pidal 109).

The probable dates for the composition of this ballad, between 1480 and 1550 (Drayson 64), coincide with the period in which the shepherd becomes a figure increasingly identified with the Old Christian. It is therefore not surprising that in Miguel de Luna's *Historia verdadera del rey don Rodrigo* (first part 1592, second part 1600) [*True History of King Rodrigo*], the same shepherd that encountered the fleeing King Rodrigo is depicted very differently. Unlike previous historiographies of Spain, Luna's history was intended to confront the predominance of *goticismo* [Gothicism] and to create a sympathetic perception of the Islamic past, aiming at the defence of the Moriscos, as they were the victims of markedly increased repression by the end of the sixteenth century.[29] In his version, after Rodrigo flees from the battle, the captain Tarif Abenziet offers a reward for his capture and both Muslims and Christians alike pursue him, but the only trace they are able to find of Rodrigo is "un pastor, el qual estava vestido con los vestidos del Rey Don Rodrigo" (43) [a shepherd dressed in King Rodrigo's clothing].

The shepherd is brought into the presence of Tarif Abenziet and the ruse is soon discovered:

> y examinando al pastor (como buen rustico que era, de pocas palabras, y menos razones) les dixo, que no sabia mas de que estando apacentando su ganado en aquella sierra llego a el un hombre cauallero en un cauallo muy fatigado, y cansado, al parecer, con aquel vestido que el traia encima, el qual con el gesto ayrado le mando que se desnudasse sus çamarros, y los tomo, y auiendose el desnudado, se los vistio, y le mando al pastor que le vistiesse aquel vestido suyo: y le pregunto si tenia algun bastimento, y el pastor le dio de lo que al presente tenia, y tomandole el cayado de la mano, le mando que le guiase al camino; y guiado, tomo una ladera arriba, y subio por ella hasta que le perdio de vista, y que no sabia mas otra cosa. (43–4)

> [interrogating the shepherd (who was, as any other rustic, of few words and fewer arguments), he told them that he knew only that he was grazing his livestock in those mountains when a rider came with a very tired horse; and, seemingly tired of the clothing he was wearing, he ordered angrily that he

(the shepherd) take off his *çamarros* (shearling coats); and the king took them, and, having undressed, put them on (the shepherd's clothing). And he asked him whether he had something to eat, and the shepherd gave him everything he had, and taking his crook from his hand, he ordered the shepherd to show him the way. And when he saw it, he went up a hill until he disappeared from sight. And that was all he (the shepherd) knew.]

Luna's rewriting of this episode focuses on the shepherd rather than on King Rodrigo. There is not even complicity between them, since the shepherd complies only under the demands of King Rodrigo, and not out of his own will to help him. Furthermore, by depicting the shepherd as an inarticulate "rústico," the *Historia verdadera* deprives the character of any agency he might have had in previous versions of the apocryphal episode, thus counteracting the perception of shepherds as an idealized community of Old Christians.

While the shepherd whom King Rodrigo encountered functions in all these versions (excepting Luna's) as the symbol which stands for the survival of traditions that predated the Islamic period, the same logic posits a similar figure when Christians finally managed to achieve political and military control over the Iberian Peninsula. This is clearly reflected in the Marian apparitions to shepherds along frontier territories, as we have seen in chapter 2, but these legends worked mostly at the local level; the major military event that included the participation of a shepherd, as emphasized by medieval and early modern historians, was the battle of Las Navas de Tolosa (1212). Although the definitive end of Islamic political power takes place with the fall of the Nasrid kingdom of Granada in 1492, the balance of power was thought to have shifted in favour of the Christian kingdoms in the years following the battle of Las Navas de Tolosa, led by the Castilian King Alfonso VIII. As it was construed in medieval sources, this Christian victory opened strategic access to most of what is nowadays Andalusia, and subsequent Castilian and Aragonese conquests over the following three decades marked the end of al-Andalus as a military and political rival.[30]

The defeat of the Muslim army at Las Navas de Tolosa was not achieved solely by Christian military prowess, at least according to earlier sources. The mountainous topography benefited Muslim defenders, and the supply lines for the Christians were so stretched out that they soon realized the need for a retreat. According to medieval sources, many of which were written by alleged first-hand witnesses of the battle, the course of events was dramatically reversed thanks to the intervention of a rustic who guided the Christian army through a secret path in the mountains.[31] While the medieval sources sometimes mention in passing that he might have guarded livestock at some point in life prior to his participation in the battle, it is only in the sixteenth century that his status as shepherd takes prominence.

He is mentioned in most of the early modern historiographical works, such as Esteban de Garibay's *Compendio historial* (1571) (738) and Juan de Mariana's *Historia general de España* (1601) (1:337). In Cristóbal de Mesa's epic poem *Las Navas de Tolosa* (1594), the shepherd attains full prominence, not only by leading royal troops through the mountains, but also by delivering a long speech encouraging the Christian monarchs and reminding them of their Visigothic ancestry (191v–205v). Argote de Molina, in his *Nobleza del Andalucía* (1588) [*Nobility of Andalusia*], introduces him when the Christian kings are hesitating over whether to attack Sierra Morena or to withdraw from the battlefield:

> Estando en este trabajo llegó a ellos un Pastor de ganado (quales son oy los grandes Pastores de la tierra de Soria) preguntando a gran priesa por los Reyes, y como fuesse llevado en su presencia, dixoles con gran determinacion, que no estuviessen en cuydado, que el los passaria sin peligro por el camino, onde repastava su ganado, que era lugar de mucha yerva, y de buenas aguas. (1:28r)

> [While they were dealing with this difficulty, a herdsman (just like the great shepherds in nowadays Soria) came asking for the kings. And as he arrived in their presence, he said with great determination that they should not worry, and that he would guide them without danger through the place where his livestock was grazing, abundant in grass and good water.]

By remarking that this shepherd, who had lived more than three centuries ago, is no different from any other shepherd of his own time ("quales son oy los grandes Pastores de la tierra de Soria"), Argote de Molina ensures that the reader does not miss the cultural continuity implicit in sheep herding.[32]

For the sake of constructing a genealogy of blood purity, all the historical accounts, and even most present historiographers, seem to presume that the shepherd was a Christian, or rather a Mozarab – a Christian living under Muslim rule. Only María Dolores Rosado Llamas and Manuel Gabriel López Payer point out that the chroniclers of the Catholic kings conflated the shepherd of Las Navas de Tolosa with a different, preexisting character named Martín Alhaja, and they suggest that he might be a Muslim shepherd as well, a plausible possibility that is rarely considered (256).[33]

Thus sheep herding as a performance of collective identity haunts accounts of both the beginning and the end of the *reconquest*, the historical axis around which early modern historiography turns. For these authors, shepherds become the piece of evidence that shows that the Iberian Peninsula before the Islamic period was the same realm which allegedly re-emerges after 1492.

For the same reason, a similar rhetorical movement of inclusion and exclusion is also set in motion in historiographical works about contemporary events related to the Moriscos. Luis del Mármol Carvajal's account of the Morisco uprising of the Alpujarras, *Historia del rebelión y castigo de los moriscos del reino de Granada* (1600) [*History of Rebellion and Punishment of the Moriscos from the Kingdom of Granada*] systematically avoids portraying the Moriscos as shepherds. From the very beginning of his work, Mármol Carvajal mentions the Moriscos as owners of livestock, but not as herders: "por todas partes tenían los moros muchos lugares poblados de gente rica por la cría de la seda y del ganado" (128) [All over (the kingdom of Granada) the Moors had places populated with people rich from the cultivation of silk and the raising of livestock]; the Moriscos inhabiting the lands of the Marquis of Cenete were likewise described (217), as were those living on the banks of the river Almanzora (275), just to mention a few.

The minute descriptions of each place inhabited by Moriscos in the kingdom of Granada invariably include a reference to the quality of pastures and livestock.[34] Mármol Carvajal also describes all of the movements of the Moriscos wandering with their livestock trying to escape the warfront nearly a dozen times, and the Morisco livestock seized as booty by the Christian army some sixteen times. However, in spite of this overwhelming amount of information about Morisco herding, there is not a single passage in which Moriscos are described as shepherds. In contrast, Mármol Carvajal automatically applies the word *pastor* when he is referring to Christian shepherds, even though they are only mentioned on two occasions (211, 242). Like many of his contemporaries, as we have seen in the introduction, Mármol Carvajal seems to imply that livestock and pastures exist in Spain without any direct relation to the Moriscos who take care of them. The Moriscos are always described in terms of their ethnic identity and almost never for their profession when they engage in herding – or any other profession attributed to Old Christians in the imaginary. Mármol Carvajal's text exemplifies how the phrase "pastores moriscos" becomes unthinkable even when faced with all the evidence available to early modern Spaniards.

In many instances, "proving" that shepherds inhabited the Iberian Peninsula immediately before the Islamic period is insufficient, and therefore the clock has to be set back to the very origins of "Spain," far before such a name even existed. Baltasar de Echave's *Discursos de la lengua cantabra vascongada* (1607) [*Discourses on the Basque Language of Cantabria*] represents such an extreme and alternative theory of pastoral beginnings. He traces the birth of the "Spanish people" back to biblical times, to the arrival of Tubal, grandson of Noah, who "partió con toda su familia, y ganados de vacas y ovejas y otros animales domésticos que le pareció bastarían, para dar principio a la nueva población de la región de España" (7) [left with all his family and herds of cows and sheep and other domestic animals that he deemed would be sufficient to repopulate the region of Spain].[35] He mentions this activity several times at the beginning of his *Discursos* (10, 13–14, 27r, 34v–5v), and even presents Tubal giving detailed instructions to those who accompany him both on how to raise livestock and on how to dress with its byproducts: "repartiéndoles vacas y ovejas a los que se aplicavan a la pastoría en los campos acomodados ... vistiéndose de la lana, de que hazían su modo de vestido, que no era más que un sayo con que se cubrían" (13) [distributing cows and sheep to those devoted to herding in the appropriate fields ... and dressing in wool, of which they made their kind of clothing, which was but a tunic they used to cover their bodies]. This commentary is followed by a detailed account of the sartorial and economic similarities between ancient Iberians and early modern Basques, demonstrating through the central role of sheep herding how every aspect of the material culture of the original Spaniards has been preserved by the Basque people.

Echave's shepherds are at the same time both biblical figures and Old Christians, proto-Basque and proto-Spaniards. They are everything at once, but for Echave there appears to be no contradiction. To be sure, Tubal does not evoke Jewishness, because he would predate Abraham.[36] Rather, Echave mentions him to demonstrate that Spain is the oldest country in Europe, within an ideological context which honours antiquity as a political value. There is no incompatibility either in the simultaneous Spanish and Basque identities of his shepherds, since in the early modern period the interrelation between these two collective identities was very different than from what it is today. Basques managed to attain a rather privileged status within Castile, which granted them "hidalguía universal," an exceptional legal concession of blood purity based on territory, while other Spaniards had to prove their "limpieza de sangre" on an individual basis. As a result, early modern Basque historiographers and jurists usually regarded themselves as the "true Spaniards" (Juaristi 18–25). This is why Echave is so interested in demonstrating that these ancient Basques were the initiators of the prestigious stockbreeding which so characterized the Iberian Peninsula. Their legacy would not only be cultural, but also economic in nature so that Spain would owe much of its prosperity to their ancient efforts. For Echave, the first action of the successive waves of European and African conquerors on the Iberian Peninsula since prehistoric times was "tomarles sus ganados que hera su particular y suprema riqueza" (39v) [taking them from their livestock, which was their own, superb wealth]. Consequently, after having emphasized how much proto-Basque stockbreeding influenced ancient Iberia, the text stops mentioning livestock as soon as the first invaders seize it. This sudden silence about who holds the herds shows how Echave is creating an image of ancient Basques as "true shepherds," and therefore the guarantors of cultural purity, while denying participation in sheep herding to other Iberian peoples who might claim different origins as markers of identity, be they Roman, Gothic, Jewish, or Arabic.

Etymology and Sheep Herding

In the sixteenth century, the existence of sheep herding in ancient Iberia could not be demonstrated by the nascent discipline of archaeology, and early modern historiographers were obliged to rely on different methods in their efforts to unveil it.[37] One of the means available to them for retrieving the past was etymology, which early modern historians regarded as a valuable historiographical tool that, in its supreme malleability, could be easily manipulated to fit their needs.

This tendency is best exemplified in Bernardo de Aldrete's *Del origen y principio de la lengua castellana* (1606) [*On the Origin and Beginning of the Castilian Language*], which is primarily conceived as a linguistic treatise that aims at connecting Spain with classical antiquity.[38] When Aldrete deals with the term "Betis" and the different names that were historically given to the river Guadalquivir, he is careful to relate every one of them to ancient sheep herding, sometimes with painstaking effort. Thus he attributes the ancient name "Perces" to the Greek verb meaning "to change colour" [mudar de color], and hence deduces that the ancient inhabitants named it in this way "por la variedad, i mudança de color, que en la lana del ganado causauan las aguas deste rio" (354) [because of the variety and change in colour that the water of this river made in the wool of sheep]. Even when he is confronted with a different reading in his sources, Aldrete manages to shift it towards his own interpretation: "Stephano tratando deste rio, dixo, que los naturales lo llamauan Pescos, i no Perces, como se lee, siendo facil trocar la S en R ... i Pescos quiere dezir Vellocino" (356) [Stephanus, talking about this river, said that the inhabitants called it *Pescos*, and not *Perces*, as we can read, being so easy to switch an S by an R ... and *Pescos* means fleece].

Regarding the most well known name for the Guadalquivir, "Betis," he starts by rejecting the hypothesis that it might derive from Hebrew, and proposes instead other etymologies, all of them related in one way or another to livestock. The first one is a purported Latin word: "Iulio Pollux dize de la Baeta, que era vna tunica larga de pellejos. Los çamarros, o pellicos de los pastores de inuierno abrigan, i en el estio defiende, i guardan del sol" (355) [Julius Pollux says that the *Baeta* was a long tunic made of skins. The *çamarros* or *pellicos* (shearling) of the shepherds warm them during the winter, and during the summer they defend them and protect them from the sun]. He goes on with a long list of classical quotations that praise the livestock of ancient Hispania, and makes use of the proverbial golden colour of Andalusian wool in antiquity to identify it with the mythic Golden Fleece (355). Aldrete continues in this way, trying to relate every ancient name of the river Guadalquivir to sheep herding, creating the hyperbolic image of an ancient Andalusia abundant with flocks of sheep. His readers – especially livestock owners – would feel at home in such a picture of the past, created not through archaeological research, but through the mastery of etymology as a historical tool that stresses the continuity between past and present.

Aldrete's aim of tracing the origins of Iberian sheep herding through the analysis of Latin and Greek words reveals the parallels and interrelations between the pastoralization of historiography and the pastoral literature in fashion in early modern Spain, because both tried to refigure the Iberian Peninsula by reinforcing both its classical legacy and its sheep herding culture. At the same time that the pastoral romances borrowed from historiography the repertoire of Graeco-Latin place names to replace those of recognizable Arabic origin, historiography borrows from the literary pastoral the iconographic prestige embedded in its community of shepherds. The main difference between these discourses is that, while the pastoral romance as a fictional idealizing genre can freely depict a Spain that has always remained "classical," a space where eight centuries of Islamic presence simply never occurred, historiography cannot fail to acknowledge it. What historiography could, and did, do was deny Muslims their rightful place in Spain's history by refiguring Christian and ancient Iberia as an idealized community of shepherds, and by depriving Muslims and Jews of any participation in sheep herding. Whereas historiographical accounts like that of Aldrete are complicit in the classicizing of sheep herding implied in the pastoral romance, there is no homogeneous project in early modern Spain for creating a unified narrative that comprises all the Iberian realms, at least not until the publication of the *Historia de rebus Hispaniae* (1592) by Juan de Mariana (Kagan 122). Therefore, competing projects of nation building created alternative historiographical narratives which sought different pastoral etymologies, each of them resorting to non-classical languages such as Basque or Arabic.

Echave, as discussed above, participates in a historiographical project which aims at proving that the original inhabitants of the Iberian Peninsula were all Basques. He shares with other historians their confidence in etymology as an indispensable tool for unburying the past and retrieving information about ancient sheep herding. Thus he writes a whole passage deriving many Iberian place names from Basque terms related to herding (33v–5v). What seems to be most significant is that, when explaining the etymology of Betis, Echave feels forced to offer a linguistic alternative to the classicizing version made by Aldrete, but at the same time agrees with him in relating it to the raising of livestock: "Guadalquivir, que antiguamente se llamó Betis, y Bética la Provincia que riega ... por ser muy apropiada para ganados de vacas; de que siempre fue abundante, como el mismo nombre en mi lenguage da bien a entender" (34r) [Guadalquivir was called Betis in ancient times, and Betica was the province that the river waters ... because it is very suitable for the raising of cattle, in which it was very abundant, as the name itself reflects in my language]. Echave does not clarify which is the Basque word that he has in mind. It is probably *behi* [cow], which is transformed to *bet* in compound names, such as *betalde* [herd of cows] and *betzain* [cowherd].[39] As the competition between Aldrete's and Echave's etymologies indicates, it does not matter which language we take into consideration in tracing the linguistic origins of "Betis" – they infallibly lead to prehistoric Iberian livestock.

Castilian and Basque identities, although competing against each other in many aspects, would still complement each other for many years to come. A different story is told when we come to consider other cultural legacies that were being excluded from the representation of Spain, as is the case for the Islamic period. I have found no evidence that the Moriscos themselves were trying to establish their own pastoral version of Spanish history.[40] Rather the insistence on the participation of Muslims in the creation of modern sheep herding seems to be a rhetorical argument that circulated primarily among Christians considering both the fate of the Moriscos and the place that al-Andalus and Arabic played in the reshaping of early modern Spain, as is clear in the case of Diego de Guadix.

The Franciscan Diego de Guadix was a prominent Arabist in early modern Spain. Because of his knowledge of Arabic, he was appointed as translator of the Inquisition of Granada in 1587 and was summoned to Rome in 1590, where he wrote and obtained in 1593 the licence to publish a lexicographical work entitled *Recopilación de algunos nombres arábigos que los árabes pusieron a algunas ciudades* (Cabanelas Rodríguez 238–40; García-Arenal and Rodríguez Mediano 367–71). Despite its title, apparently limiting it to toponymy, his work, which circulated only in manuscript form until its recent publication, constituted an encyclopedia of Arabic cultural influence on Spanish culture. Guadix, while unmistakably condemning Islam, tried to divorce religion and language: "para que ... salga del error que algunas personas tienen ... pensando que siempre anduuieron a vna la maldita seta de Mahoma y la lengua arábiga" (150–1) [so that ... some people do not make the mistake ... of thinking that the sect of Muhammad and the Arabic language were always the same thing]. Although he had a good knowledge of Arabic, which seems to have been his native tongue, his etymologies are highly imaginative, since, like most of his peer lexicographers, he relied mainly on phonetic similarities (Bajo Pérez and Maíllo Salgado 95–100).

What is striking about Guadix's text, though, is the number of places whose names are derived from Arabic terminology of sheep herding, a proportion that outnumbers any other semantic field in his dictionary. To cite some examples, he states that the place name "Aldana," in Aragon, "Consta de *al*, que, en arábigo, significa las, y de *danna*, que significa nuestras ouejas; de suerte que, todo junto, *aldanna* significa las nuestras ouejas" (245) [is composed of *al* which in Arabic means "the" and *danna*, which means "our sheep," so that, altogether, *aldanna* means "our sheep"]; or "Roa," which he derives from "*ro*^{c}*a*, que, en arábigo, significa pastores" (923–4) [*ro*^{c}*a*, which in Arabic means "shepherds"]. Even more significant for his ideological goal is the frequent place name "Mora": "Siempre entendió el mundo que este nombre significa muger mahometana, y es grande engaño, porque es *mara*^{c}*a*, que, en arábigo, significa dehesa, i., pasto o hervage" (826) [People always thought that this name means Muslim woman, but it is a big mistake, because it is *mara*^{c}*a*, which in Arabic means field, that is pasture or grazing land]. This etymology is particularly useful in de-Islamizing and re-pastoralizing the Arabic legacy in Spain, transforming the signifier for "Muslim woman" [*mora*] into the signifier for "pastures," the basic food source for livestock.[41] Guadix seems to be suggesting that, instead of regarding Moriscos suspiciously as crypto-Muslims, they should be considered as peer shepherds.

There is one more lengthy etymology which indirectly reinforces the link between sheep herding and Christianity in the Arabic language, when he deals with the place name "Çafarraya," in the kingdom of Granada:

> Consta de *fahç*, que, en arábigo, significa campo, y de *al*, que significa de las, y de *ru*ᶜ*aya*, que significa péccoras o ganados; de suerte que, todo junto, *fahçalrra*ᶜ*aya* significa el campo de los ganados o de las péccoras, porque este nombre *racaya* significa, en arábigo, lo que, en latín, este nombre *grex*, i., ganados o pécoras; y, trocándole las sílabas, hazen esta corrupción *Çafarraya*. Parecer a sido de grandes arábigos qu'este nombre consta de *fahç*, que, como queda dicho, significa campo, y de *ro*ᶜ*aya*, que significa pastorcicos o pastorcitos, porque es el diminutivo de este nombre *ro*ᶜ*a*, que significa pastores; assí que, todo junto *fahçalro*ᶜ*aya* significará el campo de los pastorcicos o de los pastorcitos ... porque e visto en el capítulo 26 del Sancto Euangelio de sant Matheo, en aquellas palabras latinas: *percutiam pastorem et dispergentur oues gregis*, en el texto arábigo está este nombre *ra*ᶜ*aya*, que corresponde a el nombre *gregis* del texto latino. (464–5)

> [It is composed of *fahç*, which in Arabic means "fields," and *al*, which means "of the," and *ru*ᶜ*aya*, which means sheep or livestock; in such a way that, all together, *fahçalrra*ᶜ*aya* means the field of livestock or the field of sheep, because this name *ra*ᶜ*aya* means in Arabic the same as Latin *grex*, that is, livestock or sheep; and hence, switching syllables, they say corruptedly *Çafarraya*. Some wise Arabians say that this name is composed of *fahç*, which as I have just said means "field," and *ro*ᶜ*aya*, which means little shepherds, because this is the diminutive of the term *ro*ᶜ*a*, which means "shepherd"; thus altogether *fahçalro*ᶜ*aya* would mean the field of the little shepherds ... because I have seen in chapter 26 of the Gospel of Matthew that those Latin words *percutiam pastorem et dispergentur oues gregis*, in the Arabic text there is this term *ra*ᶜ*aya*, corresponding to the name *gregis* in the Latin text.]

Although Guadix recognizes here that there are some uncertainties in the etymology of "Çafarraya," what seems clear is that it derives either from the equivalent Arabic word for "shepherds" or for "sheep," so the apparent debate leaves little room for doubt that the word belongs to the semantic field of sheep herding. Furthermore, in order to defend his preference, he points out the existence of an Arabic translation of the Bible. His uncertainty about the etymology is a strategic rhetorical move. Suddenly, the real point is no longer whether "Çafarraya" derives either from "shepherd" or from "sheep," but, rather to remind his reader that the Bible has been translated into Arabic and, therefore, that the Arabic language and culture is in no way incompatible with Christianity, which is exactly the same argument the Moriscos and their defenders were trying to make by the end of the sixteenth century.[42]

The emphasis Guadix gives to the etymologies that relate Spanish place names to Arabic words from the semantic field of sheep herding is driven not only by apparent or real play with homophonies.[43] Rather, it is conditioned by the deliberate aim to counteract the hegemonic historiographical project that underscores the sheep herding culture of ancient Iberia only to deprive the Moriscos of their proper place within it. Guadix takes a strategic twist that does not deny the allure of historiographical pastoralization, but appropriates it in order to reintegrate the Islamic period within the writing of Spanish history.

Without denying the importance that sheep herding has had throughout Iberian history, the appropriation made by early modern historians shows that the pastoral runs parallel to the inclusions and exclusions inherent to the crafting of an essential Spanish identity, both by the cultural homogenization imposed on different kingdoms and historical times and by the glossing over in silence of Muslim sheep herding – or, inversely, as in Guadix's case, by emphasizing it. As sheep herding informs the search for the origins of the "Spanish people," becoming in itself the visual sign for cultural continuity from a distant, inapprehensible past, it was also used to explain the origins of religious identity of early modern Spain.

Paganism and Ethnic Identity

Along with the use of place names, the other element that indirectly reflects the historical identity of the community of shepherds in pastoral romances is paganism. Despite the abundant bibliography on the importance of classical mythology in early modern Spanish literature in general, and the pastoral romance in particular, still to be explored is how it served as a mark of ethnic identity by reinforcing the Gentile nature (and therefore non-Jewish and non-Muslim) of the shepherds who invoke pagan gods.[44] At first sight, it seems natural that Pan is elected as the god of the community of shepherds in pastoral romances, as he is found in the classical and Renaissance models used by these authors.[45] In Eclogue 10, Virgil speaks of "Pan Deus Arcadiae" [Pan the god of Arcadia], and the shepherds go to visit his temple in Sannazaro's *Arcadia* (1504) (79–82).[46] The god

Pan presides over almost every pastoral fiction, such as Antonio de Torquemada's *Coloquios satíricos* (1553) where the characters refer to Pan as "dios de los pastores" (1:487) [god of shepherds].[47] While several mentions of the god Pan may sound merely formulaic, some of them enter muddy theological waters.

In Alonso Pérez's *Segunda parte de la Diana* (1563) the shepherd Sireno asks Parisiles to recount "lo del sacrificio de nuestro dios Pan, y como antes se le hazia, y desde quando se tiene en veneracion" (91r–v) [about the sacrifice to our god Pan, and how it was made before, and since when does this worship exist]. Parisiles answers:

> mi principio auia de ser declarando, quien este Dios sea, pero como el no le tenga, no se lo podre yo poner sino en si mesmo. Porque dezir que es todo uno Pan y Fauno, segun todos los autores muestran, no lo creo ... Pues dizen Fauno ser hijo de Pico, y padre de Latino: Pan hijo de Demogorgon Dios de la tierra. Dezir, como tambien affirman, que Pan y Syluano es lo mesmo, es falso. (91v–2r)

> [I should begin by declaring who this God is; however, since he does not have any (beginning), I would not be able to declare any other (beginning) than he himself. That Pan and Faun are one, as all authors show, I do not believe ... because everybody says that Faun was Pico's son, and Latino's father: Pan the son of Demogorgon, god of the earth. Also, saying that Pan and Sylvano are the same, as many state, is false.]

The shepherdess Selvagia interrupts Parisiles to warn him that his version does not match their theological knowledge: "tu nos has destruydo lo que en nosotros juzgauamos estar bien fundado, porque siempre hemos reputado a los tres por uno ... Una duda se me offrece acerca desto. Affirmaste que Pan no tiene principio y despues dixiste que era hijo de Demogorgon, como es esto?" (92r) [you have ruined what we thought was well-founded, because we have always considered the three of them as only one ... A doubt comes to me about this: you stated that Pan has no beginning, and then you said that he was Demogorgon's son: how is that possible?]. The mention of a potential Trinitarian reading of Pan suddenly brings to the social reality of early modern Spain what until now was merely a demonstration of mythological knowledge. When confronted by a question that mimics the dogma of early modern Spanish Catholicism, Parisiles responds that "tu me metes en cosa que no se como salga della con la honrra de mis dioses, o la mia" (92r) [you are dragging me to a subject from which I will not be able to leave with the honour of my gods, or my own]. Parisiles develops a confusing metaphysical explanation of Pan's genealogy, only to finally conclude that "basteos honrar a este dios Pan debaxo de que es dios de todo" (93r) [it should suffice for you to honour this god Pan under (the idea) that he is god of everything]. The rhetorical prowess of the wise Parisiles proves finally frustrating, because, as the fiction shows, he is not able to convincingly explain how Pan lacks a beginning but is the son of Demogorgon simultaneously; this inability and the final resort to dogma is in some ways analogous to the situation of Christian preachers who tried to explain the dogma of Trinity to Jews and Muslims in medieval and early modern Spain.[48]

Even more conflicting is the discussion found in Jerónimo de Covarrubias Herrera's *Los cinco libros de la enamorada Elisea* (1594). The shepherd Fontano, frustrated by the ill fortune of his love, decides to leave his town and bids farewell to his livestock:

A Dios a Dios que me voy
do no se si boluere,
voy en busca de mi fe,
ouejas si vuestras soy
a visitar os vendre. (43v)

[Goodbye, goodbye, I am going where I don't know if I will return. I go in search of my faith. Sheep, if I am yours, I will come to visit you.][49]

The ambiguity of "fe" [faith] as a term that refers to both sentimental fidelity and religious identity is common to Spanish sentimental fiction.[50] Fontano later develops the pagan religious side of the trope: "Acaba ya fortuna de affligirme, / harto estoy ya cansado y affligido, / que ayer fuy de Dios Pan oy de Cupido" (48v) [Fortune, finish torturing me. I am tired and afflicted, because yesterday I belonged to the god Pan and today to Cupid]. The narrator comments on Fontano's conversion that

aunque se via enamorado, sin saber de quien no menos pena le daba el ver que ayer huuiesse sido del Dios Pan y oy del Dios de amor. Y ansi reprimiendo las lagrimas, dixo a su amigo Rusticano. No pienses pastor que estoy poco corrido de hauer negado al Dios Pan. Estoylo tanto, que certifico que no me atreuiesse a entrar en su templo, puesto que cabe el me hallase. (48v–9r)

[even though he fell in love without knowing with whom, he was also sad realizing that yesterday he belonged to the god Pan and today to the god of

> love. And so, stifling his tears, he said to his friend Rusticano: "Shepherd, do not believe that I am not ashamed of having renounced the god Pan. I am so (ashamed) that I can certify that I would not dare to enter his temple, even if I was close to it."]

Rusticano tries to convince Fontano that there is no such religious incompatibility, because all of the shepherds in the region used to worship both gods equally (49r). The conventional ambiguity of the sentimental and religious "faith" in this pastoral romance eventually triggers a religious debate in which one shepherd defends the impossibility of worshipping more than one god while the other sees no incompatibility.

Thus, while Fontano voices the post-Tridentine vision and the project of religious homogeneity in sixteenth-century Spain, Rusticano seems to invoke an equally Iberian tradition of defending religious difference. Rusticano's position is certainly not the hegemonic discourse in early modern Spain, but traditional Iberian "convivencia" had not completely disappeared.[51] The name of the shepherd Rusticano, obviously derived from *rústico*, seems to link his theological opinion with a "rustic pelagianism," as Stuart Schwartz puts it; a popular attitude of religious toleration that was being persecuted by the Inquisition but was still alive among the people (242–55). While it may appear that the name of the character Rusticano should discredit his religious views, this passage dissolves the presumed view of Old Christian religious exclusivity as a clamour of "the people."

Spanish pastoral romances abound in disclaimers about conflicting representations of their pagan world. In Torquemada's *Coloquios satíricos*, Amintas states that he holds the life of shepherds to be the best in the world, to which his companion Florián replies that the shepherds can barely claim to be Christian, since, because of their profession, they are rarely able to attend mass and lack any basic theological knowledge (1:312). In Jerónimo Arbolanche's *Las Abidas* (1564), the paganism of the main character Abido is justified from the very beginning as part of the historical setting: "Lo demas que se trata destos Dioses, es a causa que el fue Gentil, por ser antes de la natiuidad de nuestro Señor Iesu Christo verdadero Dios" (14v) [Whatever is said about these gods is because he was a Gentile, because this happened before the birth of Jesus Christ our Lord and true God].

Strikingly enough, one of the pastoral romances in which pagan allusions are more pervasive is *Siglo de Oro en las selvas de Erífile* (1608), even though it was composed by the friar Bernardo de Balbuena, who would later become bishop of Puerto Rico. As the shepherd Cloris states, commenting on the confusion of sentimental and religious terminologies, "[s]i yo tuviera nombre entre los bucólicos, / estos lenguajes de amor heréticos / quizá los compeliera a ser católicos" (292) [if I was included among the bucolic kind, maybe I would compel these heretic languages of love to be Catholic]. Gonzalo de Saavedra y Torreblanca's *Los pastores del Betis* (1633) also underlines the pagan status of the shepherds. However, a cautious disclaimer follows the "fe de erratas," in Martín de Saavedra y Guzmán's edition of his father's work: "Siempre se protesto viuiendo el Autor deste libro ... ser palabras de lenguaje pastoril gentilico Hado, Caso, Fortuna, Suerte, Dioses, y otras palabras semejantes, puesto que todo lo que en el se contiene la resigna, y pone debaxo la correcion y censura de la Santa Yglesia Catolica Romana, verdadera Madre del mas sano sentido" (xx) [When the author of this book was alive, he always claimed that words such as Fate, Case, Fortune, Luck, Gods, and the like, were taken from the Gentile language, and that he submitted all the content of this book to the correction and censure of the Holy Roman Catholic Church, which is the true Mother of any correct meaning]. Clearly the tension between generic norms and religious orthodoxy provoked anxiety, at least on the part of some authors.

Since the ubiquity of the god Pan is clearly perceived as a sign of classicism, it is also a sign of paganism and, as such, its presence seems out of place within the climate of religious orthodoxy imposed by post-Tridentine counter-reformation. However, when churchmen attacked pastoral fiction, they did so not by focusing on paganism or doctrine, but by pointing out that they promoted lust and jeopardized female chastity. Thus the friar Pedro Malón de Chaide complained in *La conversión de la Magdalena* that pastoral romances hinder women's chastity: "¿Cómo dirá *Pater noster* ... la que acaba de sepultar a Píramo y Tisbe en la *Diana*? ¿Cómo se recogerá a pensar en Dios un rato la que ha gastado muchos en Garcilaso?" (1, 26) [How will she pray the *Pater noster* ... who has just buried Piramo and Tisbe in the *Diana*? How will she retire for a while to think about God when she has been wasting her time reading Garcilaso?].[52]

Despite the cautions regarding orthodoxy on the part of the authors of pastoral romances, it seems that neither the Church nor the Inquisition cared about the pagan content of their stories. Pastoral romances were never included in any list of prohibited books and they barely attracted any attention from censorship. Only the 1590 Lisbon print of Cervantes's *La Galatea* was expurgated of several references to paganism (López Estrada, "Cotejo" 87–8), and a passage dealing with astrology was expunged from Lope's *Arcadia* in the 1602 edition of Valencia (Morby, "Páginas olvidadas" 78).[53] Montemayor's *La Diana* was prohibited in Portugal between 1581 and 1624, arguably for political rather than religious reasons (López Estrada, "Prólogo" xii and xlv–xlvi). But these cases take place outside of Castile and they do not involve a general accusation of heresy for the genre.[54]

Graeco-Latin mythology is rarely equated to Protestantism, Judaism, Islam, or Amerindian religions. As Serge Gruzinski demonstrates, the transposition of the paganism of classical antiquity to the New World reveals the paradoxes of trying to remove indigenous cosmogonies for the sake of Catholic evangelization when Christians had their own parallel pagan system (99–101).[55] In spite of sporadic ecclesiastical attacks on Graeco-Latin paganism, it continued to be a prestigious referent for artists and writers. According to Gruzinski, it persisted because paganism allowed authors to deal with issues that would have been subjected to inquisitorial scrutiny had they chosen sacred materials (104).[56] However, one possible interpretation is that the literary flirtations with paganism were treated with leniency because they were not regarded as references to real religious practices, but instead as fictitious reconstructions of the land of one's ancestors and as part of the linear progress from a foundational culture to the revelation of Christianity's dispensation.

The tension between Christianity and paganism is implicit in the prophetical mode in which most of these romances are articulated. Scholars have focused mainly on the content of the pastoral prophecies, and how they were used to glorify patrons, the monarchy, symbolic historical figures, or the circle of poets in which they were written.[57] Thus the past from which prophecies were started is conceived to be a mere excuse for introducing present glories in a proleptic way. Be that as it may, prophecies also provide information about the characters who make them; that the characters live in the past of the readers is an obvious fact that usually goes unnoticed. It is of course a highly fictitious and vague past, but this does not mean that the shepherds inhabit a timeless utopia. Almost all of the pastoral romances are clearly located in Spain, which is by itself a foundational ideological statement, since "Spain" did not exist either as a political entity or as a cultural concept in any stage of the immemorial times in which authors of pastoral romances locate their fictions.[58]

In Gaspar Gil Polo's *Diana enamorada* (1564), the river Turia sings of "la alegre y próspera ventura / que el cielo a vuestros campos asegura" (145) [the happy and prosperous fortune that heaven assures for your fields]. In Lope's *Arcadia* (1598), the shepherds find in the temple of Pan "algunos retratos que para tiempos futuros estaban puestos" (424) [some portraits hung for future times]. In Juan de Arce Solorzeno's *Tragedias de amor* (1607), this prophecy is made by the river Sil:

> Agradecido el venerable Sil (como deidad, que sabe lo porvenir) a los ilustres hijos, que espera tener en su fertil ribera, por el alto renombre y gloria que por ellos ha de ganar ... consagra a inmortal memoria en esta sala de la fama los simulacros de aquellos, que han de llenar el mundo con la suya. Y aunque son de sucessos venideros en virtud de su deydad oculta, lo tiene todo presente, como si fuera passado. Y segun la relacion que desto nos tiene hecha, os yre diziendo cosas que han de suceder en gran numero de Olimpiadas, que os causaran gran gozo, y marauilla. (101v)
>
> [The venerable Sil (as a deity who knows the future), grateful to the illustrious offspring that he expects to have on his fertile banks for the high reputation and glory that he is going to win through them ... consecrates to immortal memory this gallery of fame, with the simulacra of those that will fill the world with his own (fame). And even though they are about future events, he has everything present as if it were past, by virtue of his hidden deity. And, according to the story that he told to us about this, I will be telling you the things that will happen in a large number of Olympiads that will bring you joy and amazement.]

Possibly the best example of the prophetical mode in the pastoral romances is Bartolomé López de Enciso's *Desengaño de celos* (1586). The plot does not explicitly state the historical period in which the shepherds are living until the shepherdess Lionida asks her lover Rosano to tell her "alguna moderna historia" (96v) [any modern story]. Rosano promises then that he will tell her "una digna de perpetua memoria: la qual sera tan alta ... y tan moderna, que aun faltan muchos años para llegar al tiempo en que sucedera, la qual oy yo un dia a nuestro sacro y claro Tajo, que contandosela como cosa passada, estaua a la altiua España" (96r) [a (story) worthy of perpetual memory, which will be so high ... and modern that there are many years to come until it happens, and I heard it once from our sacred and crystalline Tajo, who was telling it to superior Spain as if it were something from the past]. The "modern story" is the death of the Portuguese king Sebastião, whose death in 1578 in the battle of Alcazarquivir eventually leads to the annexation of Portugal by Philip II in 1580. This is only the prelude to a larger prophecy that is continually promised and postponed throughout the novel.

Later on, Celida and Laureno meet the old shepherd Criseo, who asks them to accompany him in his travels so that together they can see the things promised him (183r–v). Criseo says that one day he fell asleep on the banks of the Tajo and then there came to him "una nimpha de estremada hermosura, y soberana magestad ... La qual en su grandeça, conoci que era nuestra madre España" (183v–4r) [a nymph of extreme beauty and sovereign majesty ... who in her greatness I recognized was our mother Spain]. This nymph Spain let him glimpse the glories of the early modern empire (184r–v). The river Tajo showed up then to tell him that he should look for Laureno and Celida so that they could know the end of the prophecy (185v).

When the nymph Nirea takes them again to the cave of the Tajo, she shows some images representing Charles V, Philip II, Juan of Austria, and Philip III (256r–60r), and announces the coming of the Spanish empire along with the defence of Christianity: "verna tiempo, que en solo ellos la religion christiana se sustente, con mas veras y cuydado" (255v) [the time will come, when only they will support Christian religion with sincerity and care]. The anachronism of discussing Christianity with pagan shepherds is clear in *Desengaño de celos*, and Laureno inquires about this future religion: "querria sauer, que ley es esta, en que aquestos altos y valerosos personajes han de viuir" (260v) [I would like to know what this religion is, in which these high and brave people will live]. Nirea only reveals that Philip III will be one "quien sustentara, con mayor poder la christiandad" (260v) [who will support Christianity in its heights], but the prophecy is never fully disclosed, because it is postponed to a second part that Bartolomé López de Enciso promised (321v) but never wrote or never published.

Here, vagueness has a clear advantage. On the one hand, the characters are presented as unmistakably Spanish, even though they live in a remote past, assuming a perfect continuity with an immemorial Spanish identity. When Nirea shows the image of the Spanish monarchs, she states that "son legitimamente hijos de nuestra madre España" (255r) [they are legitimate offspring of our mother Spain]. Telling the alleged ancestors of early modern Spaniards that some of them are legitimate offspring of Spain implies that there are other "illegitimate" sons of Spain who are not mentioned in the prophecy but whose identity readers need to imagine on their own. Furthermore, by failing to address Laureno's question about the new religion, the coming of Christianity to Spain is disconnected from its biblical and Jewish origins in Israel. As the fiction of the *Desengaño de celos* puts it, the local deities of Spain (the river Tajo and the nymph Spain) appear as the prefigurations of Christianity, which they bequeath directly to the early modern Spanish monarchy, so that even Christ is elided in this narrative in favour of the genealogy of their alleged supporters.

The exact identity of these shepherds remains elusive in most of the pastoral romances. But it is unlikely that early modern readers did not frame these stories about "Gentile Spaniards" within both the historiographical search for the authentic remote identity of Spain and the debates over blood purity. Indeed, the discourse of blood purity used to be framed as if the distinction between *cristianos nuevos* and *cristianos viejos* was in fact so primeval that it was merely reproducing in new terms the division between the descendants of pagans and Jews. Writing in 1586, the same year that López de Enciso's *Desengaño de celos* was published, Gaspar de Uceda's treatise against the statutes of blood purity summarizes the social debate as "la discordia que al presente ay en España entre los desçendientes de gentiles contra los desçendientes de aquellos que antiguamente se convirtieron del judaísmo" (69) [the discord that there is nowadays in Spain between the descendants of Gentiles against the descendants of those who converted from Judaism]. When complaining about the scorn of the Jewish lineage of Jesus Christ, Uceda also underlines the paradox that Old Christians boast of not descending from the same lineage as the initiators of their religion: "Han venido los christianos quanto a esto a tener a Dios menos reverencia que tenían los gentiles a sus dioses. Tenía la gentilidad por summa honrra ser parientes de los dioses" (103) [Christians have ended up venerating God even less than Gentiles venerated their gods. Gentiles held it as a supreme honour to be related to gods].[59] Even though Uceda does not mention pastoral romances, his criticism of Old Christians' claims of descending from Gentiles can be read as a refutation of the pastoral prophetical mode already analysed in *Desengaño de amor*. The comparison of the debates on the statutes of blood purity suggests that the Graeco-Latin paganism of pastoral romances served not as a layer of erudition, but rather as a sign of temporality and ethnic identity, providing Old Christians with an idealized image of their alleged ancestors.

The God Pan and the Name of Spain

Although there is no explicit evidence confirming that early modern readers of pastoral romances identified Spanish pagan shepherds as the ancestors of Old Christians, a simultaneous development in historiography shows that the search for a link between Iberian prehistory, paganism, and the religious identity of early modern Spain was part of a larger project that went beyond the genre of the pastoral romances. In the case of historiography, the efforts concentrated again on etymology. The etymological project of depicting an Arcadian ancient Iberia described earlier still needed to be connected with the fact that Spain proclaimed itself the champion of Christendom. When facing the problem that most of the alleged Roman and Greek place names belonged to pagan, pre-Christian Iberia, historians satisfied the desire to conflate Christianity, historiography, the Arcadian myth, and sheep herding by forging a syncretic etymology of the name "España," thanks to the fortuitous homophony with the name of the pastoral god Pan.

The first step in producing this kind of "evidence" is probably taken by Florián de Ocampo, who claimes in his *Corónica general de España* (1543) that, because of a Greek settler named Pan, the Iberian Peninsula came to be known as Pania and later Spania (40r). Ambrosio de Morales expands the connection in his *Las antigüedades de las ciudades de España* (1575). Following Pliny, he states that "España" took its name from Pan, a Greek captain who ruled the Iberian Peninsula after its legendary conquest by Bacchus (34v). In order to demonstrate this, he drafts the basic phonetic evolution which was then followed by later historians:

> los Griegos ... usaron un articulo, is, para denotar una cierta excelencia, y tambien para significar aquello que los Españoles dezimos, lo de. Assi juntando este articulo a la usança Griega, con el nombre de Pan, haze Ispan, y denota la excelencia de aquel Señor; y juntando el articulo con el mismo nombre en el genitivo Griego, dize Ispanos, y dize, lo de Pan, que vale tanto como dezir, lo que posseya, y de que era señor el Principe Pan. (34v)

> [the Greeks ... used the article *is* to denote excellence, and also to mean what we say in Spanish *lo de* (that of). Thus, joining together this article in the Greek way with the name Pan, it results in *Ispan*, meaning the excellence of that lord. And combining the article with the same name in Greek genitive, it says *Ispanos*, meaning "that of Pan," which is the same as "that which he possessed and of which the prince Pan was lord."]

Both Ocampo and Morales are only trying to trace a classical origin for the name of Spain, relying upon the apparent phonetic similarity of "España" with the name of a legendary Greek ancestor, but they do not exploit all of its potential.[60]

Bernardo de Aldrete, in his *Del origen y principio de la lengua castellana* (1606), however, pursues this tack further by introducing the connection between the name "España" and the importance of livestock in the Iberian Peninsula, and by deriving it from an ancient Greek colonization:

> Arcadia se llamaua Pania tomando el nombre del Dios, que reuerenciaua [Pan] ... I como estos Griegos Arcades, i los Messenios sus vezinos, que passaron a España, fuesse gente mui dada a la cria de ganado, de que ai gran comodidad en esta tierra, por los abundantes pastos, i grandes dehesas que en ella ai ... como a Arcadia llamaron Pania, assi tambien parece, que a España donde tambien auia gouernado, consagraron su memoria al que reuerenciauan despues de muerto por Dios de los pastores. (275)
>
> [Arcadia was called Pania taking its name from the god that it worshipped [Pan] ... Both these Arcadian Greeks and their neighbours the Messenians who came to Spain were people well versed in raising livestock, which is very easy to foster in this land because of the abundant pastures and fields ... As Arcadia was called Pania, it also seems that in Spain, where he also ruled, people consecrated their memory to he who was worshipped after his death as the god of the shepherds.]

Like Morales, Aldrete struggles to make his etymology fit. First, he has to amend his own source: "Tengo para mi, que esso dixo Stephano, quando dize, que España se llamaua, Iberia, i Pannonia, i que a de dezir Pania, porque aquel libro esta mui lleno de faltas" (274) [It occurs to me that Stephanus said so when he says that Spain was called Iberia and Pannonia; and it should say Pania, because that book is full of errors].[61] Second, he has to explain the odd phonetic evolution from "Pania" to "España": "De Pania, no añidiendo articulo ... sino sola la S dixeron Spania, lo qual fue ordinario de los Griegos, que para dar mas lleno sonido la añidian" (275) [From Pania, not adding the article ... but only the S, they said Spania, which was typical of Greeks, who used to add it to give it a fuller sound].

Aldrete merely suggests that the name of Pan was introduced in Spain within the context of pagan religiosity to reveal simultaneously the classical pedigree of the name "España" and the existence of Iberian sheep herding from time immemorial. However, he might be hinting as well at a Christological reading in his etymology. The way he refers to Pan as the god "al que reuerenciauan despues de muerto por Dios de los pastores" [the god who they worshipped as god of the shepherds after his death] might read as an analogy with Christ, who was also worshipped after his death, and who was also thought to lead an idealized community of shepherds, as imagined by the biblical trope of the Good Shepherd. In fact, Plutarch's account of the death of the god Pan and the cessation of the Oracles had already been interpreted in late antiquity by Eusebius of Cesarea as the supersession of pagan religion with the arrival of Christ (Merivale 12–13).

This Christological identification circulated widely during the early modern period, both in France (Lavocat, *La Syrinx* 191–211) and England (Swaim; Merivale 13–34). In Spain, it was appropriated among others by Alonso de Villegas, who in the second part of his *Flos Sanctorum* (1583) clearly links Plutarch's story with sheep herding: "Llamauan los gentiles Pan al dios de los pastores, el qual naciendo Christo murio: esto es, perdio su credito y estimacion, porque nascia el que en especie de pan se avia de dar a los pastores y sacerdotes, y ellos communicarle al pueblo" (45r) [The Gentiles named the god of the shepherds Pan, who died when Christ was born; that is, he lost his credit and reputation, because one was born who, in the shape of *pan* (bread), was to be given to shepherds and priests, and they would communicate it to the people].[62] The dissemination of the reading of Pan as the prefiguration of the coming of Christ took a specific, additional overtone in Spain because of the homophony with *pan* (bread) and therefore with the host, which turned Pan into a symbol of transubstantiation (Osma 9–14; Kurtz 75–6; Collins 318–21).[63]

Juan de la Puente recaptures all these meanings of Pan in order to saturate the etymology of "España" in his *Conveniencia de las dos Monarquías Católicas* (1612). After outlining his own phonetic evolution of "España," quite similar to that of Aldrete (184–5), he does not fail to mention how the raising of livestock fits into this etymology: "Llamaron tambien *Pania* a este Reyno, porque Pan, Dios de los pastores, tenia especial prouidencia de los pastos y ganados Españoles. Fue España celebre en las naciones del mundo, por los buenos y grandes pastos que ay en sus fertiles dehessas" (188) [They also called *Pania* to this kingdom, because Pan, god of the shepherds, had especial inclination for Spanish pastures and livestock. Spain was celebrated among all nations in the world because of the good and large pastures that there are in its fertile grazing fields].

More important, de la Puente referred to the death of Pan as the prefiguration of the coming of Christ (182), linking it explicitly to the etymology of "España," thus arguing that its name proves to be inspired by the divinity:

> Quiso la Prouidencia Diuina, que este Reyno esclarecido se llamase tambien *Pania*, que es dezir tierra de Dios, para que hasta en los nombres uviesse tambien conueniencia entre estas dos Monarquias, la de la Iglesia Romana y el Imperio Español, adonde siempre reyno Dios, y fue de su patrimonio. Esta es la razon porque el nombre de España fue siempre sacrosanto, sin que ninguna nacion Barbara se aya atreuido a alterarle. (183)[64]

> [Divine providence wanted this illustrious kingdom also to be called *Pania*, which means "God's land," so that even in its names there was agreement between these two monarchies, that of the Roman Church and of the Spanish empire, in which God always ruled and which belonged to his patrimony. This is the reason why the name of Spain has always been sacred, without any barbaric nation daring to change it.]

The happy "coincidence" between the name Pan and the signifier "España" corroborated the ideological expectations of these authors when they began researching its etymology – that Spain was a country which had always been blessed with the natural conditions for the raising of livestock; that the pagan god who lent his name was actually a prefiguration of Christ; and, I would argue, that the god of Spain, whatever his name, was the god of the community of shepherds, in other words, the god of Old Christians.

As this chapter shows, pastoral romances and historiographical discourses influenced each other in the early modern period, producing an ethnocentric vision of Spain. One reason why the pastoral romances are so resistant to categorization as a historicist construct is because, while history shows an identifiable narrative structure of a nation from time immemorial that resisted successive waves of invaders throughout its history, the pastoral Arcadia represents a distinctly static and timeless description of a vague fragment of the past. Pastoral romances are presented as primeval historical episodes that have been isolated from the chain of historical events: if they appear to gravitate atemporally it is only because the historical sequence in which they are inserted has become unrecognizable to us after the change of the historiographical paradigm in the nineteenth century, but their pseudo-historicity would have been more evident to early modern Spaniards.

Pastoral romances became fictions about the primitive people of Spain, about a territory where shepherds had been established as the bearers of the cultural and genealogical legacy of an essential identity. The exclusion of those who do not participate in this essence is implicit in the collapse of a past moment and a present from which the concept of race emerges, excluding the middle term containing the ethno-religious contamination of unregenerated historical process. As Étienne Balibar notes, one of the effects of racism is "the imaginary 'fusion' of past and present in which the collective perception of human history unfolds" (45). The discourse of blood purity is in itself a narrative trope, a timeless confrontation between "Jews" and "Gentiles" that contaminates the writing of historiography and pastoral literature with the search for the "absolute ancestor."[65]

Even though Spanish pastoral romances take their figurative language from the Italian Renaissance and ultimately from Latin and classical models, Spanish early modern historiographical discourses portrayed Romans as invaders who did not alter the identity of ancient Spaniards in a significant way.[66] Therefore, the classicism invoked in these fictions did not necessarily imply the recognition of a Latin legacy, preferring instead to recreate an anachronistic and largely imaginary native civilization that was "classical" on its own terms. Within the need to Hispanicize classicism, rustic elements conveyed the local, specific "Spanish" feature of Arcadia. Whereas rusticity served to nationalize classicism, playing down the Italian / Latin origin of the literary form and motifs, classicist elitism kept at bay the inclusion of the popular in the representation of Spain. This way, by inserting rusticity and courtly classicism in a complex dialectics of mutual exclusion, pastoral romances neutralized the undesirable connotations of its Others, representing Latinity without Romans and rusticity without rustics.

Thus the hegemonic frame of the Spanish pastoral romances reflects the emergence of a historiographical vision that sought to deny Jewish and Islamic influences from the representation of Spain, at the same time that they provide historiography with the idealizing search for pastoral purity. This intellectual interpenetration happens at the same time as the descendants of Jews and Muslims were being socially stigmatized by increasing pressure from the statutes of blood purity or even expelled from the Iberian Peninsula. However, precisely because pastoral romances were not alien to the debates over the ethnic composition of Spain, they could also become an arena for controversy and for staging competing versions of Spanish identity, as the second part of the book details.

PART TWO

Contesting Ethnocentrism within the Arcadia

4 The Moor in Arcadia

¡Válame Dios ... y qué vida nos hemos de dar, Sancho amigo! ¡Qué de churumbelas han de llegar a nuestros oídos, qué de gaitas zamoranas, qué tamborines, y qué de sonajas, y qué de rabeles! Pues ¡qué si destas diferencias de música resuena la de los albogues! Allí se verá casi todos los instrumentos pastorales.

[My God! ... What a life we're going to lead, Sancho my friend! All those shawms regaling our ears, all those Zamora pipes, all those tambors, all those tambourines, all those rebecs! And maybe amidst all these different sounds the albogues will ring out! Then we shall have nearly all the pastoral instruments.] (trans. Rutherford 942)

Cervantes, *Don Quixote* II, 67

The previous chapter explored how Spanish pastoral romances constructed an image of primeval Spain and a landscape cleansed of ethnic "contaminations." Nevertheless, it would be too deterministic to argue that all pastoral romances automatically participate in this ethnocentric vision of Arcadia. In the same way that pastoral romances became a privileged vehicle for conveying an ethnocentric refiguring of early modern Spain, they were also a site of contestation. This chapter investigates those pastoral texts which participated in the debate over the place of the Moorish legacy in Spain between the 1560s and the expulsion of the Moriscos in 1609. It begins by analysing the place of the Moorish tale of *El Abencerraje* in Jorge de Montemayor's *La Diana* (c. 1559). The second part of this chapter then explores how, contrary to most pastoral romances, aristocratic Moorish cultural practices supersede pastoral in Gaspar Mercader's *El prado de Valencia* (1600). I conclude with an analysis of how Cervantes's entire oeuvre (particularly *La Galatea*, *Don Quixote*, and *Persiles*) struggles with the contradictions of excising the Moor from Arcadia, exposing the ideological and cultural quandaries of pastoral ethnocentrism.

El Abencerraje in Montemayor's *La Diana*

Any reader conversant with early modern Spanish literature may raise an obvious objection to the interpretation that the genre promotes the cultural cleansing of everything Moorish. Jorge de Montemayor's *Los siete libros de la Diana* (1559?) was the first and by far the most successful of Spanish pastoral romances, and the model for the subsequent development of the genre in Spain. In many ways, it also praised the pastoral life and classicism that were pervasive in the Spanish pastoral romances, and it should therefore be situated as the initiator, along with works by Boscán and Garcilaso, of an idealized vision of Spanish literary space that aimed at negating its Islamic past. However, at the same time that it provided the model for recreating a classicized landscape of Spain, *La Diana* was also responsible for spreading and reviving Moorish themes in Spanish literature. From the Valladolid reprint of 1561 on, *La Diana* was published with *El Abencerraje y la hermosa Jarifa* interpolated at the end of book four, related in the voice of the Andalusian character Felismena. *El Abencerraje* was an anonymous preexistent Moorish tale that recounts the exploits of the old frontier wars with the Nasrid kingdom of Granada.[1] The Moorish noble Abindarráez, one of the last descendants of the aristocratic family of the Abencerrajes (hence the title), is captured by the Christian noble Rodrigo de Narváez during a skirmish. As it turns out, Abindarráez approached the frontier as he was travelling in secret from Cártama to Coín in order to marry his beloved Jarifa. When Rodrigo de Narváez hears this story, he lets his captive go to his wedding under oath that he will return to his prison in three days. As promised, Abindarráez returns to Álora with Jarifa, who insists in accompanying him to his prison, until Rodrigo de Narváez finally liberates them.

In spite of the apparent simplicity of the story of *El Abencerraje*, the complex relationship established between the Christian knight Rodrigo de Narváez and the captive Abindarráez has sparked myriad and diverging interpretations. While some critics conceive the idealization of the Muslim aristocrat as a defence of the Morisco population, others consider that the game of power between Narváez and Abindarráez is but "a self-flattering depiction of Christian control over the Moor and his world" (Burshatin 197).[2] Almost every scholar, however, acknowledges the polemical nature of this text and its centrality in the debates over the Moorish legacy in early modern Spain. These pages, instead of examining the content of *El Abencerraje* while trying to decipher its agency, look at how its inclusion in *La Diana* interferes with the dialectical dynamic that pits the pastoral against the Moorish.

Because of the symbolic incommensurability between the identities of shepherds and Moors, both idealizing figures are hardly intelligible without each other. Claudio Guillén, criticizing the interpretation that texts like *El Abencerraje* may have served to defend the place of Moriscos in Spain, states that "the exaltation of the Moorish knight, always a nobleman, was far from being incompatible with a profound scorn for the *morisco*, who was always a plebeian" ("Literature" 193n39). While Guillén is right in pointing to the distance between the literary Moor and the real Moriscos in sixteenth-century Spain, he fails to acknowledge, nonetheless, that the same contrast between literary idealization and social image is applicable as well to shepherds, who are no less marginal than Moriscos in Spanish reality. David Darst also asserts that *El Abencerraje* is unrelated to the debates over the place of the Moriscos, but acknowledges that its referentiality is relational when he contends that the idealizing Moor of the text "has as little to do with contemporary attitudes about Moriscos and anti-Semitism as Garcilaso's or Montemayor's characters have to do with real shepherds and the countryside" (266). Julio Baena, more suspicious of the ideological value of literalness, rephrases Darst's dictum in the opposite direction, arguing that "los moros de *El Abencerraje* ... no son más – ni menos – reales que los pastores de las églogas" ("De Belisa" 78) [the Moors of *El Abencerraje* ... are no more (nor less) real than the shepherds of the eclogues]. Thus whatever the representational status one attributes to these literary Moors and shepherds, the verisimilitude of one idealizing figure is dependent on the other, and this corroborates the interrelated and oppositional meaning of both symbolic constructions.

Because of their oppositional meaning, the coexistence of Moors and shepherds within the same text invites an exploration of how both literary figures influence each other in *La Diana*. Some scholars refer to the version of *El Abencerraje* included in *La Diana* as "*El Abencerraje* pastoril" [the pastoral *Abencerraje*], to underline its bucolic proclivities (Fosalba, *El Abencerraje* 28–33; Guillén, *El primer Siglo de Oro* 140–8). As Israel Burshatin states, "the insertion of a Moorish version of pastoral mitigates those meanings having to do with fighting" (208). Similarly, André Stoll attributes to both the Moorish tale and the pastoral mode the utopian possibility of restoring the values of medieval *convivencia* (162). Nonetheless, if we interpret "pastoral" in the strictly literal sense used in this book, talking about a "pastoral *Abencerraje*" constitutes an oxymoron, because the Muslim characters of *El Abencerraje* are not represented as shepherds nor are herds even mentioned in their story. The Spanish expression "*El Abencerraje* pastoril" is even more contradictory, since, in the imaginary of early modern Spain, the adjective *pastoril*, as noted in the introduction, has a closer relation to sheep herding than the English equivalent "pastoral."

While the bucolic characteristics of *El Abencerraje* signal the truce in the frontier war and contributes to the idealizing Neoplatonic relationships between the characters, the relative erasure of Otherness stops short of granting their entrance to the community of the shepherds. Even if the inclusion of the gallant Moor works against the "political unconscious" of the genre, undermining the classicist denial of Moorish Spain, the categorical symbolic division between Moors and shepherds is nonetheless maintained. Whatever their particular political agency, almost every pastoral romance reproduces the concept that "shepherd" is synonymous with "Christian," while the figure of the Moor has no place in either category. In *La Diana*, this separation between shepherds and Muslims takes also a narrative dimension. Since *El Abencerraje* is a story told by the character Felismena, Moors are not at the same level of reality as the shepherds and therefore their worlds do not interact. Moors are only mentioned as a parenthetical reference to the past that ultimately confirms the ethnocentric and exclusive nature of Arcadia.

Another outcome of qualifying this version as "the pastoral *Abencerraje*" is that it confines the possible influences between the two texts to only one hermeneutical option. Scholarship has focused mainly on how *El Abencerraje* is resemanticized with its inclusion in the pastoral world and how the bucolic frame alters the reception of Muslims' religious and cultural difference (Souviron López, "*El Abencerraje*"). Talking about the "*El Abencerraje* pastoril" implies that the influence occurs only in one direction, and that it is only the Moorish tale that can adopt a limited array of bucolic elements from its pastoral frame. But it is also possible to consider the opposite alternative: that *El Abencerraje* resemanticizes and "orientalizes" *La Diana*, and that the characters of the pastoral world may well be influenced by the Moorish tale, thus opening the possibility of talking about "the Moorish *La Diana*." The diegetic and temporal separation between the Moorish story told by Felismena and the community of shepherds of *La Diana* does not preclude the identification with the misfortunes of the Moorish couple. It is especially true in the case of Felismena, since the relation of power between Rodrigo de Narváez and Abindarráez has many points in common with her own sentimental story with Don Felis. In that case, *El Abencerraje* may well be Felismena's tale, not only because she might be using it as an allegory of her own relation with Don Felis, but also because she uses the tale to foster her own cultural identity within the predominantly northern origin of the characters in *La Diana*.[3] Felismena's tale reminds the other characters that the idealization of the *reconquest* and the rejection of everything Moorish might be easy in northern Spain, but not in her native Andalusia, where the Moorish cultural legacy is much more quotidian and therefore harder to erase.[4]

The insertion of *El Abencerraje* compels the reader to reinterpret several passages of the pastoral world of *La Diana*. In the main prose text of *La Diana*, which we can unmistakably attribute to Montemayor, the story ends as Felismena and Don Felis meet in Coimbra, where a Portuguese shepherdess tells Felismena the history of the city. The shepherdess recounts the siege of the king Marsilio (Almansur), relating the name "Montemayor" to this event: "el nombre que tenía en lengua Portuguesa era Monte moro vello" (366) [the name it had in Portuguese was Monte moro vello (Mount of the Old Moor)]. Thus, even though the history of Coimbra emphasizes the conflictive encounter between Muslims and Christians in the siege, it inaugurates a fetishistic moment by keeping the memory of the Moors in the place name which is also the name of the author.

The 1561 edition of Barcelona adds as a freestanding appendix the "Historia de Alcida y Sylvano," which is a poetic, alternative rendition of the story of Sylvano as it is recounted in the main text: while in the main text of *La Diana* Sylvano is from the same geographical area of León as Sireno, and is in love with Diana and later with Selvagia, in the "Historia de Alcida y Sylvano," he is Portuguese and is in love with Alcida. In this second version of Sylvano's life, Almansur's siege against Coimbra is taken up again as well, although this time the text is careful to avoid the etymology of "Montemayor" as "Monte moro vello," offering instead an etymology that underlines the Christianity of the place: "Monte mayor el viejo tan nombrado, / y monte de fe lleno y muy glorioso, / mayor por más valiente y señalado, / llámante el viejo a ti por más famoso, / antiguo, fuerte, alto, y celebrado" (381) [You, the old greater mount so renowned, mountain full of faith and greatly glorious, greater because more valiant and eminent, they call you "old" because (you are) the most famous, ancient, strong, high, and celebrated]. The later alternative elements of the "Historia de Alcida y Sylvano" may constitute a correction to the main narrative, since the second version underlines a defence of Christianity instead of the Portuguese shepherdess's story of Coimbra emphasizing its Moorish legacy via the place name "Montemayor." The paradox, however, is that the etymological Christianization of "Montemayor" does not erase the first

Moorish etymology, and so the 1561 edition of Barcelona keeps two versions that are apparently irreconcilable. It is equally important to consider who is being addressed in each version: while the Christian etymology is told by Sylvano himself, the Moorish etymology is told to Felismena, the Andalusian character who functions as the narrator of *El Abencerraje* from the 1561 Valladolid edition on. Each character seems to listen to the etymology that is more accommodating to his or her individual cultural identity, and the insertion of *El Abencerraje* amplifies the motivated nature of each option.

Another consequence of the insertion of the Moorish tale is that it reinforces the ambiguous use of "faith" typical of sentimental fiction and thus the possibility of reading certain passages as critiques of the statutes of blood purity.[5] In book two, Dórida reports word by word the farewell between Sireno and Diana. Sireno enigmatically says to Diana: "No me haces tú partir / ... ni menos lo hace mi fe; / y si quisiese decir / quién lo hace: no lo sé" (178) [You are not making me leave, much less my faith. And if I wanted to say who does: I don't know]. He later reveals a different cause for his leaving: "Mi amo, aquel gran pastor, / es quien me hace partir" (180) [my master, that great shepherd, is who makes me leave]. But the expression "gran pastor" is still rather imprecise, since it could refer to the monarch (Philip II), to Sireno's patron, or ultimately to any higher authority that has command over him.

This mysterious "great shepherd" could also be read as a religious allusion, especially since Sireno continues by saying:

Bien podría yo dejar
mi rebaño y mi pastor,
y buscar otro señor;
mas s'el fin voy a mirar,
no convien'a nuestro amor,
que dejando'ste rebaño,
y tomando'tro cualquiera,
dime tú, ¿de qué manera
podré venir sin tu daño
por esta verde ribera? (181)

[I could well leave my herd and my shepherd and look for a different lord; but, if I consider the end that works best for our love, if I leave this herd and take any other flock, tell me: how, without harming you, could I come to this green bank?]

It is difficult to understand how Sireno's decision to join a different flock could damage Diana if he hypothetically returns to visit her.[6] One interpretation is that Sireno's claim of changing his faith is to be read literally as a religious conversion. He might be considering the possibility of becoming a renegade by passing to the Ottoman empire or the North African territories, as many of his contemporary Europeans did (Bennassar and Bennassar 147–65), or he might be adopting the position of the converts who in order to revert to the faith of their ancestors migrated to other countries in which they could openly practise either Judaism or Islam.[7]

Diana's response reveals that Sireno's travel implies going abroad: "vete, pastor, a embarcar, / pasa de presto la mar" (182) [go, shepherd, to embark, and quickly cross the sea]. Sireno insists then on the threat of changing his faith during his travels, although he is even more clearly referring here to his love for her: "Y si mudare mi fe, / por otro nuevo cuidado, / caiga del mejor estado / que la fortuna me dé / en el más desesperado" (183) [And if I change my faith for another sorrow, shall I fall from the best state that fate gives me to the most desperate one]. In another episode, when the shepherds go to the palace of the wise Felicia, they find a sign at the gate that states that the condition for entering the temple is "si la fe primera no ha perdido" (260) [if one has not lost his/her first faith]. After entering the temple, Sylvano warns Felicia with what appears to be a critique of the concept of blood purity: "asaz desfavorecido de los bienes de naturaleza está el que los va a buscar en sus pasados" (266) [quite unfavoured with the benefits of nature/birth is he who looks for those favours in his forebears]. And it is precisely at the end of this passage that Felismena tells the story of *El Abencerraje*, which can textually enter Felicia's temple because, as the legend at its entrance reads, their characters are still Muslim and have not lost "their first faith."[8]

Felicia's palace is a classicist memory place, and the first sight that the characters have of its exterior indicates so, when they see "algunas figuras de emperadores, matronas romanas y otras antigualla" (259) [some statues of emperors, Roman matrons, and other antiquities]. This memory turns specifically Spanish when Felicia takes the shepherds to a patio decorated with statues of historical figures. The first statue represents the *reconquest* hero El Cid, described as "un caballero armado de todas armas con una espada desnuda en la mano, muchas cabezas de moros debajo de sus pies" (271) [a knight fully armed with a bare sword in his hand and many heads of Moors under his feet]. The other statues are also heroes of the *reconquest*, such as Bernardo del Carpio, Diego Fernández de Córdoba, Fernán González, and Luis de Vilanova. As Rosilie Hernández-Pecoraro notes "[a]n exaltation of these men may seem out of place in a pastoral novel. In Felicia's castle, however, it fits perfectly within a symbolic fantasy of Spain as a unified, powerful, and invincible Catholic nation-state" (*Bucolic* 51).

The insertion of *El Abencerraje* in *La Diana* takes place right after the characters visit this memory place at the end of book four, precisely when the text raises the issue of memory as the key element of identity, both at the individual and collective level. The placement of *El Abencerraje* exactly here suggests that it is inserted as an antidote to the perceived ethnocentrism of the portrayal of pastoral Arcadia as an idealized community of shepherds. As a corrective, it very effectively resemanticizes *La Diana*, both exploiting the inherent ambiguity of "faith" and rewriting the site of collective memory.

The effectiveness of the insertion of *El Abencerraje* in altering the image of Spain is even more evident if we look at the international reception of both works altogether. The analogy between pastoral romances and historiography that we have already seen in chapter 3 can also be extended here. Many early modern historiographical works were written not only for local consumption, but mostly to craft an international image of imperial Spain as an essentially Christian territory in spite of the centuries of Jewish and Muslim presence (Kagan 120–5; MacCormack 86).[9] This anxiety about the European perception of Spain is found in Spanish literature, as we have seen in the meeting of Boscán and Navagero in 1526 Granada, and the resulting adoption of Petrarchan poetic forms, including pastoral eclogues, to create a classicist image of Spain. Within this process, the Spanish pastoral romances in general, and Montemayor's *La Diana* in particular, could have served as the perfect cultural ambassadors for modifying European perceptions of Spain.

However, the converse appears to be true. Even though the influence of Montemayor's *La Diana* in early modern European literature is hard to deny (Fosalba, *La Diana* 187–337), its repercussion is relatively limited in comparison with the impact of *El Abencerraje*, disseminated internationally through the version included in *La Diana* (Stoll 165; Fosalba, *La Diana* 207–8; Carrasco Urgoiti, *El moro* 93–146; Matulka 378–88).[10] Paradoxically, while the pastoral frame aimed at erasing the Moorish legacy from the perception of Spanish landscape, *La Diana* became the vehicle that disseminated the story of *El Abencerraje*, and thus ended up reinforcing the Orientalist perception of Spain. While literary demand for the pastoral world of *La Diana* waned in Spain after the 1620s, *El Abencerraje*, along with Ginés Pérez de Hita's *Guerras civiles de Granada*, outlived the pastoral romance to become the source of inspiration for European authors well into Romanticism, both because it offered an exotic stage for their characters and because it provided the literary means to keep representing Spain as a comfortable "difference."

It is a commonplace in Spanish literary historiography to say that Montemayor's *La Diana* is the most successful pastoral romance, with numerous editions and translations printed throughout the sixteenth and seventeenth centuries. Yet there is something misleading in this assertion, which must be rephrased to say that *La Diana + El Abencerraje* has more editions and translations than the rest of pastoral romances as a whole (including *La Diana* without *El Abencerraje*). At first sight, the fact that Montemayor's *La Diana* became the vehicle for spreading the image of Moorish Spain seems to contradict the idea that the pastoral romance is integral to the project of cultural cleansing. However, we must revisit the conditions that allow *La Diana* to be conceptualized as the initiator of the pastoral romance in Spain. Beginning in 1562, *La Diana* (without *El Abencerraje*) disappeared from the editorial market. In fact, readers who had a previous edition rushed to "unabridge" what they now felt was an incomplete copy, binding it with *El Abencerraje*.[11]

In contrast to the maurophile proclivities of Spanish readership, the most immediate continuators of Montemayor aimed to rewrite and contest the thread of re-orientalization of the Arcadia imposed on *La Diana* by the inclusion of *El Abencerraje*.[12] Seen in this light, the profusion of pastoral romances after the 1560s may be construed not as a desire to imitate *La Diana*, but rather as an impulse to contest the cultural influence of "the pastoral *Abencerraje*." The text of *La Diana* is thus conceived as a space that needs to be "reconquered" from a literary Moorish invasion by exacerbating the classicizing elements of *La Diana* (like the use of place names) while downplaying any suspected elements. Montemayor's successors were not able to counteract the success of *La Diana + El Abencerraje*, but they were effective at informing modern perceptions of the genre as a whole and conditioning modern philological choices.

While the text of *La Diana* (without *El Abencerraje*) was supplanted by the combined version in the early 1560s, modern editors, like its early modern continuators, share the nostalgia fixated on a return to the "original" text of *La Diana*, a "return" which has many points in common with the concept of restoration of a primeval Spanish purity. Thus the de-Orientalization of this pastoral world is projected onto the genre itself, in an attempt to recreate a hypothetical genealogical and textual purity. To emphasize the success of Montemayor's *La Diana* without taking into consideration *El Abencerraje* is only possible within the modern editorial practice of publishing both works separately as autonomous entities. This distortion of the early modern editorial life of the text is tightly related to the debates about Montemayor's participation in further editions. The selection of the 1561 edition of Barcelona as the textual basis for modern editions is justified as "la más completa, pues introduce todos aquellos fragmentos (como la 'Historia de Alcida y Sylvano') que fueron añadidos en vida del autor" (Rallo Grus, "Introducción" 93) [the most complete, since it introduces all of the fragments, like the story of Alcida and Sylvano, that were added during its author's life].

However, looking at the changes published during Montemayor's life may not be a valid, universal criterion. On the one hand, sixteenth-century editors seldom considered authors when they edited, compiled, amended, or abridged their texts, and knowing that Montemayor was alive when the 1561 Barcelona edition was published does not demonstrate that he was de facto responsible for the insertion of the "Historia de Alcida y Sylvano." On the other hand, due to the delays in printing, the fact that he died before the publication of the 1561 Valladolid edition does not conclusively discard his participation in the interpolation of *El Abencerraje*. Marcelino Menéndez Pelayo denied Montemayor's rewriting of *El Abencerraje* strongly, arguing that the Moorish tale "breaks the harmony" of the pastoral tone (273) and therefore discarding it as a spurious interpolation, which he attributes to "greedy editors" (129).[13] The stylistic argument that *El Abencerraje* "breaks the harmony" of the pastoral tone of *La Diana* is equally unconvincing, since this criterion would be also applicable to the "Historia de Alcida y Sylvano," which develops a parallel world that contradicts the main plot. Taking a different approach, Maxime Chevalier has argued that *La Diana* is not based on literary harmony, but on the technique of heterogeneous accumulation (*cajón de sastre*), thus inviting miscellaneous additions to the main text, and that the interpolation of *El Abencerraje* was in fact a way to make the text of *La Diana* more accessible to Spanish readers (49).

When confronted with such a mutable text as *La Diana*, there are in principle two philological options: either to opt for the "original text," publishing the most primitive version of *La Diana*,[14] or to choose the most complete version, including all the subsequent additions independent of whether several fragments can be reliably attributed to Montemayor or not, thus including both the "Historia de Alcida y Sylvano" and the interpolation of *El Abencerraje*. Selecting the 1561 Barcelona edition, which is the last one without *El Abencerraje*, is a compromise between both of these legitimate philological criteria that reinforces the perception of both works as two incompatible genres.[15] In this way, while philological editing is theoretically driven by the attachment to the concepts of "textual purity" and "authorial intention," in the case of Montemayor's *La Diana*, philologists tend to subvert their own theoretical postulates.

While many editors, such as Francisco López Estrada, Julian Arribas, and Miguel Teijeiro Fuentes, acknowledge the insertion of *El Abencerraje* in the 1561 edition of Valladolid, others such as Moreno Báez and Rallo Grus do not even note the place in the text where *El Abencerraje* was interpolated, even though they note every minor variant in successive editions. Furthermore, with the exception of López Estrada's 1946 edition, which includes *El Abencerraje* in a long footnote,[16] no editors feel the need to include the Moorish novel as part of the text. The influentially negative opinion of Menéndez Pelayo may be the reason why most editors still do not consider *El Abencerraje* to belong to *La Diana*. Notwithstanding the necessary and invaluable contributions of modern editors, such editorial practices are still conditioned by the implicit conception of pastoral romance as a genre that runs parallel to but remains incompatible with Moorish fictions. The exceptions to the philological logic that editors introduce reveal how previous ideologies of difference from the early modern period are preserved in present editorial practice. While both the subsequent pastoral romances written after *La Diana* and modern scholarship take Montemayor's work as the model that theoretically delineates the characteristics of the genre, it is nonetheless viewed with suspicion, anachronistically regarded as a deviation from the genre that it is supposed to inform, and as an imperfect model of de-Orientalization, itself in need of a cultural expurgation.

Reluctant Shepherds: Gaspar Mercader's *El prado de Valencia*

In contrast to the cultural expurgation that most of Montemayor's continuators inflicted in their *Arcadias*, Gaspar Mercader's *El prado de Valencia* (1600) was probably the first pastoral romance to exploit the other possibility offered by *La Diana* (with *El Abencerraje*), by creating a version of the pastoral Arcadia heavily influenced by Moorishness, more in tune with the promotion of aristocratic identity in Valencia. This work has been analysed as an academic pastoral romance, where the pastoral setting is but the vague narrative frame that serves for the gathering of the poetic verses of the "Academia de los Nocturnos," the aristocratic literary circle of Valencia of which Mercader was an integral member.[17] Equal attention has been paid to the significance of the accounts of aristocratic and royal pageantries held in 1599 to celebrate the double marriages of Philip III to Marguerite of Austria and the Infanta Isabel Clara Eugenia to Archduke Alberto (Ferrer Valls, "El duque" 260–3). The other characteristic that sets *El prado de Valencia* apart from other pastoral romances is that it shows that the mandate of cultural cleansing was not homogeneously adopted. While Mercader adopts some of its strategies, such as the practice of replacing Arabic place names by calling the river of Valencia "Turia" (instead of Guadalaviar), overall *El prado de Valencia* displaces sheep herding culture in favour of Moorish cultural practices, exemplifying how classicism and maurophilia were not necessarily incompatible representational projects.

The position of Muslims in Arcadia, systematically negated in other pastoral romances, is granted in this Valencian pastoral world, even if only in passing. At the beginning of the second book, strategically situated in the midst of the fiction, *El prado de Valencia* recounts the festivity of Saint John (San Juan), which, as the text acknowledges, was a common Mediterranean celebration shared with Jews and Muslims:

> La vispera del mas alegre dia que muestra el Sol su cara, sin poder alumbrar cosa que reguzijada no este, festejando al primo del hazedor del mundo; quando los Moros ponen de verde haya los azogados remos, las tembladoras flamulas, los gallardetes vistosos, las banderolas largas en los barcos ligeros, cubiertos ya de arrayanes verdes, y en las dos bandas dulçaynas y tamborines; quando los Iudios visten sus infames personas con mas cuydado que los Sabados, y quando toda la Christiandad entiende solo en alboroços, bullicios, juegos, danças, contentos y alegrias. (91)
>
> [On the eve of the most happy day when the sun shows its face, without being able to light up anything that is not cheerful, celebrating the cousin of the maker of the world; when the Moors decorate with green beech branches the silvered oars, shimmering streamers, colourful pennants, long flags on agile vessels, covered with green myrtle, and on both sides flutes and tambourines; when the Jews dress their infamous persons with more care than on Saturdays, and when all Christianity only busies itself with joy, bustle, games, dances, and happiness.]

The Festival of Saint John, which was likely of pagan origin, was already celebrated by Muslims in al-Andalus. Known by the name of *ᶜansara*, it was often criticized by Islamic religious authorities, who perceived it as a Christian influence (Granja 2–5; Fuchs, *Exotic Nation* 106–7). Due to the awareness in early modern Spain that the Festival of Saint John showed a great degree of cultural and religious syncretism, it is telling that Mercader acknowledges so openly that Muslims, Jews, and Christians participate simultaneously in the same celebration.[18] What is striking in this literary rendition of the trans-religious celebration is that it was no longer possible to talk about Jews, Moors, and Christians as present in 1600 Spain. The indeterminacy of the passage indicates that Gaspar Mercader is either referring to the wider Mediterranean context or to a nostalgic past of medieval *convivencia*, instead of going back to the primitive Spain untouched by the arrival of "non-Spaniard" people, as was customary in the genre.[19] Even if they are only glimpsed in the distance, this is the first time that the Moors are represented as participants in the pastoral world – but yet not as shepherds.

Mercader's recognition of the presence of Muslims goes hand in hand with the display of Moorish cultural practices. The explicit aristocratic origins of the characters may serve to explain why the pastoral background is constantly dismissed. Courtly pastoral performances had to compete with the nobility's fondness for the game of canes [*juegos de cañas*], a more local public spectacle that was specifically Spanish and in which riders dressed up as Moors and rode in the popular equestrian style called *jineta*, as seen in chapter 2. The other reason for the aristocratic refusal of pastoral performances is that the nobility did not usually identify with the values of blood purity that the figure of the shepherd came to represent. In the account of the festivities for the wedding of Philip III, the arrival of Francisco de Sandoval y Rojas, recently appointed as Marquis of Denia (later known as Duke of Lerma), gathers a hesitant community of shepherds: "pues ni quedo choça con gente, ni cayado alguno que no saliesse con alguna nueua gala hermoseado, o con listones cubierto ... y aun algunos no torcidos y tan cubiertos de borlas, que parecian mas ginetas de capitanes que cayados pastoriles" (15) [since no hut had people, nor was there a shepherds' crook that didn't appear decorated with new fineries or covered with trappings ... some of them not so twisted, and some of them covered with so many fineries that they truly looked more like *ginetas* (batons) of captains rather than shepherds' crooks]. Sebastián de Covarrubias defines *gineta* as "una lança corta con una borla por guarnición, junto al hierro dorado, insignia de los capitanes de infantería. Púdose dezir assí por ser corta y recogida, y no porque sea arma de los ginetes, cuyas lanças son muy largas" (640) [a short spear decorated with a ribbon close to the golden metallic part, which is the banner of infantry captains. It was so named because they are short and compact, and not because they are weapons typical of riders (ginetes), whose spears are very long]. Apparently, *gineta de capitán* has nothing to do with the homophonous *gineta*, the riding style that is widely recognized as being of Moorish origin and is one of the main features of the game of canes. However, even if Covarrubias was right in distinguishing two different etymologies for the two meanings of *gineta*, both signifiers are evidently similar. While it is a conventional feature of the pastoral romances to highlight the distance that separates the characters from real, rustic shepherds, among all the potential comparisons that Mercader could have chosen for the pastoral crook, he picked up precisely the one that was most easily associated with the Moorish cultural legacy in Spain and with the tradition of the game of canes, which initiates the romance's process of abandoning sheep herding pastoral practices and privileging Moorish cultural practices.

What is striking in the *Prado de Valencia* is how this distance is systematically created by supplanting material sheep herding with Moorish cultural practices and how often these so-called shepherds devote their leisure to the game of canes. The Moorish origin of the game is mentioned on the occasion of the arrival of the Marquis of Denia, when the shepherds ask their damsels "si querian ver jugar vnas cañas a lo morisco en ligeros, atreuidos y veloces caballos Andaluzes, con cifras en las adargas y alas en los bohordos" (43–4) [whether they would like to see how they played the game of canes in the Moorish fashion, with agile, daring and quick Andalucian horses, with crests on their *adargas* (Moorish shields) and wings on their *bohordos* (Moorish spears)]. It is true that the game of canes is a fully institutionalized public entertainment, and not necessarily a public display of maurophilia. Nevertheless, although pastoral romances usually reflect courtly life, the vast majority omit the omnipresent game of canes, thus emphasizing the priority of constructing a classical representation of the nation that is projected into the pastoral Arcadia.[20]

The opposition between pastoral performances and the Moorish game of canes is made explicit by the shepherd Fideno, one of the main characters of Mercader's romance, and, according to scholars, his alter ego. When the rest of the shepherds decide to celebrate a festival in his honour, they let Fideno choose three different performances. Fideno is quick to accept the invitation, and proposes in the first place a pastoral naumachia (mock naval battle):

> La primer jornada digo que sera bien que tengamos una justa de quien yo sere mantenedor, en el rio con barcos que tendre yo preuenidos para quantos quisieren, con trages y con empresas diferentes como cada uno quisiere; yo

> saldre en vn barco hecho a forma de cauallo, con quatro remos a modo de seys pies, con cuello, pecho y espaldas, proporcionado todo con paramentos naranjados sobrepuestos de coraçones de plata, y en la silla de armas ... yre yo con pellico y montera naranjada, cubierto todo de coraçones de plata, y en la montera plumas naranjadas y blancas. (140–1)
>
> [I say the first day we will have a contest in which I will be the arbitrator, in the river, with boats that I will provide to anyone who wants one, with different disguises and insignias for everyone; I will appear in a horse-shaped boat, with four oars as if they were six legs, with neck, breast, and back, all fitted with orange trappings with silver hearts over them; on the throne ... I will be there wearing a *pellico* (shearling) and an orange beret, all covered with silver hearts, and with white and orange feathers in the beret.]

Although the *pellico* invented by Fideno refers unmistakably to a pastoral garment, the spectacle is anything but rustic.

The other "shepherds" praise the selection but express their concern about the expenses that such a ceremony would entail. Due to its high costs, they would limit the three promised performances to just two: "Muy contentos quedaron Olimpo, Lisardo y Cardenio de las galas, inuencion, empresa y mote que Fideno auia de sacar, aunque les parecio que el gasto passaua de mucho, y asi le dixeron a Fideno que solo traçase otra fiesta, porque bastauan dos, auiendo tantas cosas que preuenir para ellas" (141) [Olimpo, Lisardo, and Cardenio were excited to see the fineries, ingenuities, insignia, and mottoes that Fideno should bring, although they deemed that the expenses would be too much, and they asked Fideno to propose only one more celebration, since two were enough, especially with so many things to prepare for them]. The second performance proposed by Fideno is a game of canes: "Yo saldre en mi yegua blanca, que tendra aquel dia las vñas doradas y vn jaez de monte ... y mi persona vestida a lo morisco, con marlota y capellar naranjado, con infinitos coraçones de plata y oro, turbante blanco y naranjadas plumas en la blanca adarga" (141) [I will ride my white mare, which for that day will have golden hooves and a hunting harness ... and my person dressed in Morisco clothes, with a *marlota* (Moorish tunic) and an orange *capellar* (Moorish hood), with infinite silver and golden hearts, a white turban, and orange feathers on the white *adarga* (Moorish shield)]. By describing the garments of the second game with the same patterns and colours (silver hearts and orange colour), Fideno places both the pastoral *pellico* and the Moorish *marlota* within the context of personal heraldry.[21]

As shown in chapter 2, these two garments cannot be more opposite in early modern Spanish imaginary. An additional irony in this scene in *El prado de Valencia* is that the pastoral spectacle imposes a budgetary restriction on the number of games to be celebrated in the first place, precisely when the *pellico* is the garment that is supposed to symbolize austerity and simplicity. After all, this is a fictitious romance, a genre where budgetary limitations should not even be an issue. Obviously, the element that would render the pastoral festival so onerous is not the *pellico* itself, but the naumachia that Fideno associates with it, posing a burdensome condition for the pastoral festival that can be read as an oblique parody of the pastoral project.[22]

As it turns out, the pastoral naumachia is nothing but an aquatic version of the game of canes. Since Fideno's ship is horse shaped, the colours of his garment are the same as those of his *marlota*, and the battle consists of fighting with spears; this is, in fact, a game of canes in which the only pastoral element is the name *pellicos*. Sheep herding culture fades away and the *pellico* becomes but an empty name for "dress" without the specific woollen, rustic essence of the pastoral garment, actually inhabiting the same Moorish culture it aimed at dispelling. Furthermore, the naumachia mimics the description of the Moors celebrating Saint John in their lavishly decorated vessels. Therefore, the *pellico*, the pastoral garment par excellence appears throughout the text only as a potentiality, as a pastoral signifier that is not even fulfilled as such, devoid of its oppositional values against Moorish culture.

These fake shepherds spend most of their time playing the game of canes dressed in Moorish garments, but still call themselves "shepherds," arguably because they try to have it both ways: on the one hand, they ensure their inclusion in the construction of Spanishness that seems to revolve around the "community of shepherds"; on the other hand, they keep their identification with the culture of material sheep herding at a formal, nominal level, rejecting it in favour of the most attractive and flamboyant Moorish fashion.

For these shepherds wearing *marlotas*, and arguably for many of their real aristocratic contemporaries, the pastoral genre is rather exemplified in Montemayor's *La Diana* as it was read during the sixteenth and seventeenth centuries: with the tale of *El Abencerraje* interpolated within it. According to Chevalier, the nobility displayed an acute preference for disguising as Moors (and not so much as shepherds) during the sixteenth century, and the inclusion of *El Abencerraje* in *La Diana* was an editorial strategy trying to appeal to them as an audience (45–9).[23] As Barbara Fuchs argues, the force of idealized Moors in Morisco tales such as *El Abencerraje* lay not in claiming the right to cultural difference, but in underlining instead how many cultural practices both Muslims and Christians had in common in early modern Spain (*Exotic Nation* 114). *El prado de Valencia* follows Arcadia as it is found in *La Diana*, but, in opposition to other Spanish pastoral romances, it emphasizes the material culture described in *El Abencerraje* and downplays the elements borrowed from sheep herding culture.

Arguably, Mercader's social and economic position plays a decisive role in his avoidance of the rustic, levelling values of blood purity implied in the *pellico*, in favour of the *marlota*. While he was writing his romance, he had to be aware that, as the firstborn, he would eventually take charge of the properties that his family had owned since 1425 in the barony of Buñol, which comprised the townships of Buñol, Yátova, Macastre, Alborache, and Dos Aguas, with a combined total population of 760 Morisco households in 1609 (Lapeyre 50).[24] The lords of Moriscos [*señores de moriscos*] were naturally one of their most impassioned protectors, since they profited from their labour in exchange for protection (Carrasco Urgoiti, *El problema morisco* 25–37; Lea 217–18; Colás Latorre, "Los moriscos aragoneses" 156–61). In some instances, they even defended a certain form of religious freedom that very often took the form of a nostalgic revival of medieval *convivencia* and open confrontation with the Inquisition. The most notable case is that of Sancho de Cardona, Admiral of Aragon, prosecuted in 1569 for having promoted the Islamic practices of his Morisco vassals and for planning to contact the pope and the Ottoman emperor to put pressure on Philip II, trying to revoke their forced conversion to Christianity (Boronat y Barrachina 1:445–7).[25] Consequently, ecclesiastical authorities soon complained that the main obstacle for the Christianization of the Moriscos was the opposition of their aristocratic landlords (Boronat y Barrachina 1:638–56). In the case of Mercader's family, there is evidence that the successive barons of Buñol made no effort to evangelize their Morisco vassals during the sixteenth century, allowing the practice of Islam almost openly (Ehlers 30–3). Soon after the publication of *El prado de Valencia*, following the death of this father in 1603, Gaspar Mercader became Baron of Buñol, and in 1604 Philip III granted him the title of Count (Mérimée lix).

Mercader's fate is paradigmatic of the fulfilment of the project of ethnic and cultural homogenization for the *señores de moriscos*. Even though his pastoral romance is probably the least "pastoral" and the most Moorish of the genre, Mercader actually became an active participant in the expulsion of the Moriscos only six years after he took over the family possessions in

Buñol. In September 1609, Philip III appointed him to protect one of the areas of the city of Valencia, in order to prevent any potential uprising by Moriscos during the expulsion. His participation in the expulsion is recorded in several documents.[26] He engaged as well in the anti-Morisco propaganda that followed the expulsion, with a sonnet praising Philip III for his decision, included at the beginning of the epic poem *Expulsión de los moros de España* (1610), by Gaspar Aguilar (120). The expulsion of the Moriscos had an immediate economic impact on the welfare of the lords of Moriscos, like Mercader, who suddenly lost most of their vassals and consequently their wealth. Out of the 214 households that Buñol had in 1602 there were only 91 in 1646 (Lapeyre 108), which indicates the bankruptcy of Mercader's lands. Philip III had promised economic compensation to lords who had Morisco vassals, but it seems that Mercader was not among the fortunate ones: in 1613 he travelled to the court in Madrid because of Philip III's failure to keep his promise of economic help and the lack of success in repopulating his territories, but all he received was a ludicrous tax exemption (Mérimée lxiii–lxviii).[27]

The expulsion of the Moriscos, like the previous expulsion of the Jews in 1492, constitutes the further materialization of an Arcadian dream that promotes ethnic and cultural homogeneity within clearly defined national boundaries. The only collective identity in *El prado de Valencia* is aristocratic corporatism, followed by praise of their own local identity, with no indication of forming a national identity based on "racial" criteria that work against Mercader's own class interests. In contrast to the ethnocentric worlds of most Spanish pastoral romances, which were based on the recreation of a remote mythic past or on the announcement of a forthcoming age of completion, *El prado de Valencia* depicts a local paradise of courtly festivities and literary academies. This is of course an aristocratic Arcadia based on the protection (and economic exploitation) of the Moriscos. The interpenetration of the Moorish and the pastoral is, however, not confined to the nobility, and can be found in the work of other authors, most notably Cervantes.

The Limits of Cultural Cleansing in Cervantes's Pastoral

Of all the writers studied in this book, Miguel de Cervantes devoted the most continuous analysis to pastoral, managing to explore the "political unconscious" of the genre and its pervasive denial of the Moorish legacy. Indeed, one the first works of Cervantes's literary career was the pastoral romance *La Galatea* (1585), published after his return from captivity in Algiers between 1575 and 1580.[28] Although the scholarly bibliography on this pastoral romance is abundant, there are few political readings. For Edward Dudley the success of the figure of Galatea in the troubled political environment in Italy and Spain was due to its lack of specificity, which allowed artists and writers to keep at bay courtly disputes and avoid political compromise (30–2). However, as Carroll B. Johnson ("Cervantes"), Benjamin Nelson (163–6), and José Montero Reguera (340) demonstrate, the annexation of Portugal in 1580 and the subsequent anxieties of Castilians, who feared losing leadership in the Spanish empire, constitute the political background of Cervantes's *La Galatea*. Yet, while the political confrontation between Castile and Portugal captures the overarching plot of *La Galatea* within that larger frame, conflict with the Islamic past surfaces in a very dramatic moment for the collective identity of this community of shepherds.[29]

Overall, *La Galatea* is no exception in the representation of a culturally cleansed Arcadia, showing the same use of allegedly ancient place names to replace Arabic names, as shown in chapter 3. The shepherd Lisandro begins telling his own story to Elicio and Erastro by stating that he was born "en las riberas del Betis, caudalosísimo río que la gran Vandalia enriquece" (188–9) [on the banks of the Betis (Guadalquivir), plentiful river that enriches the great Vandalia (Andalusia)]. Similarly, Teolinda introduces Tirsi and Damón by saying that the former was born in Compluto (Alcalá de Henares) while the latter was raised in Mantua Carpetana (Madrid) (251), adding that Damón's lineage comes from the mountains of León (251). In addition to classicist place names and the evocation of the geographical imaginary of blood purity (the mountains of León), the space inhabited by these shepherds on the banks of the Tajo river is equally important in the imaginary of Spanish identity.

The main setting of *La Galatea* is the river Tajo, which serves as a metonymy for Toledo. The shepherd Timbrio praises this space by comparing it with other Spanish and Italian rivers:

> No poca maravilla me causa, Elicio, la incomparable belleza de estas frescas riberas y no sin razón, porque quien ha visto, como yo, las espaciosas del nombrado Betis y las que visten y adornan el famoso Ebro y al conocido Pisuerga, y en las apartadas tierras ha paseado las del santo Tíber y las amenas del Po ... sin dejar de haber rodeado las frescuras del apacible Sebeto, grande ocasión había de ser la que a maravilla me moviese a ver otras algunas. (540–1)
>
> [Elicio, the incomparable beauty of these fresh banks causes me no small wonder, and not without reason, because for anybody who has seen as I have the spacious (banks) of the famous Betis and those that decorate the celebrated

> Ebro and the well known Pisuerga, and who has walked in far-away lands the (banks) of the sacred Tiber and the pleasant (banks) of the Po ... having wandered as well the coolness of the pleasant Sebeto, it would have to be a great event that would cause me to marvel on seeing any other banks.]

Timbrio's bucolic description of the banks of the Tajo places Toledo in a position of preeminence within the Spanish empire, while at the same time he inserts himself in a bucolic tradition of pastoral idealization of the city comparable to that of Garcilaso de la Vega.

Yet Toledo was one of the most overcharged spaces in early modern Spain. Even though the city's importance began to fade in the sixteenth century, it remained the main episcopal see of Spain and managed to keep much of its symbolic prestige (Dumora). It also occupied a central place in the imaginary of the "loss of Spain," since it was the ancient capital of the Visigothic kingdom. According to legend, it was on those same banks that Rodrigo, the last Visigothic king, raped the daughter of Count Julian, thus provoking the treachery that permitted entrance of Muslims into the Iberian Peninsula. Rodrigo's venality served as a way to conceive of the Islamic period as expiation for the monarch's sin, and the nostalgic vindication of the Visigothic monarchy in the early modern period was used to justify the imposition of Catholicism as the religion of state and the core of Spanish identity (Álvarez Martí-Aguilar 563–6; Wulff 36–41).[30]

The ineluctable association of Toledo and the Tajo, predilect space for literary shepherds after Garcilaso, with the "loss of Spain" to Muslims, requires a repression of memory of King Rodrigo and the results of his actions, in order to produce an Iberian space where there had never been an Islamic presence, and in *La Galatea* this memory surfaces only to be immediately repressed by the community of shepherds. The character who adduces Rodrigo's name is Lenio, the "desamorado pastor" (love-disillusioned shepherd), whose disdain for love places him in a marginal position within the community of shepherds and incites the opposition of the rest of the characters (Hernández-Pecoraro, *Bucolic* 113–20; Trambaioli 48). In his vituperation of love (416–30), Lenio goes over all the mythological, biblical, and historical figures for whom love caused a tragic outcome, both for individuals and for entire communities that were dragged into disaster because of the passions of their rulers.

While this conventional repertoire of classical Graeco-Latin motifs provides a comforting feeling of historical distance for debating abstract sentimental issues, Lenio concludes with an example that is much closer to home when he blames love for the loss of Spain and breaks the indulgent historical identity of his audience: "[Amor] en fin, entregó nuestras Españas a la bárbara furia agarena, llamada a la venganza del desordenado amor del miserable Rodrigo" (430) [Finally, love handed our Spain over to the barbarian Saracen fury, which was called in revenge because of the insane love of the despicable Rodrigo].[31] Dispelling the illusion of the nostalgic simulacrum and the performance of a past long gone, the mention of King Rodrigo seems out of place in Lenio's list of classical examples. It becomes clear that Lenio's attack goes beyond the individual effects of love, pointing instead to the fictitious historical imagination upon which the shepherds base their identity, when he finishes his speech with a song that calls love: "Circe engañadora que nos muda / en varios monstruos, sin que humana ayuda / pueda al pasado ser nuestro devolvernos" (431) [Deceiving Circe who changes us into varied monsters, without human help able to restore us to our past being].

Lenio's attack on the identity of the community does not go without response. The shepherd Tirsi retorts with the acquiescence of his fellows, who "deseaban que la opinión desamorada de Lenio no prevaleciese" (433) [wished that Lenio's loveless opinion should not prevail]. Tirsi refutes one by one all the points that Lenio made in his speech (434–48), but he significantly omits the last example about King Rodrigo:

> En fin, oh, Lenio, este amor es el que, si consumió a los troyanos, engrandeció a los griegos; si hizo cesar las obras de Cartago, hizo crecer los edificios de Roma; si quitó el reino a Tarquino, redujo a libertad la república. Y aunque pudiera traer aquí muchos ejemplos en contrario de los que tú trujiste de los efectos buenos que el amor hace, no me quiero ocupar de ellos, pues de sí son tan notorios. (448)
>
> [Finally, oh Lenio, love is what, if it ruined Trojans, also exalted Greeks; if it caused Carthage's works to come to a halt, it also raised the buildings of Rome; if it took the kingdom from Tarquin, it restored liberty to the republic. And even though I could adduce many other examples against those that you mentioned about the positive effects that love causes, I do not want to attend to them, because they are so notorious in themselves.]

Tirsi glides over the most uncomfortable reference, that of King Rodrigo and the loss of Spain, even though it was the one ultimately emphasized by Lenio in his speech. Thus he avoids taking the debate about love to a closer, contemporary case that is the only one that really affects sixteenth-century Spaniards.

At the same time, Tirsi justifies his neglect in broaching this case by claiming that there is no need to do so, since the effects are "so notorious in themselves." However, it is not at all clear what the self-evident effects are for the community of shepherds in Rodrigo's story. It is possible that Tirsi ends his response abruptly because he realizes the difficulties in the logic he has previously developed with the set of classical cases. If in all the cases he recognizes that a side that is damaged by the effects of love (Trojans, Carthage, Tarquin) is compensated for by the benefits that others reaped from it (Greeks, Rome, the republic), it would follow that the misfortune of King Rodrigo and Visigothic Spain was balanced by the parallel supersession of the Islamic period. Few early modern Spaniards would have endorsed the idea that there were beneficial effects of love in the fall of the Visigothic rule, unless they understood the positive outcome of the story to mean that, because of the *reconquest*, Spaniards came to better appreciate Christianity as a key element of collective identity.

Both interpretations are speculative, but the ambiguity is raised precisely by Tirsi's refusal to broach the case of King Rodrigo explicitly. In fact, the debate is suddenly brought to an end. Even before the discussion began, the entire community of shepherds wanted Tirsi to prevail over Lenio, and therefore they rushed to grant Tirsi the last word and to silence the unheard reply of the loveless shepherd: "[Lenio] ya iba dando muestras de querer responder y replicar a Tirsi, si las alabanzas que a los dos daban Darinto y su compañero, y todos los pastores y pastoras presentes no lo estorbaran" (451) [Lenio looked like he wanted to reply and refute Tirsi, and he would have done so if Darinto and all the shepherds and shepherdesses present did not prevent it with their praises to both]. Tirsi's elusiveness, along with the shepherds' haste to celebrate their eloquence, throws the issue into oblivion anew.

However, their silencing reveals that, at the level of plot, theirs is a bucolic world built upon the denial of their own history. The Arcadia of *La Galatea* (and of most of the pastoral romances) is based on the repression of cultural and ethnic heterogeneity and of the history that may hinder their concept of purity. Lenio's impropriety threatens the illusion of purity and historical vagueness: as much as the characters of *La Galatea* remove Arabic names and fill their primitive-like pastoral communities with classical references, the mention of Rodrigo reveals the imposture, since it becomes suddenly clear that this pastoral world takes place *after* King Rodrigo's fall. However, even when the cultural denial is exposed and the shepherds are confronted with the evidence that they cannot inhabit the world in which they pretend to be living, the collective self-deception is nonetheless restored. Lenio's admonition that love cannot take one back to a past state ["al pasado ser nuestro devolvernos" 431] is unheard by the rest of the characters. Thus while the debate over the effects of Rodrigo's love remains inconclusive, *La Galatea* brings to the forefront the collective delusions that support the imagining of the pastoral Arcadia.

Similar pastoral denials persist as an echo throughout Cervantes's oeuvre. Even though Cervantes never wrote another pastoral romance as such, he never abandoned the project of writing the second part of *La Galatea*, which he promised in the preliminaries to the compilation of his plays, *Ocho comedias y ocho entremeses* (1615), in the prologue to the second part of *Don Quixote* (1615), and in the prologue to *Los trabajos de Persiles y Sigismunda* (1617), only a few days before his death (López Estrada and López García-Berdoy 101–3). The pastoral reappears throughout the rest of his literary work, such as in the play *La casa de los celos*, in numerous chapters of *Don Quixote* and *Los trabajos de Persiles y Sigismunda* (1617), and as parody in the novella "El coloquio de los perros."[32] Because of the relevance that pastoral elements have in *Don Quixote*, Baena wonders whether it should take the place of the never written second part of *La Galatea* (*Discordancias* 170–1). Similarly, Jean Louis Flecniakoska argues that the goal of *Don Quixote* was to critique books of both chivalry and pastoral romances, and that this work is therefore a simultaneous parody of both genres (378).

Following these arguments, we can understand that the parody of pastoral in *Don Quixote* exposes not only the aesthetic formalism of the genre, but also the ethnocentric representation of the Spanish Arcadia.[33] While the pastoral surfaces throughout *Don Quixote*, the exclusivist component of the community of shepherds is not confronted until the end of the second part, when Sancho and Don Quixote find a group of fake shepherds who stage their own Arcadia, as the characters themselves explain:

> En una aldea que está hasta dos leguas de aquí, donde hay mucha gente principal y muchos hidalgos y ricos, entre muchos amigos y parientes se concertó que con sus hijos, mujeres y hijas, vecinos, amigos y parientes, nos viniésemos a holgar a este sitio, que es uno de los más agradables de todos estos contornos, formando entre todos una nueva y pastoril Arcadia, vistiéndonos las doncellas de zagalas y los mancebos de pastores. (II, 58, 1022)

> [In a village half a dozen miles from here, where many people of rank and *hidalgos* and rich people live, a group of friends and relations agreed that we should come with their sons, wives and daughters, neighbours, friends and

relatives, to enjoy ourselves in this place, one of the pleasantest spots in the whole area, creating a new pastoral Arcadia, with all us youngsters dressing up as shepherds and shepherdesses.] (trans. Rutherford 879)

In this passage the conventionality and artificiality of the pastoral disguise is laid bare, as well as the problematic social identity of the fake shepherds. The participants of the pastoral performance constitute the ruling classes of their village, *hidalgos* and rich people related by networks of friendship and kinship. By retiring to the wild, they set forth a fictitious idealized community that is an exact reproduction of the relations they had in their village of origin, with the differences being that they are in the midst of nature, and, more significantly, that they have left behind the presence of commoners that would contaminate the idealization of their performance. Yet, through their use of shepherds' garments, these characters appropriate the identity of rustics at the same time that they deny their presence. In order to be able to stage their imaginary community, this group needs to force itself into exile from the reality that they seek to idealize, thus showing that the pastoral exclusion of the Other is also an exclusion of the self.

This self-destructive feature of ethnocentrism is later emphasized when Don Quixote and Sancho plan their own pastoral project. When Don Quixote returns from Barcelona, after having been defeated by the Knight of the White Moon, he recalls the fake Arcadia and decides to switch from the imitation of chivalric adventures to the imitation of the pastoral life. He proposes that they change their names to "pastor Quijotiz" and "pastor Pancino," fantasizing about every detail of the pastoral romances that they would imitate (II, 67, 1093). He even describes the pastoral instruments that they would play:

> ¡Válame Dios ... y qué vida nos hemos de dar, Sancho amigo! ¡Qué de churumbelas han de llegar a nuestros oídos, qué de gaitas zamoranas, qué tamborines, y qué de sonajas, y qué de rabeles! Pues ¡qué si destas diferencias de música resuena la de los albogues! Allí se verá casi todos los instrumentos pastorales. (II, 67, 1094)
>
> [My God! ... What a life we're going to lead, Sancho my friend! All those shawms regaling our ears, all those Zamora pipes, all those tambors, all those tambourines, all those rebecs! And maybe amidst all these different sounds the albogues will ring out! Then we shall have nearly all the pastoral instruments.] (trans. Rutherford 942)

Sancho interrupts this enthusiastic description of Quixote's dreamed pastoral Arcadia by showing his ignorance of the word *albogues*: "¿Qué son albogues ... que ni los he oído nombrar, ni los he visto en toda mi vida?" (II, 67, 1094) [What are albogues? ... I've never heard of them or seen one in all the days of my life] (trans. Rutherford 942). Don Quixote, happy to enlighten his squire with his erudition, describes the instrument as "unas chapas a modo de candeleros de azófar, que dando una con otra por lo vacío y hueco, hace un son, si no muy agradable y armónico, no descontenta, y viene bien con la rusticidad de la gaita y del tamborín" (II, 67, 1094–5) [thin plates rather like brass candle-holders, and when you strike them against one another on the concave side they make a sound which, if not very agreeable or harmonious, is not altogether unpleasant, and blends well with the rustic tones of the pipe and the tambor] (trans. Rutherford 942). Immediately Don Quixote qualifies *albogues* as an Arabic name: "este nombre *albogues* es morisco, como lo son todos aquellos que en nuestra lengua castellana comienzan en *al*" (II, 67, 1095) [this word *albogues* is of Morisco origin, as are all the words in our language that begin with *al*] (trans. Rutherford 942). Don Quixote provides then a quite conventional list of easily recognizable Arabic words in Spanish.

Yet the apparent simplicity of the passage encodes a complex negotiation between Don Quixote and Sancho about the way Spanish culture is being imagined. To begin with, Don Quixote's description of the *albogues* is highly defective. Most of the scholars who analyse this passage note that this musical instrument appears in all medieval and early modern sources as a wind instrument, and that Don Quixote's definition of *albogues* as a percussion instrument is unique (Diego 18; Salazar 54–5; Arco y Garay 256–7; Haywood 140–1). Strikingly, in deference to the authority of Cervantes even the *Diccionario de la Real Academia* has given credit to Don Quixote's ludicrous definition. However, Miguel Querol Gavaldá questions the authority of Don Quixote as a musicologist: "Estamos por pensar que don Quijote, excepto el nombre de *albogues*, no sabía más de ellos que Sancho. Casi nos atrevemos a decir que la explicación que don Quijote da de los *albogues* es un rasgo humorístico de Cervantes, que puso en boca de nuestro caballero una respuesta disparatada a la pregunta de su escudero" (148) [We are led to think that Don Quixote, except for the noun *albogues*, knew no more than Sancho about them. We almost dare to say that the explanation that Don Quixote gives about the *albogues* is a humorous note of Cervantes, who put in the mouth of our knight an absurd answer to his squire's question]. This caution of Querol Gavaldá allows the reader to analyse what is going on in this scene.

This is an unlikely error on Cervantes's part, since he mentions the *albogues* in several of his works.[34] It is also unlikely that Sancho does not know the *albogues*, even if he proclaims his ignorance. In an episode in part I, Sancho alludes to a time when he worked as a shepherd (I, 20, 195–6), and *albogues* have been typical of shepherds until recent times (Lizarazu de Mesa 245–6). Furthermore, Sancho's way of saying that he has never heard of them and that he has never seen them is an immediate self-contradiction – if he did not know what the *albogues* were, he could not know that he had never seen them. It seems implausible as well that Don Quixote, a voracious reader of pastoral romances, in which the *albogues* are ubiquitous, would not know that this was a wind instrument.[35] Yet, following on the heels of Sancho's implausible admission of ignorance, is Don Quixote's equally misleading erudition. Taking into account the cultural competency of both characters, their ignorance appears as an imposture rather than as lack of knowledge. The game of questions and answers between Don Quixote and Sancho does not aim to reach the "truth" about the object, but rather to stage the pretence that they do not know something that it would be impossible to not know in early modern Spain. Baena argues that, within a Spain dominated by the wool trade, "[l]a 'falsedad' ... de la pastoril radica en aparentar, en afectar que no se sabe nada de ovejas" ("De Belisa" 79) [the insincerity of pastoral resides in pretending that one does not know anything about sheep]. It can equally be argued that the insincerity of Don Quixote and Sancho's Arcadian project is pretending that they do not know anything about the *albogues*.

The Arabic name of the instrument may justify the ludicrous reaction of both characters. Right after the expulsion of the Moriscos, finished in 1614, it is not advisable to include a musical instrument with an obvious Arabic name in their pastoral Arcadia. From the initial moments of sixteenth-century cultural cleansing, music was a significant target. In 1526, Charles V attempted to ban Morisco *zambras* (Gallego Burín and Gámir Sandoval 89–98). Hernán López de Yanguas's *Farsa Nuevamente compuesta sobre la felice nueva de la concordia y paz*, composed around 1529 to celebrate the Peace of Cambray, already reflects this cultural pressure on Moorish music and its substitution by rustic equivalents, when Tiempo, disguised in pastoral garments, requires of Placer: "No nos tañas la morisca / sino el villano de antaño" (*Obras* 119) [Do not play the Morisco dance, but the peasant music of yesteryear]. Philip II reinstated the prohibition of *zambras* in 1566, regardless of their content, mandating that "no hagan zambras, ni leilas con instrumentos moriscos en alguna manera, aunque ellos no canten cosa que sea contra la Fe, i Religión" (*Los códigos españoles*

11:241) [(Moriscos) shall not make *zambras* nor *leilas* with Morisco instruments in any way, even though they do not sing anything against (Christian) faith and religion].

One plausible reading of the interchange, then, is that Sancho employs his implausible ignorance to warn Don Quixote. Realizing the impropriety of including the *albogues*, Don Quixote initiates a complex ploy to cover for his lapse: on the one hand, he displays his encyclopedic erudition to show that he knows how to differentiate the Arabic elements in Spanish, and is therefore able to eradicate them from his discourse. On the other hand, both Sancho and Don Quixote collaborate in their performance of denial as exculpation, displaying an ignorance that they lack in order to artificially widen the cultural boundary that separates them from Moriscos and from accusations of maurophilia.

Cervantes's selection of this particular musical instrument may well be motivated more deeply. What is truly dramatic in the whole situation is that, in spite of the Arabic name, *albogues* were very likely an instrument typical of Christians, not of Moriscos. Fuchs notes that Don Quixote "never calls the objects he describes Moorish: all that is Moorish is the *nombre* – the word" (*Exotic Nation* 26). Sebastián de Covarrubias, in his *Tesoro de la lengua castellana* (1611), attributed a Moorish origin to the musical instrument: "Es cierta especie de flauta o dulçaina ... de la qual usavan en España los moros especialmente en sus çambras" (67) [It is a certain kind of flute or clarinet ... which was used in Spain by the Moors, especially in their *zambras*]. Although Covarrubias is considered one of the most authoritative lexicographers of early modern Spain, his encyclopedic knowledge is not free of ideological bias, especially with regard to the Moriscos (Lezra).

In this case, Covarrubias stands alone in attributing the *albogues* to the Moriscos (like Don Quixote and Sancho), while the rest of the documentation of the period suggests that the instrument belonged to Christians. Adolfo Salazar quotes the testimony of the Andalusian Abou Said el Nisaburi, who contends that the *albogue* was used by Christians in medieval al-Andalus, and that the word was an Arabization of a romance *voce* (voice, sound) (54n16). Christian sources also attest that the *albogue* was not an instrument used by Muslims, at least in medieval Spain. According to Juan Ruiz's *Libro de buen amor*: "Albogues e bandurria, caramillo e çampoña / non se pagan de arávigo quanto d'ellos Boloña" (1517a–b) [*Albogues* and *bandurria* (a type of mandolin), flute and panpipes do not care about Arabs as much as Bolonia does not care about them]. In a previous stanza, he attributed its use to shepherds: "El pastor ... taniendo su çanpoña e los albogues" (1213a–b) [The shepherd … playing his panpipes and his *albogues*].[36]

Diego de Guadix's *Recopilación de algunos nombres arábigos* (c. 1590), one of Covarrubias's sources, provides one of the more detailed definitions of the *albogues*:

> Llaman en España a un cierto género o suerte de flauta que su siluo se ceua y sustenta del aire de una votilla o cuerezuelo hinchado que el que lo tañe trae deuajo del braço; tiene una fístula gruesa y larga que siempre, a un tono, lleua un vaxo, con el qual concierta y acuerda los puntos de la otra flautilla que el dicho tañedor tiene y trae en las manos. (221)
>
> [In Spain the name is given to a certain kind of flute whose cane is fed through a pouch or hide bag that the one who plays it carries under his arm; it has a long and thick lever that always gives a bass tone, and with it he tunes it and harmonizes the notes with that of the other small flute that the performer plays in his hands.]

The alleged Arabic etymology is rather tautological: "Consta de *al*, que, en arábigo, significa el, y de *buq*, que significa este dicho instrumento" (221) [It consists of *al*, that in Arabic means *the*, and *buq*, which means the aforementioned instrument]. Guadix never says that it is a Morisco instrument; on the contrary, he ends the entry illustrating the linguistic substitution that underlies the absurdity in this passage of *Don Quixote*: "An deseado, en España, dar de mano a esta algarabía y llamar y nombrar a esta dicha flauta por nombre castellano y la an baptizado por este nombre: *gayta çamorana*" (221) [In Spain, some have wanted to remove the Arabic name and call this aforementioned flute by a Castilian name, calling it *gayta çamorana* (bagpipe from Zamora)].

Guadix's equation of the *albogue* and the *gaita zamorana* is not necessarily accurate.[37] Nonetheless, both instruments must have shared a strong similarity in order for Guadix to be able to state that they were the same. Furthermore, there is another case in which Covarrubias manipulates Guadix's definitions. Guadix defines *zambra* in two separate entries. He first states that "zamara ... significa, en arábigo, flauta, gayta o çampoña" (1031) [*zamara* … means in Arabic flute, pipes, or panpipes], later adding that "zambra ... significa música" (1034) [*zambra* … means music]. The definition that Covarrubias offers for *zambra* is based on Guadix's, but shows significant exclusions: "Dança morisca. En rigor zambra vale tanto como música de soplo o silvo, porque se dança al son de dulçaynas y flautas" (1017) [Morisco dance. Strictly speaking *zambra* means wind music, because it is danced with the sound of clarinets and flutes]. Covarrubias's omission of Guadix's *gayta*, and the translation of his *zampoña* for the most recognizable "Morisco" instrument *dulçaina*, indicates that he is expunging his source in order to construct a neat cultural differentiation between Old Christians and Moriscos.

If Guadix is right in denouncing the linguistic substitution taking place in Spanish society, it means that Don Quixote's list of pastoral instruments goes one step farther in its absurdity, since he mentions both the *gaita zamorana* and the *albogue* as if they were two different instruments – instead of the same instrument with two different names. That is, once the signifier *albogue* is repressed in the social imaginary, it returns renamed in fantasmatic form. In this case, *albogue* surfaces in Don Quixote's pastoral dream as a ghost signifier, whose meaning has been displaced by the *gaita zamorana*. The arbitrary definition improvised by Don Quixote is but a strategy to alleviate its threatening appearance by reinscribing it within the ordered form of an encyclopedic taxonomy. Thus the passage turns out to be emblematic of how the project of cultural cleansing eventually is turned against Christian cultural practices, as happened in many other cases in which the zeal to extirpate so-called Morisco cultural practices led to the repression of purely Iberian or Mediterranean ones (Milhou, "Desemitización" 44–50; Stallaert, *Etnogénesis* 65–9). This reflects what Christiane Stallaert identifies as the final phase of ethnocentrism, which does not stop with the expulsion of the Other, but, on the contrary, the expulsion of the Other is only the beginning of a never-ending search for the Other in the self (*Ni una gota* 251–60).

The end of this pastoral project is at best ironic. As they are likely dreaming of the flocks of sheep that they will put to pasture, Don Quixote and Sancho are awakened by a terrifying noise. Unable to escape in the dark, a herd of six hundred pigs runs over them, ruining their pastoral plans when they have not even begun them (II, 68, 1099). It is not hard to establish the connection between the pigs and the Old Christian perception of New Christians as a differentiated "flock," usually identified by the animal that Muslims and Jews are not supposed to eat and that is signified by the term *marrano*.[38]

The irruption of the "wrong" livestock in Don Quixote's Arcadia reveals the illusory nature of the ethnocentric search for purity. This scene can be read also as a punishment inflicted upon Don Quixote for his failure to craft a cultural cleansing in his own pastoral dream. The connection between the selection of the musical instrument and the arrival of a determined ethno-religious "flock" is already found in Juan de Ávila's sermon "¡Dichosas ovejas que tienen tal pastor!" that glosses the biblical verse of John 10:16 "oves meae vocem meam audiunt" [My sheep will hear my voice]. According to Juan de Ávila, the swineherd and the shepherd use very different musical instruments to call their herds:

> Vi una vez, yendo por un camino, que apacentaba un hombre unos puercos y que estaba tañendo una trompetilla de muy mal sonido, una música muy mala, y estándola tañendo venían los puercos saltando y corriendo con gran regocijo al hombre que traía la trompetilla. Y preguntando qué fuese aquello, dijéronme que los puercos venían a aquel sonido, porque aquélla era la voz con que llamaba aquel pastor a su ganado. ¡Oh oveja! ¡Oh cristiano! ... ¿y no conoces tú la voz de tu Señor? (262–3)
>
> [Once I saw, walking down a path, a man who was grazing some pigs and was playing a small trumpet with a very horrid sound, and very bad music, and while he was playing it the pigs came hopping and running with great joy towards the man with the small trumpet. And when I asked what that was, they told me that the pigs came to that sound, because that was the tune which that herdsman used to call his livestock. Oh sheep! Oh Christian! And don't you know the voice of your Lord?][39]

The textual contiguity of the debate about the inclusion of the *albogues* in the Arcadia and its collapse with the irruption of the herd of pigs signals that the Spanish pastoral world rests on the fiction of both cultural and genealogical purity, and that therefore neglecting to carry out cultural cleansing allows the Other to burst into Arcadia, causing its destruction.

While the whimsical connection of ethnocentrism and the pastoral ultimately wrecks Don Quixote's pastoral project, there are other moments in *Don Quixote* in which the bucolic appears precisely in the ethnic and cultural heterogeneity denied by pastoral romances. Alban Forcione suggests that the "true pastoral" should be found in the convivial conversation between the Old Christian Sancho Panza, the Morisco Ricote, and a group of German Protestants (1040). Similarly, Johnson points out that the encounter between Sancho and the Morisco Ricote close to a beech tree seems to be marked by the pastoral tradition of Virgil and Garcilaso (*Cervantes* 52).

These observations on whether the *albogues* was Morisco or Christian and whether it should be included or not in Arcadia ultimately touch on the authorial role of Cide Hamete Benengeli, who is unexpectedly silent precisely at a moment in which his opinion would have mattered most. As Monika Walter shows, as far as *Don Quixote* is a parody of books of chivalry, it becomes yet another book of chivalry itself, and, furthermore, Cide Hamete Benengeli should be credited as an author of books of chivalry. Incidentally, Walter asserts in passing that, since pastoral romances constitute part of *Don Quixote*'s subtext, then Cide Hamete Benengeli should be considered as well "autor moderno de novelas pastoriles" (40) [modern author of pastoral romances].[40] As Frederick de Armas observes, it is Cide Hamete Benengeli who combines Moorish and pastoral elements in his tale (*Don Quixote* 59–77), noting that the pastoral episode of Marcela and Grisóstomo comes right after his authorial voice surfaces in *Don Quixote* I (*Don Quixote* 68). Paradoxically, Grisóstomo is criticized precisely because "mandó en su testamento que lo enterrasen en el campo, como si fuera moro" (I, 12; 119) [he ordered in his will to be buried out in the wilds, like some Moor] (trans. Rutherford 89). The accusation against the Moriscos for wanting to be buried outside the city walls as a proof of Islamism was common (García Pedraza 2:598–603; Cardaillac 38–9). This does not mean that Grisóstomo is a Morisco (the text never suggests that), but rather, and more important, that the signs of cultural difference are equivocal, and that Morisco cultural practices and pastoral imagery may have unexpected points in common.

The role of Cide Hamete as writer of pastoral romances is reinforced in the final words of *Don Quixote*, in which he says goodbye to his pen: "Aquí quedarás, colgada desta espetera y deste hilo de alambre, ni sé si bien cortada o mal tajada péñola mía" (II, 74, 1138) [Here you shall rest, hanging from this rack on this length of brass wire, O quill of mine – whether well trimmed or not I do not know] (trans. Rutherford 981). Here, Cide Hamete Benengeli poses as Sannazaro, who at the end of his *Arcadia* also says goodbye to his *sampogna* (bagpipe), with which he claims to have composed his work (Clemencín 6:463; Avalle-Arce, *Las novelas* 224).

This final act of Cide Hamete posing as Sannazaro rewrites the latter's *Arcadia* inasmuch as it rewrites as well the diegetical containing of Moors within a pastoral frame, as found in Montemayor's *La Diana*. While in *La Diana*, *El Abencerraje* was the tale of Felismena, separated in time and space and contained within the frame of the pastoral romance (if not subjugated to it), inversely, in *Don Quixote*, the Morisco narrator is the authorial voice that frames, and therefore contains, the pastoral interludes of the Christian characters.[41] Yet Cide Hamete's authorship is rarely acknowledged, nor is the significance of an Arab from La Mancha allegedly writing a parody of pastoral romance.

Cervantes continues this exploration in his last work, the posthumous *Los trabajos de Persiles y Sigismunda* (1617), in which the connection between pastoral, ethnocentrism, and the neo-Gothic myth reappears. When the main characters Periandro and Auristela arrive in their pilgrimage at the city of Toledo, the view of the river Tajo inspires Periandro's praise and remembrance of Garcilaso de la Vega's first eclogue:

> No diremos: *Aquí dio fin a su cantar Salicio*, sino: Aquí dio principio a su cantar Salicio, aquí sobrepujó en sus églogas a sí mismo; aquí resonó su zampoña, a cuyo son se detuvieron las aguas deste río, no se movieron las hojas de los árboles, y parándose los vientos, dieron lugar a que la admiración de su canto fuese de lengua en lengua y de gente en gentes por todas las de la tierra. (327)
>
> [We won't say, 'Here Salicio brought his song to an end,' but, here Salicio began his song; here he outdid himself in his eclogues; here his shepherd's flute trilled, the waters of this river paused to hear it, the leaves of the trees ceased fluttering, and the winds grew calm to give news of his astonishing song a chance to go from tongue to tongue and from one nation to another through all the world.] (trans. Weller and Colahan 234–5)

Periandro emphasizes the capacity of Garcilaso's eclogue to halt time, to describe an unchanging picture of the central landscape of Spanish identity.

However, unlike the shepherds of *La Galatea*, Periandro immediately makes the connection between the bucolic banks of the Tajo and the city of Toledo as both the ancient capital of the Visigoths and the episcopal see of the Spanish Church:

> Y poniendo la vista en la gran ciudad de Toledo, fue esto lo que dijo:
>
> ¡Oh, peñascosa pesadumbre, gloria de España y luz de sus ciudades, en cuyo seno han estado guardadas por infinitos siglos las reliquias de los valientes godos, para volver a resucitar su muerta gloria, y a ser claro espejo y depósito de católicas ceremonias! ¡Salve, pues, oh ciudad santa, y da lugar que en ti le tengan estos que venimos a verte! (327)
>
> [Then turning to look at the great city of Toledo, this is what he said. "Oh, massive rock, the glory of Spain and brightest of its cities, in whose heart of countless centuries the relics of the brave Goths have been kept to bring their dead glory back to life and to be a shining example and safe treasure house of Catholic ritual! Hail, then oh holy city, and grant that those of us who have come to see you may find a place within!"] (trans. Weller and Colahan 235)

What Periandro's enthusiasm reveals is that the bucolic of Garcilaso and the restoration of the neo-Visigothic myth belonged to the same project of turning the clock back to the largely mythical pre-711 Spain.[42] The narrator points out that, since Periandro is a foreigner, his knowledge of Spain is mediated by his readings: "como es uso de los setentrionales ser toda la gente principal versada en la lengua latina y en los antiguos poetas, éralo asimismo Periandro" (327) [since it's customary among the northern nations for the most important people to be well-versed in the Latin language and the classical poets, Periandro, being one of the principal people in his country, certainly knew those things] (trans. Weller and Colahan 234).

In contrast to Periandro's bookish understanding of the Spanish landscape, Antonio, a Spaniard who had been exiled for many years in the Barbarian Isle, is not able to perceive it in the same way:

> Esto dijo Periandro, que lo dijera mejor Antonio el padre, si tan bien como él lo supiera; porque las lecciones de los libros muchas veces hacen más cierta experiencia de las cosas, que no la tienen los mismos que las han visto, a causa que el que vee con atención, repara una y muchas veces en lo que va leyendo, y el que mira sin ella, no repara en nada, y con esto excede a la lección la vista. (327–8)

> [It was Periandro who said this, although the elder Antonio might more appropriately have done so, had he known all that. But lessons learned in books often teach us more about things than is known by people who have actually seen them, since a person who reads with care thinks over and over again about what he's reading, while a person who looks without paying attention observes nothing, and so reading can, indeed, surpass seeing.] (trans. Weller and Colahan 235)

As the narrator's intervention underlines, one must be trained in the explicit discourses of history and myth that modulate the perception of landscape in order to be able to apprehend the spatial symbolism of early modern Spain. Thus, even though Periandro is a foreigner, his intellectual background allows him to quickly frame the view of Toledo and the Tajo within its proper ideological interpretation better than a native not trained in them.

The mention of Garcilaso triggers the timely appearance of a group of shepherds on the Tajo's banks (328), immediately preceding the character's arrival at a coastal town close to Valencia inhabited by Moriscos (353–9). The pastoral instruments of the shepherds on the Tajo's banks are similar to those listed in *Don Quixote*: "uno tocaba el tamboril y la flauta, otro el salterio, éste las sonajas y aquél los albogues. Y de todos estos sones redundaba uno solo, que alegraba con la concordancia, que es el fin de la música" (328) [One was playing the finger drum and flute, another the psaltery, this one the tambourine and that one the reed pipes. These sounds flowed together as though they were all one and lifted everyone's spirits with their harmony, the goal of music] (trans. Weller and Colahan 235–6). Although the inclusion of the *albogues* is not questioned in this passage, it is the last instrument enumerated and the narrator notes right away that there is no lack of harmony. Still, after having seen how much fuss Don Quixote and Sancho created concerning this instrument, the final mention of the *albogues* may be interpreted as the residual signifier that prevents the perfect closure of a pastoral world that gravitates around the centre of the Spanish empire, the neo-Gothic myth, the ethnic interpretation of the Christian community, and the institutionalization of Castilian pastoral poetics.

Thus, throughout his entire oeuvre, Cervantes unveils the deep connection that pastoral had with an ethnocentric vision of the history of Spain and the project of ethnic cleansing through the voices of individuals who are not fully integrated into the community of shepherds. While in *La Galatea* it was the "desamorado pastor" Lenio, in *Don Quixote* it is revealed through the ludicrous knight and squire and their short-lived pastoral project, and in the *Persiles*, it is properly a foreigner that only knows Spain through his readings. They are the only ones who, out of ill-will or pretended incompetence, reveal the underlying assumptions of the Spanish pastoral world built upon the historical and cultural denial of the legacy of Islamic Spain.

The literary texts analysed in this chapter are significant for different reasons. In the case of Montemayor's *La Diana*, the interpolation of *El Abencerraje* serves to resemanticize the ambiguity of sentimental faith and the cultural identity of each character depending on their geographical origins. Simultaneously, analysing how both the continuators of *La Diana* and modern editors have reacted to this interpolation shows that the model of the genre was created, not on the basis of *La Diana* by itself, but rather in reaction to it when combined with *El Abencerraje*. Mercader's *El prado de Valencia* exemplifies how the genre is rewritten by a lord of Moriscos in order to craft an alternative Arcadia more in line with his own class interests. In his version of the pastoral genre, the elements of sheep herding are displaced in favour of Moorish cultural practices best suited for the performance of an aristocratic identity and the Moriscos are recognized their place within it – even if subordinated to the protection of their feudal lord. Meanwhile, Cervantes shows in his entire oeuvre how the presence of the most minimal element of Iberian Moorish legacy (or even the suspicion of its presence) threatens the pastoral fictions of characters, destabilizing their idealized sense of community. What these literary works have in common is that all reveal in different ways the impossibility of containing the Moorish legacy that the genre as a whole was trying to suppress. If, with Moriscos living in their own communities, containment and suppression were impossible, we must explore what was the effect of their expulsion in 1609 for authors of pastoral romances.

5 Imagining the Spanish Arcadia after 1609

fue por el río Betis conocida
la expulsión de los Moros deseada ...
Al son de varios dulces instrumentos
los ríos comen y las Ninfas cantan.

[The river Betis knew about
the desired expulsion of the Moriscos ...
At the sound of some sweet instruments,
the rivers dine and the nymphs sing.]

Gaspar Aguilar, *Expulsión de los moros de España* (1610)

Son pocas las novelas pastoriles del siglo XVII que
conservan un elemento
bucólico fuerte y auténtico ... *El premio de la constancia* ... y la
Soledad entretenida ... han sido invadidas por elementos moriscos.

[Only a few pastoral romances in the seventeenth century
keep a true and authentic bucolic component ... *El premio de la constancia* ... and the *Soledad entretenida* ... have been invaded by
Morisco elements.]

Willard King

The previous chapter analysed those pastoral romances that questioned the fundamental ethnocentricism of the Spanish Arcadia. After 1609, however, there is an important feature that differentiates the writing of pastoral. Philip III's decree that year expelling the Moriscos radically altered the imaginary of cultural and racial purity. Rather than placating ethnocentricism, the measure set the precedent that entire ethnic communities could be expelled from Spain, even if Christian – and Moriscos had been formally Christians since their forced baptism at the beginning of the sixteenth century (Tueller 153).[1] This alteration is foreshadowed in Cervantes's oeuvre: the second part of *Don Quixote*, published after the expulsion, is much more critical of the cultural underpinnings of the pastoral Arcadia, and Cide Hamete Benengeli mimicking Sannazaro's voice already points to the radical rewriting of the genre that some pastoral romances will undertake.

As we shall see, this moment of exacerbation of early modern Spanish ethnocentrism is reflected in Lope de Vega's *Los pastores de Belén* (1612), in which Hebrews and Gypsies – but no Moors or Moriscos – attain sudden prominence in a pastoral world the orthodoxy of which has never been questioned. In contrast, the disappearance of real Moriscos suddenly imprinted a nostalgic tone to pastoral fictions, causing a progressive and yet hesitant inclusion of the Moor in Arcadia, as seen in Jacinto de Espinel Adorno's *El premio de la constancia* (1620) and Juan de Barrionuevo y Moya's *Soledad entretenida* (1638). Each of these three texts has been identified by scholars as the last pastoral romance marking in one way or another the decadence of the entire genre. Analysis of each of these works shows how the genre was transformed after losing in 1609 the element that pastoral was aiming to counteract, suddenly turning suppression into nostalgia for a waning ethnic heterogeneity.

Jews and Gypsies in Lope de Vega's *Pastores de Belén*

Lope de Vega, although mostly known as a playwright, cultivated the pastoral genre as well in various fictions. In 1598, he published *La Arcadia*, a pastoral romance that is largely a compilation of poetic compositions. In 1621 he published the novella "Las fortunas de Diana"; and in the novella "Guzmán el Bravo" (1624) he promised another pastoral romance, *El pastor de Galatea* (*Novelas* 337), a promise that likely went unfulfilled. In contrast with these more conventional pastoral texts, Lope's *Pastores de Belén* (1612) is a pastoral romance "a lo divino," a rewriting of the genre that replaces the alleged sensuality and paganism of the courtly pastoral literature with religious materials.[2] The content is certainly atypical of the genre and at first sight looks like merely a pious rewriting: some Hebrew shepherds meet to celebrate the birthday of John the Baptist and the imminent birth of Christ. The work ends with the shepherds heading off to worship the newborn Christ in Bethlehem, and with the flight of the Holy Family to Egypt to avoid the slaughter of all newborns ordered by Herod.[3]

Lope made the change in tone explicit in the dedication to his son: "será bien, que cuando halléis *Arcadias* de pastores humanos, sepáis que estos divinos escribieron mis desengaños, y aquellos mis ignorancias" (81–2) [it will be good that, when you find *Arcadias* of human shepherds, you realize that my disenchantment composed these divine ones, and my ignorance the aforementioned]. At the end of the book he renews the attack against profane pastoral fictions, which he brands "aquellas fábulas vanas, inútiles, copiosas de mentiras y lisonjas, halagadoras de hermosuras, que en tan breve tiempo feas, han sido la luz de mis engaños" (589) [those vain fables, useless, abundant in lies and flatteries, so quickly ugly, those have been enlightenment from my illusions].

Lope's shift in register has traditionally been read as the outcome of a sudden religious vocation (Avalle-Arce, *La novela pastoril* 271; Carreño, "La otra *Arcadia*" 142–5; López Estrada, "Los libros de pastores" 193–6). Consequently, Antonio Carreño considers it an orthodox counter-reformist didactic text ("La otra *Arcadia*" 150). However, in spite of the romance's biblical content, Lope's Hebrew shepherds are not so different from the shepherds in other pastoral romances: they entertain themselves with the same courtly games, intercalating biblical tales with the celebration of aristocratic festivities, music, and riddles. Indeed, as Carreño notes, several individual copies of *Pastores de Belén* were severely expunged by their owners of the episodes with erotic content, such as those dealing with Susanna, the attraction of David for Bathsheba, or the rape of Tamar ("La otra *Arcadia*" 147).[4] The presence of such erotic content and courtly games suggests that *Pastores de Belén* did not engage primarily with other ecclesiastical attacks on the sexual morality of pastoral romances. Yet, while the biblical element covers the book from beginning to end, this does not necessarily suggest that this work is a perfect embodiment of post-Tridentine orthodoxy.

It is interesting to note that scholars rarely question Lope's religious belief; they confirm his orthodoxy while they rush to find signs of heterodoxy as soon as the same topic is approached by a different individual. This is the case for Felipe Godínez, who composed two short plays inspired by Lope's pastoral "a lo divino": *Coloquio primero y segundo de los Pastores de Belén*, published in 1655 but likely written in the 1620s (Menéndez Onrubia 179). Godínez, probably a friend of Lope (Pamp 23), was prosecuted by the Inquisition for Judaizing in 1624 (Carrasco Urgoiti, "*De buen moro*" 546–8; Menéndez Onrubia 179). It seems to be this biographical knowledge that leads Carmen Menéndez Onrubia to conclude that Lope's *Pastores de Belén* is an ingenious model of religious orthodoxy (177–8), while Godinez's version is quite the opposite, a polemicizing work full of heterodox passages (180–5). Instead of creating such a stark divide based on biographical assumptions, it is possible to see why a *converso* prosecuted by the Inquisition may have been attracted by Lope's text, which already contained the same controversial elements that are regarded as heterodox in Godínez's text, arguably the counteracting of an ethnocentric appropriation of both the pastoral Arcadia and scripture.

By 1612 the process of ethnic engineering seemed unstoppable: it had started with the expulsion of the Jews in 1492, continued with the deportation of the Moriscos from the kingdom of Granada in the early 1570s, and reached a climax with the expulsion of the Moriscos between 1609 and 1614. As Lope was writing *Los pastores de Belén*, between January 1609 and November 1611 (Carreño, "La otra *Arcadia*" 141), several waves of expulsion, directed by the Marquis of Santa Cruz, had officials searching for Moriscos throughout Spain. As soon as these initiatives started, voices were also raised asking to continue with the expulsion of Gypsies. By the 1610s the dream of a purified ethnic Spain seemed endless.

Pastores de Belén positions itself as a rewriting of the pastoral romance from its opening lines. If the beginning of Jorge de Montemayor's *La Diana* was "Bajaba de las montañas de León el olvidado Sireno" (109) [Sireno the forgotten was descending from the mountains of León], Lope rewrites it as "Bajaba de las montañas de Judea a la torre de Belén ... el pastor Aminadab" (93) [the shepherd Aminadab was descending from the mountains of Judea to the tower of Bethlehem]. While the intertextuality with *La Diana* has been amply acknowledged (Avalle-Arce, *La novela pastoril* 271; López Estrada, "Los libros de pastores" 194), its ideological connotations have yet to be explored, since, in the early modern Spanish imaginary, one can hardly think of more antithetical mountains than those of León and those of Judea. Lope's rewriting of the mountains of León, genealogical guarantors of blood purity as we have seen, actually confirms that he (and with him many of his contemporaries) read the beginning of *La Diana* as a statement about the doctrine of blood purity. Thus, if the beginning of *La Diana* associates Sireno with the imaginary space of blood purity, *Pastores de Belén* introduces Aminadab as unmistakably Jewish.

The Jewishness of these biblical shepherds is a central issue of *Pastores de Belén*. Right after contrasting Sireno's space, the narrator moves to an explicit reflection over Aminabad's ancestry:

> descendiente del Tribu y casa de Jacob, y deudo del santísimo José, a la sazón esposo de la hermosa María, criada ... para Madre de Dios; aunque no entonces padre putativo suyo, si bien cerca de tan sublime nombre. Iba el pastor dichoso revolviendo en la memoria aquellas antiguas historias de la creación

> del mundo, tapices que por la ancianidad del tiempo intentaban los años cubrir de olvido; contando, pues, desde el primer padre de las gentes hasta el segundo, que pasada la mayor tempestad vio su nueva regeneración y principio. Y prosiguiendo por el largo proceso de sus descendientes, causábale notable alegría la memoria de aquellos antiquísimos pastores y patriarcas, antecesores suyos. (93)
>
> [descendant of the tribe and household of Jacob, and relative of most holy Joseph, who was by then the husband of Mary, raised ... as the Mother of God; he was not by then his presumed father, even though he was close to deserving that sublime name. The fortunate shepherd was pondering those ancient stories of the creation of the world, tapestries that the years were trying to put into oblivion with the ancientness of time; and he was recounting from the first father to the second, who after the greatest tempest came with renewed generation and beginning. And going on with the long progression of [Noah's] descendants, it made him very happy to recall the memory of his ancestors, those very ancient shepherds and patriarchs.]

In spite of the explicitness with which these shepherds display their Jewishness, *Pastores de Belén* is careful to follow Catholic orthodoxy closely, by stressing repeatedly that the birth of Christ implies the coming of a new age and the supersession of Judaism: "ya estaban cumplidas sus esperanzas, que aquél era el Mesías prometido a nuestros padres Abraham, Jacob y David, y el que tantos Profetas habían predicho" (541) [their hopes were already fulfilled, that he was the Messiah promised to our fathers Abraham, Jacob, and David, the one that so many prophets had predicted].[5]

However, the religious supersession is not interpreted as an abrupt event, but rather an internal development derived from the Old Testament's typological announcement on the arrival of the Messiah: "De todos era la verdad este Señor recién nacido ... Todos los institutos de la mosaica ley, ceremonias, ritos y sacrificios eran su sombra, y todas las promesas hechas a nuestros padres iban previniendo su venida hasta que se cumpliesen" (414) [This was truly the recently born Lord of everyone ... All the institutions of Mosaic Law, all the ceremonies, rites, and sacrifices were his shadow, and all the promises made to our parents were foreseeing his coming until it was brought to fruition]. This vision of Christianity is strictly orthodox, but at the same time would be more acceptable to *conversos*, since it privileges the evangelic message over ethno-religious differences.

The *hidalgo* [noble] lineage of Christ and the Virgin throughout the divinized pastoral fiction, even if Jewish, is one of the privileged topics in the debates of this community of shepherds. One of their songs about the Virgin praises her as *hidalga*: "¿quién dudará, que el sol salga / de una Virgen tan hidalga?" (122) [who would doubt that the sun will rise from so *hidalga* (noble) a Virgin?]. Another song by Damón and Fabio borders on irreverence as it suggests that the nobility of the Virgin would be immemorial: "Virgen, la nobleza vuestra / hoy vuestra patria averigua, / que sois más que el cielo antigua ... ¡Qué más antigua hidalguía, / que haber cuando os hizo Dios, / uno solo y ese Dios, y que en sí mismo vivía!" (232–3) [Virgin, your fatherland comes to know your nobility today, since you are older than heaven ... What a very old *hidalguia* could you have had when God created you, only one God, that God who lived in himself!]. Bato wonders about the lineage of the Virgin in the same terms as the inquiries into blood purity [informaciones de limpieza de sangre] in early modern Spain: "si se hiciese una información en que jurase el Príncipe, ¿qué fe podrían hacer los demás testigos? Y así no era necesario examinarlos. No hay quien pueda deponer en la hidalguía de la Virgen, de tiempo inmemorial ... en la probanza de su antigüedad, sólo puede jurar el mismo Dios" (233) [if an investigation was held in which the prince swore, what kind of references would the other witnesses be able to give? And so it would not be necessary to examine them. Nobody can testify about the *hidalguía* of the Virgin, since time immemorial ... To prove its antiquity, only God can swear it].[6]

As Ronald Surtz points out, to talk about the *hidalguía* [nobility] of the Virgin Mary is not the same as to talk about her blood purity; the latter was less common because her lack of blood purity is obvious ("Sobre hidalguía" 610). However, even if they were two different legal and social concepts, *hidalguía* and blood purity were closely related and usually overlapped in discourse. Since one of the seventeenth-century prerequisites for reaching the status of *hidalgo* was having blood purity, the individual who acquired *hidalguía* could retroactively claim to have blood purity as well.[7] In *Pastores de Belén*, there are two references to the lack of *mancha* in the lineage of Christ and the Virgin that suggests that Lope's target was blood purity and not just nobility. About Christ, it is first said that he is "[e]l Cordero blanco / de la ovejuela, / que nació de Adán / sin la mancha negra" (537) [the white Lamb from the ewe who was born from Adam, but without the black stain]. Shortly after, the Virgin is called "aquella Cordera / sin mancha alguna" (558) [that ewe with no stain]. The context suggests that *mancha* should be interpreted as a reference to original sin, but the mention of sheep is reminiscent of the zootechnical language of blood purity already analysed in chapter 1.[8]

When faced with this issue, the defenders of blood purity argued that Christ and Mary were from a different bloodline than the rest of the Jewish people in order to avoid questioning their lineage (Sicroff 214–15). However, they tended to avoid this debate, and it was the detractors of the statutes of blood purity who raised the Jewishness of Christ (Nirenberg, "Race" 343n51; Stallaert, *Ni una gota* 241–4). By emphasizing the ties of kinship between the Jewish shepherds and the Holy Family, *Pastores de Belén* attacks precisely the line of argument that leaves aside questions about the lineage of Christ and the Virgin Mary. This is clear in Aminadab's aim to inform his fellows about "el origen y principio de nuestro pueblo Hebreo, del cual tanto tiempo después ha nacido el divino Mesías, Salvador del mundo" (553) [the origin and beginning of our Hebrew people, from which the divine Messiah was born, the Saviour of the world].

A similar debate surfaces in relation to the Adoration of the Shepherds. Several scholars have pointed out that, between the second half of the fifteenth century and the beginning of the sixteenth, *conversos* favoured the Adoration of the Shepherds (which emphasized the evangelic message) over the Passion of Christ (which is the basis of the accusation of deicide against the descendants of Jews) (Gitlitz, "Lucas Fernández" 504; Surtz, "Pastores judíos" 334–7). The critics of the statutes of blood purity emphasized the Jewish identity of the shepherds of the Adoration, as did Alonso de Cartagena in his *Defensorium Unitatis Christianae* (1449–50) (181) or Hernando de Talavera. In his *Tratado de la misa* (c. 1480), Talavera explicitly stated that the birth of Christ was announced to Jewish shepherds:

> cuando la denunciaron a los pastores judíos e hovieron singular gozo por la mucha paz que la Encarnación hizo ... de los judíos con los gentiles, haciendo de entramos linajes un pueblo ... Esta paz, muy necesaria a nuestra salvación, demandemos al Señor con todo corazón, especialmente aquella ... paz de buena voluntad con todo cristiano, de cualquier seta e linaje que haya venido. (84)
>
> [when it was announced to the Jewish shepherds they had singular joy because of the great peace that the Incarnation brought ... between Jews and Gentiles, making of both lineages one single people ... This peace, which was very necessary for our salvation, we should ask of the Lord with all our heart, and especially that ... peace made out of good will with every Christian, no matter from which sect or lineage they come.]

As shown in chapter 2, while it is initially impossible to establish a deterministic identification between the figure of the shepherd and either the Old Christian or the *converso* across exemplars of the genre,[9] with the emergence of the discourses on blood purity later in the sixteenth century, the association between the figure of the shepherd and the Old Christian became hegemonic. Even if the setting of the Nativity in the city of Bethlehem should emphasize the Jewishness of the shepherds, it was appropriated as well as a metaphor for the community of Old Christians by abstracting both the specificity of the location and the original ethnicity of its inhabitants. By emptying Bethlehem of its historicity, Nativity performances treated it as what Michel Foucault calls heterotopia, a real place which is made to enact a utopian space where a society sees a perfected reflection of itself ("Of Other Spaces" 24).

In contrast, while actual Bethlehem was effaced in the imagery of the Nativity in order to ritualize the bonds and boundaries of the imaginary Old Christian community, Lope first reconstructs a historical Jewish space, cancelling its absolute virtuality at the same time as he recovers the ecumenical message of the Adoration. In a second move, however, *Pastores de Belén* restores the function of Bethlehem as a heterotopia of Spain, although no longer as a spatial reflection of utopian ethnic purity, whether Old Christian or Jewish, but instead as a meeting space where the heterogeneity of early modern Spain is suddenly represented, as I shall now explore.[10]

Lope's first move that turns *Pastores de Belén* into a heterotopia of a diverse early modern Spain is his use of a repertoire of ethnic engineering selected from the Old Testament, which goes from a blending of the cultural and genealogical to genocide. One of the first such biblical stories, taken from Genesis 33–4, tells how Shechem raped Dina when Jacob established his family in Canaan. Shechem, regretting his act, offered to marry Dina. Jacob's sons replied that they could not accept him until he was circumcised, saying that, if he agreed to do so, "recibiremos vuestras hijas, y os daremos las nuestras en contracambio; viviremos con vosotros y seremos un pueblo" (107) [then we will mutually give and take your daughters, and ours: and we will dwell with you, and will be one people]. Even though Shechem agreed to the conditions imposed by Jacob's family, he and his lineage were exterminated within the next three days, as the offer of integration turned out to be only an excuse to prepare revenge for Dina's rape.

A different kind of ethnic engineering appears in Lope's recounting of Herod's aim to erase the collective memory of the Hebrew people by destroying their books of lineages: "Herodes ... ha hecho quemar essos libros de las genealogías de los mayores; porque como por ellos parecía la claridad de su sangre, y la de este cruelísimo Rey es tan oscura, no quiso que se pudiesen alabar los Hebreos, a quien gobierna, de su nobleza antigua, y de la bajeza y novedad de la suya vituperarle" (329) [Herod ... has ordered those books on the genealogies of the elders burned, because they show the illustriousness of their bloodlines, and that of this King is so obscure that he refused to allow the Hebrews, over whom he rules, to be praised for their ancient nobility and to scorn him for his low and parvenu lineage].[11] The end of *Pastores de Belén* is the Massacre of the Innocents, taken from Matthew 2, when Herod orders the slaughter of all the newborns and young males up to the age of two in Bethlehem to avoid the birth of the Messiah. Lope's shepherds here criticize Herod openly, debating cases of cruelties committed by tyrants against their own people, most of them telling of the extermination of entire lineages (187–8).

Even though scholars have pointed out the persistent presence of violence in the Spanish pastoral romances (Mujica; Castillo Martínez, "La violencia"), *Pastores de Belén* stands out as a space where much more typical individual violence is displaced by state actions against entire communities. This critique of the ruler, when raised in a biblical pastoral romance, counters the biopolitical concept of the king as the incarnation of the good shepherd, as it was developed around the reign of Philip III, as we have seen in chapter 1.[12] Lope has carefully selected the biblical passages dealing with ethnic engineering to show that the ruler who is prone to commit such acts is more similar to a tyrant than a good shepherd, and his readers would probably have recognized his contribution to this political controversy of their own time.

The repertoire of ethnic and racial heterogeneity characterizing Lope's Bethlehem may at first seem an exotic account of ancient folklore, but is actually an allegorical chart of early modern Spain, and, as such, a heterotopia of Lope's own society. The final chapter describes the Adoration of the Shepherds, emphasizing that conversion to Christianity erases all previous differences, since Christ guards "todo el ganado del mundo" (481) [all livestock in the world]. Again, this vision of Christ as a good shepherd who takes every flock under his care contradicts the ethnocentric image of Philip III as a good shepherd whose task is to separate the good sheep from the bad.[13] Yet, this ecumenical message, rather than being a general abstraction, includes the people from different parts of the world who converge in Bethlehem. For introducing ethnographic descriptions of these people, music takes a preeminent place. First, the Jewish people are praised by Pireno for their musical skills: "cuán natural es a los Hebreos la música, entre todas las naciones del mundo, pues aun los Babilonios, cuando los llevaban cautivos, les pedían que cantasen, que es indicio de la fama que por las naciones extranjeras tenían sus voces" (175–6) [music is so natural for Hebrews, among all the nations in the world, since even the Babylonians asked them to sing when they were being taken as captives, which is an indication of the fame that their voices had among all the foreign nations]. When other peoples from the entire world come to worship the newborn Christ, some are praised for the same reason, such as Gypsies and Ethiopians.[14]

The inclusion of the Gypsies in this biblical world may seem anachronistic to modern readers but not to early modern historical knowledge, which held that the Gypsies who arrived at the Iberian Peninsula in the fifteenth century came from Egypt (whence their ethnonym).[15] In their first appearance in *Pastores de Belén* the Gypsies, who come to worship Christ in Bethlehem, are represented as a local population: "los Gitanos, / que en Jerusalén vivían" (522) [the Gypsies who were living in Jerusalem]. They are also described, in their dress and their music, like the Gypsies of early modern Spain.[16] In their second intervention, when the Holy Family arrives in Egypt on their flight from Bethlehem (583–4), these Gypsies clearly represent ancient Egyptians. This early modern stereotype in which Gypsies were usually presented as musicians and sorcerers is apparently confirmed in this passage. In it, they warmly welcome the Holy Family with their instruments and a Gypsy woman, Marandra, addresses the Virgin to tell Christ's fortune: "Parad el niño bendito, / Señora llena de gracia, / porque la buena ventura / la diga aquesta Gitana" (585) [Stand your blessed child up, oh gracious Lady, so this Gypsy woman can tell his fortune]. Marandra prophesies the events that happen in Christ's life through the Passion, thus introducing his tragic fate and momentarily breaking the joy of the moment. While her fortune-telling constitutes one of the reasons why Gypsies were regarded with growing suspicion in early modern Spain, the fact that her prophecy is in agreement with New Testament accounts dispels the potential threat that her magical abilities may pose against Christian doctrine.

It is significant that the warm reception that the Egyptians (Gypsies) give the Holy Family is placed at the end of *Pastores de Belén*, which turns the passage into a clear counterargument within contemporary debates on measures against them. Shortly after Philip III decreed the expulsion of the Moriscos, Gypsies became the next target of Spanish ethnocentrism (Stallaert, *Ni una gota* 203–5; García-Arenal). In August 1610, the Duke of Lerma proposed in the Council of State to carry out the expulsion of the Gypsies following that of the Moriscos. On 24 December 1610, the Courts of Madrid debated the question, suggesting that Gypsies be confined to big cities and prohibiting the use of distinctive clothing and cultural practices. On 23 July 1611, the Council of State recommended that expulsion of the Gypsies be undertaken as soon as that of the Moriscos of Murcia was complete (Pym 61–4). Even the Mesta, the association of herdsmen, raised its own petition to the Court of Madrid on 16 June 1610, accusing the Gypsies of stealing livestock and asking for their expulsion (Pym 173n21). Over a decade after the expulsion of the Moriscos, Pedro Fernández de Navarrete's *Conservación de monarquías* (1626) suggested that Gypsies should have been expelled instead of Moriscos, who were industrious subjects (467–8). Since one of the arguments in favour of the Gypsies' expulsion was the accusation that they had failed to welcome the Holy Family during their flight from Bethlehem (Pym 8; Charnon-Deutsch 24), Lope's emphasis on the warm welcome is a clear statement in their favour, though neither they nor the Ethiopians belong to the community of the shepherds, a symbolic status reserved only for Hebrews in *Pastores de Belén*.[17] As with the case of the Jewish shepherds, scripture is elaborated to contest the theological underpinnings of the doctrine of blood purity and to critique the justification of ethnocentric policies.

Within the climate of never-ending ethnic purification of the national body in the 1610s, the similarity of the racial map of *Pastores de Belén* to that of early modern Spain omits one significant group, as if the Moriscos were left out in this representation precisely as they are being expelled. Their omission is of course historical, as Islam was not yet in existence.[18] Moriscos are the void that mediates the genocide of the past (the expulsion of the Jews in 1492) and the latent genocide of the future (the expulsion of the Gypsies). Might Jews and Gypsies indirectly represent the Moriscos here?[19] Might Moriscos serve instead as scapegoats? By 1612 their expulsion was largely a fait accompli, and standing against the institutional propaganda campaign that tried to justify the measure would be to defend a lost cause.[20]

Precisely because the Moriscos were now gone, they could be used in *Pastores de Belén* as a symbolic scapegoat.[21] The only moment that may allude to the Moriscos is found when the characters, talking about Herod's cruelties, mention his mixed genealogy: "Pasó últimamente el Reino a un extraño, porque aunque es verdad que Herodes judaíza, al fin es hijo de padre Idumeo y de madre Árabe" (305) [The kingdom finally came to a foreigner, because, even if it is true that Herod practises the Jewish religion, he is the son of an Edomite father and an Arabic mother]. This genealogy is mentioned again when the Magi arrive in Judea asking for "the king of the Jews," which causes Herod to order the massacre of all the young males in Bethlehem (496). In both cases, Herod's miscegenated, half-Arab lineage is quoted as a possible explanation for his cruelties. We may read this as the final irony of *Pastores de Belén*, which rearticulates the motifs of the discourse of blood purity in such a way as to make them unrecognizable.

Concerns over blood purity are located in the Jewish victims of Herod, while the perpetrator of genocide is an individual with mixed origins and a convert to Judaism – even though, as the same shepherds who criticize him acknowledge, he is a true believer. In 1612 making a half-Arab the genocidal tyrant against Jews seems like a clear warning for Philip III that, if he follows the actions of such a counter-model of political action, he will be in the position of Herod. Thus the missing Moriscos of Lope's *Pastores de Belén* are paradoxically present as a kind of undertone and real object of the message of this pastoral romance "a lo divino." Moriscos will return to the pastoral post-1614, bringing a sense of nostalgia that marks such works as Jacinto de Espinel Adorno's *El premio de la constancia.*

Nostalgia for the Moor: Jacinto de Espinel Adorno's *El premio de la constancia*

Jacinto de Espinel Adorno's *El premio de la constancia y pastores de Sierra Bermeja* was published in Madrid in 1620, eight years after Lope's *Pastores de Belén* and six years after the expulsion of the Moriscos had already been completed. Although there is no explicit reference to the expulsion of the Moriscos, by including Moorish characters this pastoral romance invalidates the ethnocentric instrumentalization of the pastoral Arcadia as a fictitious community of shepherds from which the descendants of Iberian Muslims are excluded.

Very little is known about Espinel Adorno. Besides this pastoral romance, he is the author of the epic poem *Descendencia ilustre de don Yñigo Manrique de Lara, alcayde de las fuerças y castillos de la ciudad de Malaga* (1626).[22] He was the nephew of the famous writer Vicente Espinel, who introduced him to the social and literary circle of the Madrid court (Haley 57). Vicente Espinel is mentioned several times throughout *El premio de la constancia*, in the introductory poem by Juan Serrano Brochero and in the narrative itself, praising him as the inventor of the poetic form known as the *espinela* (33v). Any other biographical information about Espinel Adorno has to be inferred from the main character Arsindo, who seems to function as his alter ego.[23] Arsindo describes himself as born in Manilva and raised in Ronda (Munda), both towns located in the same Andalusian Mountains of Sierra Bermeja that give the subtitle to the book.[24]

Both Vicente Espinel and Jacinto de Espinel Adorno were descendants of Christians who had settled in the kingdom of Granada by the end of the fifteenth century, and had lived in an area densely populated by Moriscos. Ronda surrendered to the Catholic Kings in 1485 with a peace treaty that respected the practice of Islam. In 1501, due to the forcible conversions mandated by Cardinal Cisneros throughout the entire kingdom of Granada, the Moriscos of Sierra Bermeja launched a bloody uprising that ended with the death of Alonso de Aguilar, elder brother of Gonzalo de Córdoba, and was only suppressed through negotiation that allowed some of the Moriscos to relocate to North Africa (Lea 101–3).[25] In spite of the migration of 1501, there were still more than 7,000 Moriscos in the area of Ronda before the uprising of the Alpujarras in 1568 (Domínguez Ortiz and Vincent 79). Even after the deportations of 1571, about 214 Morisco households managed to stay on their land, according to the census of 1609 (Lapeyre 180).[26] The possible influence on Vicente Espinel of cohabitation with the Moriscos has been widely acknowledged, especially in his picaresque novel *Vida de Marcos de Obregón* (1618) (Carrasco Urgoiti, "Reflejos" 204–16; Montoro 225–6; Aguirre de Cárcer).

Espinel Adorno's pastoral romance shows similar influences, although, in line with its generic features, the relation between Christians and Moriscos is more stylized.[27] The beginning of *El premio de la constancia* conforms to the conventionalities of the genre of the pastoral romance. A group of shepherds welcome the arrival of Arsindo, at first an enigmatic outsider, by telling him their sentimental stories and singing while they guard their livestock. The plot suddenly shifts when the shepherds ask Arsindo to tell them about his own life and inquire about the reasons that brought him to that remote hamlet. Arsindo goes back to his birth in Manilva, "a la parte del Leuante del monte Calpe" (35v) [to the Mediterranean side of Mount Calpe (Gibraltar)], and his later move to Munda (Ronda) in order to begin his studies.

The action of his tale begins with his visit to La Mina (literally, "the mine"), an artificial cave that descends from Ronda to the banks of the river Guadalevín.[28] While in La Mina, he is requested by a ghostly voice to come back alone the next day. He descends then to the cave where he is welcomed by a gallant Moor who takes him to the presence of King Zelimo, "un venerable moro, de aspecto graue y amable presencia" (39r) [a venerable Moor, of solemn countenance and pleasant appearance]. The Moorish king reveals that he is one of the first Muslim conquerors who came to the Iberian Peninsula in 711 and founded Ronda afterwards (39v). One day, he comes down to the cave, where a *morabito* [a Muslim sage], who had been bewitched since the time of the Goths, puts a spell on him and his retinue, forcing them to wait until the arrival of "un mancebo altiuo de pensamientos, fuerte contra sus enemigos" (41v) [a young boy of high thoughts, strong against his enemies].[29] His prophecy has to be accomplished by Arsindo (41v). In his next visit, the *morabito* explains the ritual that he needs to complete in order to cast them out: Arsindo has to climb Mount Calpe [Gibraltar] in order to collect some herbs that can be found only there. On his way to his task, he falls in love with Amarilis, who eventually dies.

When he returns to Ronda and descends again to La Mina with the herbs, the *morabito* explains to him that he has to kill a snake and asks for the key of the cave to "[e]l viejo Guadaliuin, deydad encerrada en aquel sitio" (79v) [the old Guadalevín, deity imprisoned in that place]. Once liberated, the *morabito*, King Zelimo, and his companions return to Africa, leaving a chamber full of treasures as a reward for him. Guadalevín warns Arsindo that he has only thirty days to take the treasures before the river takes them away. However, Arsindo suddenly recalls the death of Amarilis and is driven crazy until the deadline passes and the treasure is gone. He gets over his sorrows and falls in love with Celia. Unfortunately, he hurts one of her suitors in a fight and has to flee from Ronda. Only at this point of Arsindo's tale, close to the end of the romance, does the audience of shepherds find out the reason why he has come to the hamlet in Sierra Bermeja. King Zelimo, who still feels indebted to the young man, sends a magic storm that takes Arsindo to his African kingdom, where he offers him a place in his court. Arsindo, after considering his offer, decides to return to his hamlet and the *morabito* takes him back to Spain, secretly putting some jewels in his pockets. Again in Sierra Bermeja, he meets Celia, who was following him, and the story ends with the wedding of Felino and Grisalda, two of the local shepherds.

Thus the pastoral functions as a frame for Arsindo's tale, itself a combination of popular legends, courtly adventures, and Byzantine romance. Most critics of this pastoral romance see the inclusion of the Moorish element and the subsequent blending of different literary genres as a symptom of the decadence of the pastoral romance.[30] Certainly, the pastoral setting in *El premio de la constancia* is a mere scenario, one in which the real interest lies in telling a story about bewitched Moors. In fact, the alleged narrative justification for telling Zelimo's story turns out to be misleading: as the reader finds out only in the denouement, the story of La Mina had no direct bearing on Arsindo's arrival in the hamlet; he reached this pastoral setting in Sierra Bermeja while fleeing justice after injuring

Celia's suitor, and the story of Arsindo's love for Celia has no relation at all with the events in La Mina, serving only to provide a conventionally sentimental ending. This leads one to question Espinel Adorno's choice of precisely a pastoral setting to write a Moorish tale, and to wonder how both genres resemanticize each other. *El premio de la constancia* is really the story of Arsindo recounting to a group of shepherds how he helped a group of enchanted Moors escape their confinement in Spain. The idealized community of shepherds that forms the audience of Arsindo's tale is thus exposed to the latent desire of Spanish pastoral romances, with a perfected fantasy of ethnic cleansing that excises any feeling of guilt.

One of the most striking motifs in *El premio de la constancia* is Zelimo's insistence in rewarding Arsindo, for which at least two readings are possible. The first one would be a rather cynical interpretation in which the courtly relation between King Zelimo and Arsindo alludes in an idealized manner to the economic conditions in the history of the Moriscos. The first treasure that Zelimo leaves to Arsindo, lost with the flood of the river Guadalevín, serves as a metaphor for the temporal benefit that the monarchy obtained from the confiscation of the expelled Moriscos.[31] The benefit of this confiscation proves to be as illusory as the treasure wiped out by the Guadalevín, because the real value of the Moriscos was not their properties, but the colonial situation in which they served as a cheap labour force.[32] The second payment, now in Africa, recalls the fact that the Moriscos, after being expropriated of their real estate, were forced to defray the passage to their exile. In the decree of 12 January 1610, which ordered their expulsion from Andalusia, Murcia, Granada, and Hornachos, Philip III stated that, even though he was entitled to confiscate all the possessions of the Moriscos, he conceded thirty days (reduced to twenty by Juan de Mendoza) to sell their movable assets (Harvey 402–5). Nevertheless, they were forbidden to sell their immovable property or to take any kind of currency with them, except the amount to pay for their passage (Harvey 403–4).[33] The generosity of King Zelimo, who pays twice for his "liberation," could easily be interpreted as an idealization of the economic conditions of the expulsion, an idealization that is already contained in the legal language of the decrees.[34]

However, the text also invites the reader to look for a second reading of the Moorish treasure as cultural nostalgia. The preliminary note to the reader introduces the plot by foreshadowing the misfortunes that it contains:

> Aqui te presento (si acaso el leer desdichas gustas) unas, que si bien las consideras, podra ser que te obliguen a leerlas mas vezes, solo por tomar escarmiento en suerte de hombre tan infelize ... y si acaso los pastores te parecieren

> rusticos, y el mio que no te deleyta, no te espantes, pues solo llora desgracias, y siente desventuras, que incendios de amor llegan tales fines. (n.p.)
>
> [I am introducing here, if you like to read misfortunes, some that, if you properly consider them, will perhaps force you to read this many times, only to learn from the ill fortune of such an unhappy man ... And if the shepherds look rustic to you, and mine does not please you, don't run away, because he only mourns tragedies and feels misfortunes, because the fires of love lead to that end.]

The promised misfortunes of the shepherds remain unfulfilled, however. In spite of the prologue's gloomy opening, the book ends with the reunion of Arsindo and his lover Celia and with the wedding of Felino and Grisalda, which constitutes one of the most strikingly happy endings in a genre usually characterized by the lack of sentimental fulfilment. We must wonder then whether the announced "misfortunes" are really those of Arsindo, or rather those of Zelimo and the *morabito*, whose "liberation" could easily be interpreted in 1620 as a reference to the recent expulsion of the Moriscos.

The uncertainty of where to locate these misfortunes opens up a narrative space that blurs the difference between the shepherds on the surface and the Moors below the ground. In this sense, Zelimo's insistence on rewarding Arsindo does not symbolize a payment, but rather a metaphor of how Moorish cultural practices resist simply fading away after 1609.[35] While the description of pastoral life in the hamlet is rather schematic and dry, the underground palace of King Zelimo and the gallant ceremonies that take place there are depicted in thorough and vivid detail. From his very first visit to the cave, "obra morisca que para ser tan curiosa no pudo ser que otros la hiziessen" (37v) [a Moorish building which, because so exquisite, it was not possible that others built it], Arsindo is impressed by the chambers that he finds there:

> entramos por otra sala muy bella, adornadas las paredes de unas colgaduras de seda y brocado muy finas, cuyos suelos estauan llenos de unas alfombras mas ricas que damasquinas, tan grandes y de tanta labor que cogian de una parte a la otra. Entramos en una recámara, donde estauan en un estrado muchas almohadas de terciopelo açul, sobre alcatifas del mismo color. (39r)[36]
>
> [we came through another beautiful room, the walls embellished with very fine silk and brocade hangings, and the floors full of carpets more luxurious than Damascene drapery, and so large and with so much needlework that they stretched from one side to the other. We entered one chamber where

there were on one *estrado* (platform) many blue velvet cushions over *alcatifas* (fine carpets) of the same colour.]

Admiration of the luxurious decor leads to full integration into the cultural practices of King Zelimo, who is quick to share "his" customs with Arsindo: "le traxeron de comer a su usança, puestas las mesas en vez de bufetes sobre las propias alcatifas que en el suelo estaban y me hizo comer con el tanta diuersidad de manjares, que os puedo afirmar que dudo hauer en España quien tan bien coma" (42r) [they brought him something to eat in their way, the table cloths not on a table, but directly on the fine carpets themselves on the floor, and he made me eat a great array of delicacies, and I can assert that there is no one in Spain who eats as well].

Arsindo repeats this commensality "a su usança" [in their way] when he later visits the African kingdom of Zelimo, forgetting in both situations that he is participating in a cultural practice that was perceived as related to Islam.[37] Certainly there is a constant caution by the narrator to point out that these ceremonies happen "a su usança," trying to create a safe distance of exoticism, but these passages reveal nonetheless not only fascination, but also a great familiarity with such practices, undermining the alleged cultural impermeability between "Moorish" and "Christian" in early modern Spain. A similar fascination occurs in the way the narrator shows his admiration for Moorish clothing, as happens with one of Zelimo's knights, who is described wearing: "un paño o toalla atado en la cabeça, que le caia atras gran parte del; un almayçal blanco, puesto encima de un medio baquero de lo mismo, que traia sobre unos calçones blancos, y unos pantuflos en los pies" (42v–r) [a cloth or towel wrapped over his head, of which a big portion hung down in the back; a white veil over a short tunic of the same colour that he was wearing over some white breeches, and slippers on his feet].

The final trip to Africa seems unnecessary in terms of plot structure once the spell has been broken. The only element that justifies the reappearance of these fantastic Moors is that Arsindo, tormented by the loss of his lover Amarilis, forgot to retrieve the treasures from the banks of the Guadalevín before the flood took them. Arsindo's reencounter with the Moors in Africa is openly happy:

estauan abaxo algunos Moros vestidos gallardamente para el recebimiento del afortunado moço, que viendolos el casi le dio alguna tristeza, mas quando la nuue se quito y conocio al Morabito que del encantamiento hauia librado, y a los demás Moros que con el rey Celimo estauan, le dio un grandissimo gusto y mas como todos se le allegaron y abraçaron, dandole mil parabienes, haziendo muchas fiestas y regocijos con su venida, formando al son de algazaras, a su usança, varias danças y saraos ... lleuauale al lado derecho el Morabito con la alegria que generalmente todos gozaban. (112r)

[some Moors were beneath him, lavishly dressed for the reception of the fortunate young man, who, when seeing them, almost became sad, but the cloud lifted and he recognized the *morabito* that the enchantment had freed, and the rest of the Moors who were with King Zelimo; and he was very happy, and even more so when all of them came close to him and hugged him in congratulation, partying and rejoicing upon his arrival, organizing several parties with music and dance in their customary fashion ... And the *morabito* led him at his right side with all of the joy that all of them had.]

When he arrives in Zelimo's chamber, who was seated "con su acostumbrada seueridad y Regia pompa" (112r) [with his customary gravity and royal splendour], Arsindo addresses the Moorish king as if he were Zelimo's vassal, kissing his feet and calling him "gran señor" [great lord] (112r–v). Such an attitude of vassalage towards a Muslim king in 1620 seems highly polemical.

The entire episode of Arsindo's visit to the African kingdom of Zelimo accentuates the bond established between them. For two days, Zelimo invites him to stay in his African kingdom, promising to make him rich, until finally Arsindo asks to be taken back to the pastoral hamlet in Sierra Bermeja (129r–31r). In their last meeting, Zelimo embraces him and predicts that "tu te acordaras de mi, y de todos segun imagino muchas vezes" (132r) [you will remember me and all of us, as I often imagine]. Although Arsindo in the end decides to come back to the pastoral community in Sierra Bermeja, he hesitates long enough to show that his allegiances were divided between the two worlds. As Cristina Castillo Martínez states, Arsindo is a character who, "no adscrito a ningún mundo literario, constituye el puente de unión entre lo pastoril y este mundo mágico-morisco" ("*El premio*" 279) [not attached to any of these two literary worlds, constitutes the bridge between the pastoral and this magic Moorish world]. Even though he eventually stays with the shepherds, he is also a foreigner to that world, and one may conclude that, had it not been for his relationship with Celia, he would have seriously considered joining Zelimo's court. Once he is back in the hamlet, Arsindo finds that Zelimo hid a box full of jewels in his pockets. He then recounts to the shepherds what happened in North Africa, stating in praise of Zelimo that "estas no podian ser sino cosas de un tan gran Principe" (135v) [these things could only belong to a great prince]. Through Arsindo's praise of King Zelimo, the prestige of everything Moorish has been restored within the very literary genre that initially excluded the Moriscos from the idealized refiguring of Spain.

In order to fully understand the polemical nature of this seemingly naive romance, it is necessary to contrast it with the use of pastoral moments in the propaganda literature that celebrates the expulsion. Gaspar Aguilar's epic poem *Expulsión de los moros de España* (1610) [*Expulsion of the Moors from Spain*] ends with an eclogue which announces the fulfilment of the long awaited Arcadia as the last Moriscos leave the coasts of Spain:

> Viendo la gente alborotada
> fue por el río Betis conocida
> la expulsión de los Moros deseada,
> con tantos vaticinios prometida.
> En siendo esta verdad averiguada,
> publica grandes fiestas y convida
> para aumentar su gozo verdadero
> a Tajo, Ebro, Guadiana y Duero ...
> El coro de Diana consagrado
> vino a ver esta fiesta y juntamente
> vinieron las hermosas Semideas,
> las Dríadas, las Ninfas, las Napeas ...
> Al son de varios dulces instrumentos
> los ríos comen y las Ninfas cantan ...
> Danzan las bellas Ninfas de Diana
> cubiertas de un volante transparente,
> con que su gentileza soberana
> se cubre y descubre juntamente.
> Y el padre Betis, por el bien que gana
> con la expulsión de la Morisca gente,
> brinda los convidados a la gloria
> que España tiene por tan gran victoria. (334–7)[38]

[By seeing the tumult of the people, the river Betis knew about the desired expulsion of the Moriscos, promised by so many omens. As soon as the truth is verified, he announces grand festivities, and invites Tajo, Ebro, Guadiana, and Duero to celebrate his true joy ... Diana's consecrated choir came to see this celebration, along with the beautiful female semi-gods, the dryads, the

> nymphs and the *napeas* ... At the sound of some sweet instruments, the rivers dine and the nymphs sing ... The beautiful nymphs of Diana dance covered with a transparent veil, so their supreme grace is simultaneously concealed and revealed. And father Betis, with all the good that he earns with the expulsion of the Moriscos, drinks a toast with his guests to the glory that Spain has with such a great victory.]

Aguilar's eclogue crowns in pastoral form the violence of the expulsion. Here, anthropomorphized rivers celebrate the event with their nymphs, and even the animals that formed the pastoral scenario participate in the general laity: "Ya por saber que han vuelto las espaldas / muchas cosas se alegran y alborozan, / pues de Moncayo en las soberbias faldas / los corderillos entre sí retozan" (342) [Many things rejoice at knowing that they have turned their backs, since from the arrogant slopes of Moncayo the lambs frolic]. The erasure of both the physical Moriscos and their cultural legacy is thus conceived as the beginning of a time of completion and a restoration of the Golden Age of innocence and prosperity: "Ya vuelve el tiempo en cuyos Siglos de Oro / hubo una gente digna de alabanza, / que toda su riqueza y su tesoro / fundaba en el ganado y la labranza" (358) [Here returns the time in whose Golden Ages there were people worthy of praise, for all their wealth and prosperity was based in livestock and tilling the soil].[39] Aguilar's poem perfectly exemplifies the link between ethnocentrism and pastoral analysed by Iain Twiddy, for whom: "The pastoral vision is one of an artificial view of nature ... and ... is present in the nationalist desire to make an artificial vision of a country real. The desire to make real the disturbing illusion of ethnic or blood purity in order to establish freedom, prosperity and harmony involves genocide" (65–6). Although Twiddy refers not to the early modern period nor to Spain, but to the deconstruction of contemporary nationalism and pastoral idealism found in the Irish poet Seamus Heaney, the dynamics of pastoral and ethnic-cleansing propaganda that he describes are strikingly similar to Aguilar's poem.

It is within this propaganda combining pastoral and expulsion that the generic contamination in *El premio de la constancia* becomes controversial, by displacing the focus from the pastoral world to the cave of the Moors. The most bucolic moment occurs when Arsindo, the *morabito*, and Zelimo thank the river Guadalevín before their departure: "començamos a ver muchos corros de Ninfas bellissimas, que con adornos vistosos y coronas floridas, nos salieron a recibir por mandado de su Deydad" (83v) [we began to see many circles of very beautiful nymphs, with colourful adornments and garlands, who came to greet us as ordered by their deity]. The nymphs of the river Guadalevín appear singing pastoral songs (83v–7r). The entire passage contains all the elements of pastoral imagery – except shepherds. By contrast, the space of shepherds in Sierra Bermeja is absent of nymphs and fluvial deities. In the context of literary propaganda like Aguilar's *Expulsión de los moros de España*, making nymphs and Moors celebrate together the "liberation" of the latter and excluding the shepherds becomes a parody of triumphalist pastoral accounts about the expulsion of the Moriscos.

Beyond its complication of the relation between pastoral and expulsion, *El premio de la constancia* also questions the genealogical assumptions of ethnic purity in early modern Spain. If pastoral mobilizes tropes of "purity," it is only to create an underlying dialectical notion of contamination. Yet Espinel Adorno's fiction counters the pastoral perception that Moors are the foreigners in Arcadia. First, the cultural influence, already noted in the case of Arsindo, is not one-sided. King Zelimo offers an unexpected use of the parallel naming strategy analysed in chapter 3. When he recounts how he founded Ronda, he upends the very logic of the classicist strategy of the pastoral romances, by saying in passing that it is "la ciudad mas fuerte de toda la Vandalia, que vosotros llamays Andaluzia" (39v) [the strongest city in Vandalia, which you call Andalusia]. Thus, in Zelimo's account, it is the early modern Christian Spaniards who call the territory by its Arabic version (Andalusia), while the Muslim king prefers to use the classicized Germanic name (Vandalia), suggesting at least a cultural chiasmus by which Moors trapped in time reveal the cultural influence of the Visigoths they encountered in the eighth century, while the alleged descendants of the Visigoths are culturally Arabized.

On the other hand, even the genealogy of the Moors is questioned. King Zelimo, who was not born in Spain but had to live ensorcelled underneath it for eight centuries, from the defeat of the Visigoths until sometime around 1620, certainly fits the conventional narrative of the history of Islam in the Iberian Peninsula, which implies that Zelimo and his court are a foreign "race" whose presence is but temporary. This simplification conveniently forgets that most Moriscos might well have descended from the original Hispano-Roman population that converted to Islam, thus preserving the myth that the "true Spanish line" only survived in the northern mountains. The long period of Zelimo's confinement under the earth reinforces in a fantasmatic way the opinion that a total, almost supernatural separation existed between Hispano-Romans and Moors. Magic here naturalizes as well the recent expulsion of the Moriscos, construing it as liberation and a return to the African fatherland. The *morabito* explains to Arsindo that, once they are disenchanted, they will return to "nuestra tierra" (51v) [our land]. At the end, the god Guadalevín addresses both Zelimo and the *morabito* to announce to them that they are now free to go, ordering them to cross the sea and rejoice with their remote descendants [sucessores], "que hasta oy os esperan" (87v) [who are waiting for you even now]. Thus, in spite of the eight centuries that have passed since their arrival in the Peninsula, the *morabito* and Zelimo would be welcomed by the descendants of their lineage in their African homeland. At first sight, this account is in agreement with the vision of the apologists of the expulsion, who bequeath their voice to the Moriscos only to attribute to them an unequivocal desire to return to Africa in order to rejoin a mythical genealogy uninterrupted by the passage of time.[40]

Nevertheless, there are some details in the biography of the *morabito* that complicate such a fantasy of genealogical and religious impermeability of Spanish identity, as it is revealed in passing in his story:

> muchos años antes que sucediesse aquella ruyna de los Godos y perdicion de sus Reinos, teniendo noticia que en estos lugares hallaria muchas yeruas conuenientes para exercer su arte magica, entrando por aqueste lugar halló una cueua no muy grande, donde assi como entro se hallo encantado tambien, y un letrero que dezia, que encantando el primer Rey Arabigo que aqui fundasse, saldria de aquel encantamiento y quedaria libre del. (41v)

> [many years before the downfall of the Goths and loss of these kingdoms, knowing that he would find in this place many herbs which were very convenient for practising his magic art, entering here he found a small cave, where upon entering he found himself enchanted as well. And there was a sign that said that, after enchanting the first Arabic king who settled there, he would be free of the spell.]

The passage is confusing, since it states that the *morabito* would be liberated after enchanting "the first Arabic king," but in fact it turns out that this was only the first requirement, because the *morabito* lingers with Zelimo until the arrival of Arsindo. The confusion of whether the *morabito* was free to go after enchanting Zelimo or not is raised again in his own explanation of the spell: "[e]l que ordeno este encantamiento, lo hizo con intento, de que el que una vez en el entrasse, nunca mas saliesse, si no tuuiesse la compañía de este Rey, y aunque yo entre en el tanto ha, y Zelimo que esta presente, no por eso he dexado de estudiar en mi Magica de que yo era gran Maestro" (51v) [the person who arranged this enchantment made it with the intention that, once one entered he may not leave unless he is in the company of this king. And, even though I came in a long time ago, and Zelimo is here, I have not stopped studying the magic art in which I was a great master].

Although none of the characters comment on the temporality of the *morabito*'s story, it is striking that he was enchanted "muchos años antes que sucediesse aquella ruyna de los Godos" (41v) [many years before the downfall of the Goths occurred]. Logically, if he was enchanted *before* the arrival of the Muslims, it means that he was not one of them, but rather a Visigoth or a Hispano-Roman. Nevertheless, willingly or not, he has to leave with Zelimo for Africa, and we may speculate that it is because he has converted to Islam during his stay in the cave and has therefore changed his "lineage" and even his "homeland." The case of the *morabito* is the pièce de resistance contesting one of the arguments frequently used by apologists of the expulsion, who used to ignore the *muladis*, Hispano-Romans and Visigoths who converted to Islam, thus denying the possibility that any Moriscos were genealogically linked to Old Christians.[41]

Thus *El premio de la constancia* attests that, only a few years after the expulsion of the Moriscos, the prophecy of the Golden Age has not come to fruition, and that its dream has had devastating material, cultural, and sentimental effects. By rearranging one by one all the elements of both the rhetoric defending the expulsion and the pastoral ideal of ethnocentric completion, Espinel Adorno's work suggests that the expulsion, rather than making possible the advent of Arcadia, has impoverished Spain both economically and culturally, confronting Spaniards with the darker side of the pastoral dream. Yet, while Espinel Adorno's *El premio de la constancia* constitutes a nostalgic idealization of the Moors that conveys the notion that there is no Arcadia without them, it cannot transcend the ideological parameters of the "pastoral habitus." In spite of Arsindo's transcultural allegiances, the world of the shepherds and the underworld of the Moors barely communicate with each other, and thus *El premio de la constancia* still conceives that Moors and shepherds are two distinct identities who inhabit two separate if entangled worlds, a separation broken down almost two decades later by Juan de Barrionuevo y Moya's *Primera parte de la soledad entretenida* (1638).

Pastoral Hierarchies: Juan de Barrionuevo y Moya's *Primera parte de la soledad entretenida*

Juan de Barrionuevo y Moya's *Primera parte de la soledad entretenida* was published in Écija in 1638.[42] As we shall see, this is the first fiction to break the symbolic divide between the world of shepherds and that of Moors, although it still reproduces a pastoral hierarchy of Iberian people depending on their allegiance to Christianity. *Soledad entretenida* is a heterogeneous romance set in medieval Iberia that blends almost every literary genre of its time together with a large dose of Christian moralization. Since most of its first part takes place in the Nasrid kingdom of Granada, many of the literary motifs and the materials are borrowed predictably from the maurophile literature depicting the idealized frontier wars between Moors and Christians.[43] Like Espinel Adorno, Barrionuevo y Moya is from Andalusia, and their common geographical origins may explain their attachment to Moorish themes.[44] In scholarship on Spanish pastoral romances, this work is generally mentioned in passing as merely an example of the decadence of pastoral as a genre.[45] I will not enter here into the formalistic debate on whether the *Soledad entretenida* should be considered a pastoral romance, but rather I will analyse how the first part rewrites pastoral tradition by interweaving it with Moorish themes.

The title is an obvious allusion to Luis de Góngora's most famous and polemical poetic work *Soledades*, and the literary language of Barrionuevo y Moya is profoundly influenced by this poet from Córdoba. Like Góngora's *Soledades*, the *Soledad entretenida* begins in medias res with a shipwreck, followed by a pastoral encounter with a group of goatherds. The first scene presents Ambrosio Calisandro climbing a mountain and fleeing from the Moors who pursue him. On his way to the top, he finds the shepherdess Haja, who takes him to a pastoral setting: "se hallo en una llanura apacible que dava vista a todas aquellas sierras de donde se descubria no muy lexos una pequeña aldea, avitacion propia de cabreros segun lo davan a entender algunas manadas de cabras que pacian a su vista, y se oyan muchos silvos de los que las venian espantando" (7r) [he found himself in a peaceful valley which provided a view of all those mountains and from which could be seen, not very far away, a small hamlet, a home belonging to goatherds, as some flocks of goats grazing in his sight gave him to understand, and many whistles of those who were shooing them away were heard]. The story of the Moorish shepherdess Haja is also properly set within the pastoral frame, when she tells him about her unhappy marriage to the wealthy livestock owner Adurramen and her impossible love for the shepherd Almançor:

> En lo interior de aquel valle, que de la otra parte de aquella aldea parece, componen unas pagiças casas una grande cabaña de cabreros: dulce albergue mio, y de mi tirano dueño; desta breve aldea es mayoral Ali mi padre a que en poco acierto, y mucho desamor le deve Haja su mal casada hija pues contra el gusto mio, y de toda razon me entrego por mujer a un hombre anciano. Vive al presente en la misma aldea, y vivira en mi memoria lo que yo viviere Almançor, hijo de un mediano ganadero, a quien siempre tuve por idolo de mi aficion. (8r)
>
> [Inside that valley, which appears on the other side of that hamlet, there are some straw huts that constitute a great group of goatherds; my sweet shelter, and that of my tyrant master; my father Ali is the head of this small hamlet, and due to his bad judgment, his daughter Haja, unhappily married, owes him ill-will, because against my will and against all reason, he gave me as wife to an old man. Almançor, the son of a middling livestock owner whom I have ever held as the idol of my love, lives now in the same hamlet, and he will always live in my memory as long as I live.]

The visit to the Moorish hamlet follows the conventions of the pastoral romance, when the shepherd Almançor comes to them: "siguiendo algunas cabras vieron dos çagales, y al uno dellos que venia ... cantando en Aravigo al son suave de un instrumento musico" (9r) [they saw two young shepherds following some goats, and one of them came ... singing in Arabic to the sweet sound of a musical instrument].[46]

In appearing as "cabreros" or "pastores de cabras" [goatherds], these Moors constitute one of the rare cases in which Iberian Muslims are represented as herdsmen in early modern Spanish literature.[47] As I have argued throughout this book, because of the centrality of sheep herding for refiguring Old Christian and Spanish identity, Muslims are excluded from it. There are, however, some cases in which the participation of Moors in herding is acknowledged, although the potential inclusion is usually played down by presenting them as goatherds, which constituted in the imaginary a debased form of herdsman.[48] One of these instances is found in Vicente Espinel's *Vida de Marcos de Obregón* (1618), as a shepherd from Sierra Morena recounts the story of a Morisco goatherd in sixteenth-century Ronda. The Morisco found out how to channel the waters that supplied two rival villages, Balastar and Chucar. After obtaining a reward from the inhabitants of Chucar for bringing them the water, they killed him so that he could not return the favour to the rival town of Balastar (1:193–5). Consequently, the secret of how to alter the flow of water died with him. Espinel calls this character "cabrero moro" [Moorish goatherd] or, more simply, "moro," while the Christian who tells the story is called simply "pastor," as if the act of saying that he was a shepherd made it unnecessary to clarify that he was Christian.[49] Similarly, in Lope de Vega's *La hermosura de Angélica* (1602) the Moorish sage Antito was a "pastor de cabras solitario" [lonely goatherd] before becoming a sorcerer (641).

The beginning of *Soledad entretenida* is inscribed in this stratification of pastoral belongings by describing the Moors as goatherds, to the extent that, by the same linguistic ideology that equates *pastor* with Christian, at times *cabrero* becomes a synonym for Moor. Thus Calisandro and Haja are attacked by "mas de treinta cabreros con alfanjes" (9v) [more than thirty goatherds with scimitars]. *Soledad entretenida* moves one step farther in these ideological equivalences when Calisandro recounts how he dressed as a Moor to enter Granada and boasts of his passing: "Todos me tenian por cabrero" (23r) [Everybody saw me as a goatherd].[50] Finally, Calisandro reveals the reason for his presence in Muslim Granada. His original aim, before he got enmeshed with all the Moorish lovers that he encounters on his way, was to recover the loot that the Muslim captain Muça took from his uncle Protasio Provaliano, mainly livestock and shepherds. Calisandro finds out about how the booty was distributed as it arrived to Granada: "las obejas las compró Aliçuel un rico ganadero ... las cabras, y los otros pastores fueron vendidos a otro señor de ganado" (23r) [Aliçuel, a wealthy livestock owner, bought the sheep ... and the goats and the other shepherds were sold to a different livestock owner]. Thus the only occasion when the text mentions sheep instead of goats is because they have been stolen from Christians. Even though *Soledad entretenida* seems to reflect this dichotomy, so that Moors are called goatherds and Christian shepherds, this opposition is not systematic and we can find many cases in which the opposite term is also applied to each category. For example, the same passage that describes the booty that the Moors gained in Christian territory also reveals that Christians possessed goats as well, and goatherds are also found among Christians (61r). Inversely, Moors are called shepherds in a few instances, most commonly when they are perceived as sympathetic characters that collaborate with Calisandro (11r, 27v, 35r, 45v).

Calisandro's ability to pass as a *cabrero* makes him undistinguishable from Moors themselves. The text emphasizes several times that Calisandro speaks perfect Arabic and he shows a high degree of transcultural competence, as when he serves the king of Granada "in their way" (37r). As the soldiers of the captain Alihabencier observe, Calisandro "era Christiano ... aunque no lo parecia en el avito y lengua" (12r) [he was a Christian ... although he did not look like it in his dress and language]. Similarly, when he is about to be executed by the king of Granada, the Moor Ali Abenciet pretends that Calisandro is his nephew Dragut, "el mas valiente soldado que se crio en Africa y Europa" (217r) [the bravest soldier raised in Africa and Europe]. Calisandro himself boasts of his ability to pass as a Moor, taking the name Habenragel:

> Todos me tenian por cabrero; todos me hazian buen pasage, sin aver moro de pequeño, ni de alto estado que me pidiesse quenta a donde iva, ni de a donde venia, porque parlando tan diestramente como aprendi su lengua y haziendome en aquella saçon mas alentado de lo que tengo uso, no por adversario Andaluz; mas por su connatural del Reyno, ya que no por patriota era tenido de quien mas me via y me tratava. (23r)[51]

> [Everybody regarded me as a goatherd; all of them welcomed me warmly, without there being a single Moor, of low or high status, who asked me where I was going or where I was coming from, because speaking their language so perfectly as I learned it, and becoming more vivacious than I usually am, anybody who saw me and talked to me held me not as an Andalusian foe, but as a fellow of their kingdom, if not their own neighbour.]

In some instances, the passing goes beyond an instrumental deception by becoming a fully inhabited identity, as when Calisandro translates for his Christian fellows the pastoral ballad in Arabic that he sang while alone in the forest (23v).

Further, Calisandro's doubles abound among the Moors. First, when he enters the city of Granada he listens to the soldiers proclaiming a physical description of himself: "un pastor de edad como de treinta años, alto de cuerpo, y jarretudo de color trigueño, barbinegro, vasto de facciones" (39v) [a thirty-year-old shepherd, tall and with long legs, olive-skinned, black beard, and hard features]. Consequently, Calisandro changes his appearance, but he is nonetheless momentarily arrested because he is now identical to the Moor Tarfe (41v). Later on, when Calisandro returns again to the wild, he sees a Muslim shepherd who is being arrested because of his similarity to himself: "un pastor muy gentil hombre ... que quien lo viera y me viera dixera con mucho acierto ser un retrato mio" (46r) [a very gentlemanly shepherd ... whoever saw him and saw me would say that he was truly my portrait]. Significantly, the narrator degrades this shepherd from *pastor* to *cabrero* as soon as his vile personality is unveiled. As it turns out, the name of this Moorish "herdsman" is Abenhamet and he had been the slave of Calisandro's relatives. Being able to recognize Calisandro as a Christian, Abenhamet later denounces him in order to seduce Fatima (49v–50v). Interestingly, Abenhamet betrays him at the moment that Calisandro acknowledges that he was fighting against his own lust towards Fatima (50v), transforming Abenhamet into the double who incarnates Calisandro's own temptations. Thus Calisandro is confused twice with Moors of very different class status, when he dresses both as a Moorish aristocrat and as a Moorish *cabrero*.

Calisandro's cultural permeability is paired with the conversion of several Muslims during the romance, such as the couple of goatherds Haja and Almançor (15r) and Xarifa and her father Ali Abenciet (222r). Conversion to Christianity, even if promoted, is not a precondition for Calisandro establishing bonds of solidarity. The lovers Abenhuc and Fatima repeatedly refuse to convert, yet Calisandro helps them flee to Oran (52v). The permeability of the frontier is exemplified in one anonymous "çagala ... de nacion agarena" (163r) [young shepherdess ... of Saracen stock] living in Christian territory (and not as a slave, as Abenhamet did). There is one significant case of inter-ethnic marriage, when the Christian shepherd Nemesio Frasano, who had been living as a captive with the Muslims, marries Daraxa after convincing her to convert to Christianity (210r). Tellingly, when Nemesio Frasano falls in love with Daraxa, he inevitably fears "la infamia de degenerar de la sangre limpia que me entregaron mis padres" (201r) [the infamy of degenerating from the pure blood that my parents bequeathed me]. While his concerns seem to reflect at first the doctrine of blood purity, it is a rather particular understanding of it. Frasano expresses his concerns when he fears that, if Daraxa does not convert to Christianity, he might be tempted to convert to Islam in order to marry her. Therefore, for him blood purity is an individual adherence to Christianity, and not an idea about lineage and ancestors, since he is more than willing to marry a Moorish shepherdess should she convert.

This cultural permeability also affects the way the characters describe the space that they inhabit. In opposition to other pastoral romances, Arabic place names are widely acknowledged as such, not only in the Nasrid kingdom of Granada (202r), but also in Christian Andalusia. Thus for the river Guadalimar, the author gives both the alleged Latin name (Salfo) and the etymological Arabic Guadalhamar [red river] (82v). By acknowledging every cultural legacy of the territory, the text claims an accumulation of historical layers rather than a selection and a hierarchy among them.

While emphasizing the superiority of Christians, the text also acknowledges the collaboration between them and Muslims, even by rewriting key historiographical myths. In book seven, Calisandro leads the Christian army in defence of the territory against the military incursion of Hali Abençulema. The Christians, aware of the disadvantage they have in not being familiar with the terrain, capture several Muslims taken as spies, although they turn out to be but some poor villagers who were fleeing from the war front (197v). One of them recognizes the Christian military weakness and offers his help:

> El que era de mas crecida edad entre estos tenia por nombre Hamete, hombre sagaz aunque barbaro: este, sabiendo la confusion y duda en que al presente estava puesto Calisandro, se le ofrecio servir en aquella necessidad, prometiendole, como hombre que sabia bien aquellos terminos y estancias de aquellos asperos montes, de guiarle con todos sus soldados por buen camino a un cierto lugar seguro y fuerte, donde pudiese defender la entrada de la sierra. (198r)

> [The elder among them was an astute man, even if a barbarian, whose name was Hamete: he, recognizing Calisandro's confusion and bewilderment, offered to help him in that necessity, promising him that, as a man who knew well the surroundings of those rough mountains, he would guide all of his soldiers through a good path to a certain place that was secured and protected, where they would be able to defend the entrance to these mountains.]

The Christians are hesitant about whether or not to trust the rustic Moor, until they finally follow his advice, thus gaining a tactical advantage over their enemies which is crucial for their victory.

This episode is a reminiscence of the historical accounts of the battle of Las Navas de Tolosa in 1212 (already analysed in chapter 3), which was emphasized as the Christian victory that allowed for the conquest of large territories in Andalusia, and in which the key to their victory was the intervention of a rustic who showed the Christians a secret path through the mountains. By identifying a rustic Moor as the equivalent helper of the Christian army, *Soledad entretenida* emphasizes that the Castilian territorial advances were only possible with the collaboration of the Muslims themselves, and not by their own military prowess alone. Further bolstering this echo, the alleged location of Las Navas de Tolosa falls right in the centre of the geographical area where this romance is located (as well as the geographical area where Barrionuevo y Moya lived). This rewriting suggests that Barrionuevo y Moya's heterogeneous romance went beyond concocting a literary pastiche, and that he was aiming at contradicting both the literary and historiographical image of the Moor as alien to sheep herding and thus to Spanish identity.

Similarly, the pastoral hierarchy of inclusion in *Soledad entretenida* accentuates the potential assimilability of Muslims by introducing the Jew as the third element that is definitely excluded. The reluctance of the Spanish imaginary to acknowledge that other ethno-religious identities were also shepherds is even more prominent in the case of Jews than in that of the Moriscos. If there was some reality behind the stereotype that Jews did not participate in Iberian sheep herding, it would be due largely to their specific legal status during the Middle Ages. As Joseph Pérez points out, in medieval Castile Jews probably devoted themselves to agriculture and sheep herding as much as Christians until they were forbidden to own land by the end of the thirteenth century (24–5). Andrés Bernáldez's *Memorias del reinado de los Reyes Católicos*, written at the very beginning of the sixteenth century, claims that the Iberian Jews had never devoted themselves to sheep herding, which seems to justify their expulsion in 1492: "Nunca quisieron tomar oficios de arar ni cavar, ni andar por los campos criando ganados" (98) [They never chose to accept the duties of tilling the land, nor going into the fields to raise livestock]. *Soledad entretenida* reflects this perception when, during his stay in the mountains of Andalusia, Calisandro encounters the Jew Manases. Manases addresses Calisandro disrespectfully, asking him about Fatima, the fleeing sister of the king of Granada, and threatening to hang him if he does not tell the truth (29r–v). Calisandro replies and, after a short fight, he leaves Manases alive in exchange for 3,000 *escudos* (29v). The Jew Manases stands out in *Soledad entretenida* as the only character from whom Calisandro feels entitled to ask for money without questioning his own aristocratic values. During the rest of his journey he entertains himself by rehearsing all the alleged affronts Jews have committed against Christians (29v).

There is something odd in the way Manases insults Calisandro: "Dime pastor, o salvage" (29r) [Tell me, shepherd or savage], and the potential subjective positions articulated around this insult are more complex than it seems at first sight. From Calisandro's perspective, it is the act of a Jew insulting an Old Christian. The use of *pastor* as insult here recalls the insult that fifteenth-century *conversos* raised against Old Christians who claimed superiority because of their blood purity, an insult that, as we have seen in chapter 2, is probably reenacted in the early Castilian drama. From the perspective of the Muslim knight Mulahacen who accompanies Manases, however, a Jew is insulting a Muslim. The perspective of the Jewish Manases blends with that of the Muslim Mulahacen and that of the Christian Calisandro: on the one hand, Manases, like Mulahacen, believes that Calisandro is a Muslim; on the other hand, he calls him *pastor* – a term that in this narrative is mostly reserved for Christians, instead of *cabrero*. Thus, within the hybrid gaze of the Jew, Muslims become "shepherds," while they are mostly "goatherds" for the Christian narrator. The perspective of the outsider (the insult of the Jew) dilutes, even if only momentarily, the symbolic opposition between Christians and Muslims.

Begoña Souviron López regards *Soledad entretenida* as a "proto-racist" text, because of the negative stereotyped image of Muslims, Jews, and Blacks (*La mujer* 179). I am inclined to agree with her with one important qualification: all pastoral romances are in some way or another "proto-racist" texts. What stands out in Barrionuevo y Moya's work, in comparison to the pastoral romance as a genre, is its too explicit racial hierarchy in which the core of identity are shepherds (Christians) and the potentially assimilable Other are goatherds (Muslims), while those outside of the pastoral community are considered the not-assimilable Other – the Jew and the Black. The message of this fiction is thus the opposite of Lope's *Pastores de Belén*, whose main point was precisely to remind readers that biblical Hebrews were shepherds. In its pairing of the Jew and the Black as the most alien element to Spanish identity, *Soledad entretenida* also points to the reformulation of Spanish ethnocentrism and therefore of the exclusions operated by the pastoral metaphor, as we will see in the conclusion.

In the pastoral romances analysed in this chapter, the inclusion of Muslims, Jews, and Gypsies in Arcadia is contextualized within a wider negotiation about their place within Spanish identity, which did not end with the expulsion of the Moriscos in 1609–14.[52] In fact, the proclamation of its completion showed an acutely laconic tone. A note issued by the Council of State on 20 February 1614, urged the monarch to conclude the never-ending expulsion, stopping any further inquiries about Moriscos returning from or staying in Spain. Against official triumphalism, the document does not state that ethnic cleansing has been completed, but rather that this is an impossible and never-ending project "porque si esto no se ataja, es cosa que nunca tendrá fin" (qtd. in Lapeyre 329) [because if this is not stopped, it will never have an end]. Subsequent rewritings of the pastoral genre echo such inconclusiveness and, strikingly, in what seems to constitute a nostalgic move, fully integrate the Moor within an Arcadia that becomes a place of nostalgia rather than a prophecy of completion.

In Lope de Vega's *Pastores de Belén* the emphasis on the ties of kinship that biblical Jewish shepherds have with key figures of Christianity, such as Christ, Mary, and Saint John the Baptist, counters the theological arguments that underpinned the discourse of blood purity and the stigmatization of the *conversos*. Furthermore, by concluding his pastoral romance with the alleged protection that Gypsies provided to the Holy Family, precisely at the time when their expulsion seemed imminent, Lope's text becomes an intervention against the theological arguments used to promote the measure. This explicit rewriting of pastoral romances shows that Lope understood the role that the Arcadian dream played in imagining and promoting an ethnically cleansed landscape. In Espinel Adorno's *El premio de la constancia* and Barrionuevo y Moya's *Primera parte de la soledad entretenida*, ancient frontier wars are gradually transformed into the site of a nostalgic Arcadia. Although these two texts had a minor impact on a genre that was fading away by the time they were written, and have been largely ignored within the formation of the Spanish literary canon, their exclusion is clearly emblematic of the implicit ideological connotations that have conditioned the evolution and reception of the genre in contemporary scholarship.

Willard King comments that both *El premio de la constancia* and *Soledad entretenida* are pastoral romances that "han sido invadidas por elementos moriscos" (147) [have been invaded by Morisco elements], a choice of words that echoes the militaristic discourse of the *reconquest*. The purity of the genre is thus conceived within the same historical view of Spain as an essential identity that needs to be protected from those influences construed as foreign. Simply speaking, Moors do not belong to pastoral inasmuch as they do not belong to Spanish identity, or rather, they do not belong to Spain inasmuch as they do not belong to its idealized image. In this metaphoric language, Espinel Adorno and Barrionuevo y Moya are regarded as some kind of Don Julián, the traitors that allow the Muslims to invade anew the imaginary space of the Spanish Arcadia precisely as their last remnants have been forced to leave the country.

Such attitudes on the part of critics of works like Espinel Adorno's *El premio de la constancia* and Barrionuevo y Moya's *Soledad entretenida* are part of a larger rejection of literary history, as shown in the study of the editorial practices of Montemayor's *La Diana* analysed in chapter 4. Thus the symbolic expulsion of the Moor from the idealized representations of

Spain in the pastoral romances is not a phenomenon limited to the early modern period, but persists nowadays in the curriculum of literary historiography that reiterates uncritically a stark division between the pastoral and the Moorish.[53] More important, this unarticulated assumption so persistent in literary historiography indicates that the "pastoral habitus" has lingered in the Spanish imaginary well into the twentieth century, as the conclusion outlines.

Conclusion: Pan's Labyrinth

Media España ocupaba España entera
con la vulgaridad, con el desprecio
total de que es capaz, frente al vencido,
un intratable pueblo de cabreros.

[Half of Spain occupied all of Spain
with the vulgarity and the total disdain
of which it is capable, in the face of the defeated,
a disrespectful country of goatherds.]

Jaime Gil de Biedma

Canto me gustaría facer o meu belén
Solimán e Mustafá facendo de pastores.

[How much I would like to do my own Nativity scene,
with Solimán and Mustafá acting as shepherds.]

Os resentidos, "O belén da caixa"

Guillermo del Toro's film *El laberinto del fauno* (2006) tells the story of pockets of rural resistance (also known as *maquis*) against the Franco regime in the mountains of Aragon, in the aftermath of the Spanish Civil War. Half historic film, half horror movie, it explores the ghosts that haunt Spain's more recent history. The film does not bear a direct relation to this book: the faun that inhabits the underground is taken from fairy and gothic tales rather than from the literary pastoral tradition, there is no representation of sheep herding, and the rural setting is anything but an idyll.

And yet, it is a felicitous coincidence for this book that the title of the film was translated into English as *Pan's Labyrinth*, arguably because it sounded more evocative than the word "faun." Most likely, Guillermo del Toro was unaware of the story described in chapter 3 of how early modern historians traced the name of Spain to the god Pan, but *Pan's Labyrinth* could well describe the serendipitous history of Spanish ethnocentrism vis-à-vis the pastoral and sheep herding traditions.[1]

The Pastoral Habitus: Early Modern to Present

Up until what point does pastoral remain a conceptual frame that defines instances of racial purity in Spain? While developing a thorough analysis of the uses of pastoral from the seventeenth century until present-day Spain is beyond the scope of this book, I would like to conclude by analysing several documents showing that it is still a persistent category in the Spanish imaginary. The importance of sheep herding well into the twentieth century, the weight of the discursive tradition of pastoral, and the contemporary reshaping of an ideal of the nation inherited from the early modern period (along with its repertoire of exclusions), invite an exploration of how the "pastoral habitus" is still felt as a remnant of the past.

Many idiomatic expressions reflect that the logic of exclusion inherent to sheep herding is still alive at least as a cultural residue. Expressions such as "cada oveja con su pareja" ["birds of a feather flock together"; literally, "each sheep with its spouse"] and "ser de mal pelo" [to be of bad fur/fleece, i.e., bloodline], as well as the prejudice against black sheep are part of a larger gamut of semiotic fossils derived from the doctrine of blood purity.[2] At the same time that these linguistic remnants reverberate with old systems of social exclusion, pastoral idioms can be reworked to dispel that ideology, such as the proverb "no con quien naces sino con quien paces" ["it's not where you come from that counts, but what you make of your life," literally, "not with whom you were born, but with whom you graze"], which displaces the emphasis on lineage to the social set of relations outside the family as the relevant criteria for individual status and social identity.

Literary texts in which the pastoral and Islam are pitted against each other as two opposite categories are still found between the end of the nineteenth century and the first half of the twentieth, such as Miguel de Unamuno's *En torno al casticismo* (1902), analysed in the introduction, and the short novel "En la guerra [Episodios de Melilla]" (1909), by

Carmen de Burgos (originally signed under the alias of "Colombine"), set in the context of the Spanish colonization of Morocco. In Carmen de Burgos's novel, while Spaniards are celebrating a military victory, a group of Moroccan boys arrive to entertain them with their instruments. The narrator compares one of these instruments to Pan's flute and finds it a very bucolic scene. The magic of the moment is only altered by the narrator's final remark that "[s]in sus vestiduras árabes, aquellos tres muchachos hubieran recordado las pastorales de Longo" (186) [without their Arabic clothing, these three youngsters would have brought back memories of Longus's pastorals]. Carmen de Burgos is not alluding to the Spanish pastoral romances, but to the classical tradition through the memory of Longus's *Daphnis and Chloe* (second century CE).[3] Nevertheless, the importance attributed to clothing as the mark of a distinct identity is not found in Longus, but is filtered through the Spanish imaginary, which posits pastoral clothing as the mark of Spanish identity vis-à-vis the Moor, as we saw in chapter 2.

The survival of the imaginary divide between the Moor and the shepherd is also reenacted in other encounters with North African Muslims, not only in the failed colonial project in Morocco, but during the Spanish Civil War (1936–9). Miguel Hernández's last play *El pastor de la muerte* (1937) [*The Shepherd of Death*], rehearses the defence of Madrid by the Republican troops. At the beginning of the play, the shepherd Pedro tells his mother that he intends to join the cause of the Republic by invoking the pastoral tradition: "quiero / dejar ganado y redil, / trocando por un fusil / mi cayado ganadero" (1832) [I want to leave the livestock and the corral, swapping my shepherd's crook for a rifle]. When she complains, Pedro replies with an enigmatic statement that conflates class struggle with a reference to racial difference:

> Hay muchas malas personas
> Dentro de nuestra nación.
> Son de la casta de aquel
> de quien aún llevamos huellas,
> y hemos de acabar con ellas
> como acabamos con él.
> Son los que han sido los amos
> del pastor y del gañán,
> con quienes acabarán
> si con ellos no acabamos. (1833–4)

> [There are many bad people within our nation. They belong to the bloodline of that one from whom we still bear traces, and we need to finish them as we finished him. They have been the lords of the shepherd and the peasant, and they will finish us if we do not finish them.]

The bloodline or stock [casta] to which Pedro seems to be referring is certainly enigmatic, never fully identifying the "him" from whom Spaniards still keep some trace. "He" might be a Jew, because of the recurrent imaginary that associates Jews with merchants. But "he" might also be a Muslim, since the play emphasizes the presence of Moroccan troops in the Nationalist army in order to construe the Republican resistance as a reenactment of the *reconquest*.[4] Be it the Jew or the Muslim, the same person who advocates for a crusade to eradicate them, nonetheless acknowledges that it is a legacy common to all Spaniards ["de quien aún llevamos huellas"], as if the revolutionary cause was before anything a fight against the *raça* within the self.

The references in the play to North African Muslims fighting on the Nationalist side are abundant: one soldier laments the superiority of the Nationalist army, which includes Moors (1875); Pedro promises to defeat "two battalions of Africa" (1880); a group of Moors provokes the soldiers in the trenches during the peak of the battle (1885); and, finally, one of the officers is captured and executed by "aquella piara / de asesinos africanos" (1890–1) [that herd of African assassins].[5] In one instance, the *reconquest* is explicitly evoked as an analogy of the Republican cause: "salid / a las puertas de Madrid, / de donde los africanos / salían por vuestras manos / igual que por las del Cid" (1906) [go to the gates of Madrid, from where the Africans flew away because of you, as if your hands were those of the Cid].

During the battle, Pedro, who had been mocked until that point because of his rural origins, earns the respect of his companions because of his military prowess, and a Cuban soldier gives him the nickname that is the play's title, "Shepherd of Death" (1909). Pedro himself quickly adopts the nickname, stating "soy pastor de la vida, / por ser pastor de la muerte" (1909) [I am the shepherd of life because I am the shepherd of death]. From this point on, pastoral metaphors are applied to the battlefield, describing the tanks as sheep (1909). Beyond its appropriation of these traditional pastoral tropes, the continuity between tradition and *El pastor de la muerte* resides in the set of oppositions it establishes, by simultaneously idealizing the Republican side as pastoral and emphasizing the Moorish element in the National side.[6]

The characterization of Pedro as rustic may be an element to encourage other rural Spaniards to join the Republican cause, but it is also openly autobiographical. Miguel Hernández was a herdsman from Orihuela and introduced himself as such during his first efforts to become a poet in 1931, when he travelled to Madrid seeking literary success (Ferris 111–28). While his first trip was largely a failure, he was interviewed by Ernesto Giménez Caballero, who wrote an article about him entitled "Un nuevo poeta pastor" [A New Shepherd Poet].[7] Giménez Caballero patronizingly described Miguel Hernández as a "simpático pastorcillo caído en esta Navidad, por este nacimiento madrileño" ("Un nuevo poeta" 10) [cute little shepherd who showed up during Christmas, during this Nativity season in Madrid], and making a satirical appeal to the government of the Republic (which he opposed) to provide some economic support for him: "Queridos camaradas literarios: ¿no tenéis unas ovejas que guardar? Gobierno de intelectuales: ¿no tenéis algún intelectual que esté como una cabra para que lo pastoree este muchacho?" ("Un nuevo poeta" 10–11) [Dear literary comrades: don't you have some sheep to guard? Government of intellectuals: don't you have any intellectual who is completely nuts so this youngster can shepherd him?].[8] Although not completely destructive, Giménez Caballero reacted towards the anachronistic appearance of the shepherd poet by adopting a condescending tone and by instrumentalizing his significance against his political adversaries.

But Giménez Caballero's reaction to the pastoral in the form of a real Spanish shepherd is in itself ironic when considering his own intellectual evolution and how the political turmoil of the 1930s set the two writers apart. While Miguel Hernández eventually joined the Republican side, affiliated himself with the Communist Party, and died in a Francoist prison in 1942, Giménez Caballero supported the National rebellion and the Franco regime and tried to make a political career within the Falangist movement. We can see the shifts in his attitude towards the pastoral in two separate moments, marked by two publications made before and after the war.

Roughly three years after his first encounter with Miguel Hernández, Giménez Caballero treats the traditional Christian imagery of shepherds in his *El belén de Salzillo en Murcia* (*Origen de los Nacimientos en España*) [*The Manger Scene of Salzillo in Murcia* (*Origin of Nativity Scenes in Spain*)] (1934). When analysing the different moments in which pastoral imagery has emerged in Europe, he observes that it surfaces in times of crisis, and that "[e]l tema pastoril aparece siempre en el arte europeo como síntoma romántico de una vuelta a lo elemental de la vida, a lo primitivo humano, del retorno al pueblo, a la masa" (78) [the pastoral theme appears always in European art as a romantic symptom for a return to the elemental side of life, the human primitive, the return to the common people, to the masses]. Thus, although the work is in general a laudatory essay about the tradition of Nativity representations, he adopts a distant and critical gaze towards the identity value of pastoral.

What is striking in Giménez Caballero is that, only a decade after mocking the shepherd poet and then later pointing out how the pastoral emerges in periods of crisis in search of a stable origin for collective identity, he exalts the figure of the shepherd as the core of Spanish national identity. In his manual, *España nuestra: El libro de las juventudes falangistas* (1943) [*Our Spain: The Book of Falangist Youth*], he offers his particular vision of national history and its symbols, devoting an entire section specifically to the shepherd and connecting it with imperial ideology:

> ¡Tierra de guerreros Castilla! Por eso el hombre más representativo de España no es el agricultor, sino el pastor ... El pastor y su grey se desplazan, trashuman ... Y – en un momento de peligro – el cayado o la garrocha se transforman en fusil. Su ojo, habituado a la lejanía, precisa la bala. Y sus ganados se hacen víveres de todo un Ejército, para que ese Ejército pueda siempre avanzar, moverse y vencer a las ciudades inmóviles, a las huertas comodonas, a los obreros que empuñan martillos, a los payeses que empuñan hoces. Por eso nuestra guerra–y todas las guerras de España–las ha ganado siempre Castilla, la guerrera y pastoril. Contra el resto de las tierras pacifistas, industriales, agrícolas. Y asustadas. (84)

> [Castile, land of warriors! This is why Spain's most representative man is not the farmer, but the shepherd ... The shepherd and his flock wander around, move to new pastures ... and, in the moment of danger, the shepherd's crook or the cattle prod is transformed into a rifle. His eye, accustomed to the distance, aims the bullet. And his livestock become the supplies of an entire army, so that this army can always advance, move, and defeat the immobile cities, the cushy gardens, the workers who brandish hammers and the *payeses* (Catalan peasants) who brandish sickles. That is why our war – and all the wars in Spain – have always been won by Castile, warlike and pastoral. Against the rest of the pacifist, industrial, farming – and shocked – lands.]

Here, Giménez Caballero rearticulates the symbol of the shepherd as the guarantor of Spanish identity, clearly opposing its Castilian core to a mob that is marked by both class difference ("los obreros que empuñan martillos") and by geopolitical difference within Spain – the *payeses* (instead of *campesinos*) is a term that refers specifically to Catalan peasantry.[9]

This exacerbation of sheep herding as the core of Spanish – or rather Castilian – identity explains why sheep herding is discredited and downgraded in post-Civil War formulations of Spanish identity. Anti-Francoist writers generally try to get rid of the symbolic capital of the Franco regime, specifically recognizing this pastoral element. For example, in his poem "De los años cuarenta" [On the Forties], first included in *Moralidades* (1966), Jaime Gil de Biedma describes the victorious Spain that emerged after the Civil War as "Media España ocupaba España entera / con la vulgaridad, con el desprecio / total de que es capaz, frente al vencido, / un intratable pueblo de cabreros" (49) [Half of Spain occupied all of Spain with the vulgarity and the total disdain of which it is capable, in the face of the defeated, a disrespectful country of goatherds].[10] Gil de Biedma criticizes Francoist attachment to pastoral not by denying it, but by presenting it in its debased form.

More recently, Jesús Encinar complains in his blog that the covers of the Lonely Planet travel guides align Spain with Africa, instead of Europe, by representing Spain as a semi-deserted and nomadic "country of shepherds." Here, sheep herding has lost its iconographic prestige, and has quickly switched from representing Europeanness to becoming a sign of Africanization and orientalization as the opposite of modern concepts of civility. While the blog is probably right in denouncing the stereotypes created by the rest of Europe about Spain as an underdeveloped country, it does not take into account how much Spaniards themselves have persistently contributed to creating this image over centuries, as shown herein. We may assume that early modern Spaniards would have been struck by these images for the opposite reasons: not because they align Spain with backwardness, but rather because they project onto Africa the same economic activity they were trying to establish as the basis for Spanish identity against the legacy of the Islamic period.

From Blood Purity to Whiteness

The final period of Franco's dictatorship and the transition to democracy in the 1970s and 1980s entailed an intense reworking of the tradition and within it, the transformation of old racial categories as well as the creation of new sets of oppositions. While the core of the pastoral idealization of early modern Spain was ethno-religious, physical criteria were not completely absent in Spain, although they took a secondary position, thus my focus on the primary features of this hitherto unexplored dimension of the pastoral in early modern Spain. The phenotypic opposition does, however, reappear in several texts, to which we now turn.

In Lope de Rueda's short play "Paso de Polo y Olalla negra" [Farse of Polo and Black Olalla] (published in 1567) one Black slave woman is scorned for her social-climbing pretensions. The play includes very conventional satiric images, depicting Olalla trying to whiten her skin and dye her hair; she refuses the suitor proposed by her owner, stating that she prefers to marry a shepherd instead: "yo quiere con un cagañeroz" (212) [I want a shepherd].[11] The pun in this scene derives from her confusion between literary representations, in which idealized shepherds serve to represent the concept of blood purity, and the reality of sixteenth-century Spain, where real shepherds were no less marginal than many Blacks.[12]

In the seventeenth century, the figure of "el negro heroico" [the heroic Black] attained a high degree of success in the *comedia*, with the abundant presence of Black saints, kings, or heroes.[13] As Martínez López notes, the importance of this dramatic character marks the independence of the figure of the Black from previous iconography that always associated it with the Epiphany and the Adoration of the Wise Men (14). Traces of Blacks' roles in the Epiphany survive, however, in the tropes through which most of the characters try to underline their honourable status. Thus in Lope de Vega's play *El prodigio de Etiopía* [*The Prodigy of Ethiopia*], in order to explain the social ascension of the main character Filipo, he is explicitly compared to King Melchior: "Parece que Filipo se ha soltado / del portal de Belén; Melchor pintado / en cuadro de los reyes me parece" (118) [It seems that Filipo has escaped from the manger in Bethlehem; he seems to me a painted Melchior in a painting about the Wise Men]. Because of the contiguity of the Adoration of the Magi and the Adoration of the Shepherds, they are usually represented together. In Lope de Vega's *Los pastores de Belén* (1612) Blacks (under the name of Ethiopians) are depicted also coming to worship Christ (536). Similarly, in Andrés de Claramonte's play *El valiente negro en Flandes* [*The Valient Black in Flanders*] the Duke of Alba is unable to find the brave Black soldier Juan to honour him for his military deeds, but the captain Don Agustín suggests that "[e]n el pesebre le han de hallar sin duda; / que esta noche los negros y pastores / le están diciendo a Dios sus villancicos" (73) [you will doubtless find him in the manger, because tonight the Blacks and shepherds are singing their Christmas carols to God]. They are, however, consistently characterized as two distinct identities, since Blacks are not included in the community of shepherds, and they can only benefit from contiguity with the latter. Even in this case, contiguity is not always necessarily harmonious. Felipe Godínez's *Coloquio segundo de los Pastores de Belén* (written c. 1620) ends with a confrontation between the Jewish shepherds and the Black pages of King Balthazar (146r).

By the middle of the seventeenth century, Blacks start being conceived of as a threat to pastoral Arcadia, usually cast as lustful characters that try to rape white shepherdesses. For example, in Juan de Barrionuevo y Moya's *Primera parte de la soledad entretenida* (1638), the main character Calisandro fights against two Blacks who attempt to rape the Muslim shepherdess Daraxa (207v). An anonymous ballad of 1687 recounts the rape of a shepherdess at the hands of a Black slave (Berco 73–4 and 159n49). These cases are few and marginal in the period under consideration, but suggest a growing ideological opposition towards Blacks in the pastoral imaginary in parallel with the emergence of a racial stratification based on phenotype.

As this indicates, shepherds are differentiated from Blacks until well into the seventeenth century, but only marginally opposed to them, in contrast to the articulation of the pervasive symbolic opposition of shepherds and Muslims and Jews in early modern Spanish culture analysed above. Yet, while ethno-religious difference was the focus of early modern Spanish racism, phenotype dominates racist formulations as soon as Spain is reintegrated into modernity and Western identity with the transition to democracy. Franco's death in 1975 and the subsequent democratic elections of 1977 meant a cultural and economic revamping of Spain that also entailed a rewriting of tradition coincident with the arrival of new waves of immigration. Even though the figure of the shepherd lost most of its past allure as a central symbol of collective identity, the "pastoral habitus" has kept its functionality as a semantic fossil for idealizing regimes of difference – but also for contesting them. As one of the features of "habitus" is its "successful adjustment to the objective conditions" (Bourdieu 62), so the "pastoral habitus" is always ready to provide its accumulation of cultural capital to the new scenarios and logics of Spanish racism, adapting to the ever-changing conditions to which forms of exclusion are geared.

The transposition to modern racial thought of Spain's early modern pastoral racial division is evident in José Camón Aznar's *El pastor Quijotiz* (1969).[14] This twentieth-century sequel of *Don Quixote* takes over the proposed transformation of Don Quixote into a shepherd. At the end of the novel, Don Quixote finds out about the pillage that the band of Lupercio Latrás ("Latrán" in Camón Aznar's novel) (mostly comprising shepherds from the Pyrenees) was committing in the Ebro valley against the Moriscos working in agriculture (142–6). Camón Aznar, an erudite from Aragon himself, brings to the fictitious world of *Don Quixote* actual historical events that took place in the 1580s.[15] As it turns out, Don Quixote does not actually meet the mythical Lupercio Latrás, but only one squire who passes as the bandit. Camón Aznar's rendering of the historical event is telling of how early modern pastoral ethnocentrism transposes to the present taking on new meanings:

> Los pastores del Pirineo, bajando a las llanuras del Ebro como una manada de lobos feroces, habían invadido un pueblo que censaba el caballero aragonés y habían asesinado a sus habitantes. El pleito era viejo. A la rivalidad entre los pastores trashumantes y los labradores sedentarios se unía un odio de razas. Los pastores eran de tipo nórdico, altos, de mirada azul, de recia complexión, sobrios de palabras y con la idea, que era casi su único credo, de una selvática libertad personal. En tanto que los labradores ribereños eran descendientes de los moriscos, de prieta tez, fecundos en el trabajo y en el habla, pegados a una tierra de la que habían sido expulsados sus padres. (142)

> [The shepherds of the Pyrennees, coming down to the plains of the Ebro valley like a pack of fierce wolves, had invaded a town that belonged to the Aragonese noble and slaughtered all its inhabitants. The dispute was old. To the rivalry between migrating shepherds and sedentary peasants race hatred was added. The shepherds were of a Nordic type, tall, with blue eyes, of robust complexion, rather laconic, and with the idea, for them just about their only creed, of a wild personal liberty. Meanwhile, the peasants on the banks of the Ebro were descendants of Moriscos, with dark skin, labourers in their trades and speech, attached to a land from where their fathers had been expelled.]

Camón Aznar reflects Unamuno's interpretation of the rivalry between peasant Moriscos and Christian shepherds – indeed, he had a close personal relationship with Unamuno (Aranguren Egozkue 41–54). While his description seems to praise the Morisco legacy in Aragon, he nonetheless imposes a phenotypical difference that is nowhere found in the early modern historiographical sources documenting Lupercio Latrás's activity, and not even in Unamuno's *En torno al casticismo*. Thus the more typical of the colour-based racism of modernity reformulates the pastoral imaginary of difference.

A more recent example of this phenotypic reformulation of the construction of racial difference in the "pastoral habitus," although this time in a mocking tone, is the song "O belén da caixa" (1991), by the Galician rock band Os Resentidos.[16] It describes a Nativity scene in the city of Vigo during Christmas, beginning by construing it as an immutable tradition: "Todo cambia, nada cambiará / O belén da caixa sempre vai estar" [Nothing changes, everything will change: the Nativity representation of the savings bank will always be there]. The song articulates the equalizing force of the manger scene: "Non hai ideoloxías, non hai clases sociais, / nin distinción de sexos, nin de relixións. / Todos van visitar, visita-lo belén / o belén da caixa: Vigo é Nazaret" [There are no ideologies and no social classes, nor discrimination among sexes or religions. Everybody's going to visit the Nativity scene of the savings bank: Vigo is Nazareth]. By the end of the twentieth century, Nativity scenes have lost their purely religious meaning, and have undergone a process of desacralization that transforms them into urban attractions, but they keep working as a symbolic national space by blurring social distinctions.

As the song unfolds, however, the universal inclusivity of the Nativity scene turns out to be illusory, since the sense of community is altered when two sub-Saharan immigrant street vendors show up at the scene hoping to sell their merchandise to the audience. Mostafá and Solimán bear with them a double Otherness, both because they are Blacks from Cameroon, and because their names suggest a Muslim identity. If immigration was a new phenomenon in Spain in general by the beginning of the 1990s, it was even more so in Galicia, a region from which people mostly emigrated. The Otherness of Solimán and Mustafá, their lack of belonging to the community that contemplates the Nativity scene, is also emphasized here because, for twentieth-century Spaniards, the place of Blacks was understood to be *on the other side* of representation as pages of King Balthazar.

The song then proposes to integrate them *into* the Nativity scene: "Canto me gustaría, facer o meu belén / Solimán e Mustafá facendo de pastores" [How much I would like to do my own Nativity scene, with Solimán and Mustafá acting as shepherds]. However, this apparently naive wish, which seems to reinsert the black Muslim immigrants into their "proper place" in the racial imaginary, turns out to constitute a further transgression of the iconographic conventions: the audience would expect to find Blacks among the objects of their gaze, as the pages of the Magi or in the figure of King Balthazar, but never participating as the shepherds of the Adoration. This final twist, and the surprise that it produces, reveals how the shepherds of the Adoration stand as symbols of ethnic purity and how the alleged universality of the evangelic message of the Nativity scene has very specific, ethnocentric limits.

A very similar displacement occurs in José Luis Cuerda's film *Amanece que no es poco* (1988) [*To See the Dawn is Enough*], a choral and surrealist vision of the Spanish countryside. The shepherd of the village is Nge Ndomo Álvarez Martínez, who introduces himself as a mulatto from the village, born of a black father and a white mother: "Yo heredé de mi padre el nombre, la raza y el acento, y de mi madre los dos apellidos y el lugar de nacimiento" [I inherited from my father my given name, race, and accent, and from my mother my two family names and my place of birth]. The film tries to portray him as a positive character, although it often reproduces stereotypes about Black people, attributing to Nge certain characteristics like sexual prowess and liking music and dance. His accent also marks him as Caribbean (the actor, Samuel Claxton, is of Cuban origin), thus conflating anxieties about immigration from both Latin America and sub-Saharan Africa.

Even though his name does not link him to the imaginary of Muslim Africa, he is inexplicably forbidden admittance to the church because he is a "catecúmeno" [catechumen]. According to his mother (Chus Lampreave), Nge's social stigmatization happens because of his affair with a white married woman in the town. Nge is, however, well aware that none of the inhabitants of the village is banned from entering the church in spite of their sins, and therefore complains that the true reason for his exclusion is his skin colour: "No entro porque soy negro" [I am not admitted because I am black].[17] Nge's "race" is thus perceived as an ethno-religious stigma. Yet he is the only shepherd of the village. And it is precisely this double identity as shepherd and Black that constitutes the "surrealist" feature of Nge in the film and within the Spanish imaginary, as Nge himself exploits it.

Every night, Nge takes his goats to a mountain nearby, not to graze, but merely to form bucolic picturesque scenes with no apparent audience: "Anda que no debe estar bonico esto; las cabras ahí quietas y yo aquí de perfil como un masai" [Look, shouldn't this be beautiful, with the goats quiet over there and me here, posing in profile as a Masai]. At first, it seems that Nge uses this pastoral solitude to stage for himself a fabricated African identity. Nonetheless, he does so through the filter of Spanish pastoral, with the crook and the characteristic *pellico* [shearling jacket] that he wears throughout the film (figure 16). Soon afterwards, the *guardia civil* [rural police] Pascual (Ovidi Montllor), appears in the fog, addressing Nge, visibly enraged: "¡Nge, Nge! ¡Ya te has traído otra vez las cabras, jodido!" [Nge, Nge! Here you are again with your goats, jerk!]. The scene is left in suspense with this image of potential police violence bursting into Nge's nocturnal Arcadia.

Figure 16 *Amanece que no es poco* (Dir. José Luis Cuerda, 1988). Courtesy of Video Mercury Films.

When the camera returns to the encounter between Nge and Pascual, it turns out that the same scene is repeated over and over every night, and that all Pascual does is accompany Nge back home:

PASCUAL. Bueno, pues hasta mañana. Mañana iré a buscarte otra vez.
NGE. Pero, ¿a usted le gusta la estampa que hago yo allí con las cabras?
PASCUAL. Hombre, claro que me gusta; es muy bonita, muy curiosa ...
NGE. Se lo digo porque así al menos no pierde su viaje todas las noches.
PASCUAL. No te preocupes, hombre, no te preocupes. Mañana voy a por ti.
NGE. Perdone las molestias, pero no le veía otra salida.

[PASCUAL. Well, see you tomorrow. Tomorrow I'll look for you again.
NGE. But do you like the picture that I create here with my goats?
PASCUAL. Of course I like it; it's very nice, very polished ...
NGE. I am telling you because that way at least you don't have to waste your time every night.
PASCUAL. Don't worry, don't worry. I'll look for you tomorrow.
NGE. Sorry for the inconvenience, but I couldn't see any other way out.]

The film does not clarify why Nge states that he is unable to see any other way out. What kind of "solution" is his pastoral performance, whose only audience is the local authority who comes to interrupt it? What is he trying to solve with it? While none of the two characters verbalize it, it is clear that Nge's pastoral performance is perceived as an illicit act – an infraction that needs to be repressed every night by Pascual, even though it is such a minor infraction that it merits only a slight quotidian reprimand. Both the semi-clandestine pastoral performance and the – apparently no less absurd – repression only make sense if we consider that what is going on in this scene is a transgression of the "pastoral habitus," the inclusion of the racial Other in the imagery that had implicitly and for centuries served as the core of Spanish identity.

This interpretation is supported by a Basque novel from the same year. In Bernardo Atxaga's novel *Obabakoak* (1988), skin colour appears in quite a different light. In this tale of the life of an imaginary Basque town, skin colour is a caesura within the community of shepherds, which is in turn singled out as a separate and marginalized community within the village. At first, the narrator fails to perceive any difference between the two groups, suggesting that there are no perceptible distinguishing characteristics. However, one of the narrator's informants warns him "de lo *especiales* que eran, de la diferencia que había entre ellos y el resto de la comunidad" (137–8, emphasis in the original) [of how *special* they were, of the differences that there were between them and the rest of the community]. When the narrator investigates the reasons for the marginality of the group, the native Daniel details the distinctions between them and asks wonderingly: "¿Es que aún no te has dado cuenta de que hay pastores blancos y pastores negros?" (143) [Have you not yet realized that there are white shepherds and black shepherds?].

Daniel fills in all the physical and moral characteristics that distinguish both groups, until he finally suggests that the narrator go visit the shepherds to see the differences between both groups for himself: "Ya verás, uno es de los negros y otro de los blancos" (143) [You will see, one group is all Blacks and the other is of Whites]. The information takes the form of a self-fulfilling prophecy, to which the narrator abides when acknowledging "¡Pues tienes razón! – me pareció que su clasificación era del todo pertinente" (143) [You are right! – I thought that his classification was wholly pertinent]. In these interactions the narrator is trained to perceive skin colour as difference, a classification by phenotype that is always a social construct, as Livio Sansone points out (11). What is idealized in *Obabakoak* is that it depicts the precise moment of injunction to perceive difference: the narrator was simply not seeing it until he was explicitly told to do so and what to look for. This is certainly an idealization of social practices concerning the perception of race, since this process of awareness is not so clear and easily identifiable in common socialization. It is also an idealization because the narrator, an outsider to the village, seems in fact an outsider to both Basque and Spanish society in their entirety, since by the late 1980s the social concept of race as conveying phenotypical difference was amply disseminated. It is, however, a radical rewriting of the idealization underlining early modern pastoral: while difference was an implicit feature embedded in the figure of the shepherd, *Obabakoak* presents the communities of shepherds as an explicit instance of social racialization.[18]

Bernardo Atxaga's novel, José Luis Cuerda's film, and Os Resentidos's song were all composed between the end of the 1980s and the beginning of the 1990s, precisely the period in which North African and sub-Saharan immigration became a phenomenon in Spain. All of them use the term *pastor* to refer to Black herdsmen – even in the case of Nge in *Amanece que no es poco*, who guards only goats. More than two decades have passed since then, and I was about to conclude that the racial "pastoral habitus" had faded away with the transition to democracy when I came across an article entitled "El último cabrero vino de África" [The Last Goatherd Came from Africa], written by Óscar Iglesias and published in the Galician edition of the newspaper *El País* (9/24/2009). It reports the story of João Almeida, an immigrant from Angola who works as a herdsman in the mountainous region of Ourense (Galicia). Even though the text clarifies that João Almeida guards forty-eight sheep and only five goats, and therefore it would be more accurate to identify him as *pastor*, he is nonetheless labelled as *cabrero* [goatherd]. While at first this seems to be only a news report of how traditional occupations are preserved thanks to the arrival of immigrants, it is personalized in the figure of Almeida by exploiting his Blackness and refusing to label him *pastor*. Although presented in a purportedly positive light, it is striking how, even now, the same linguistic ideologies I have analysed in these pages reemerge from latency in the Spanish imaginary.

As all these cases show, the link between Spanish ethnocentrism and sheep herding is not limited to the early modern period, although its sets of oppositions are reworked and reformulated from the late Middle Ages until the beginning of the twenty-first century. The persistence of the "pastoral habitus" suggests that early modern Spaniards were successful in construing Spain as the Land of Pan, the god of shepherds (and a prefiguration of Christ), when searching for an essential identity for the imaginary community of Spaniards, in a way that shaped the boundaries of the collective imaginary and crystallized racial and cultural purity through the language and tropes of sheep herding.

However, the Land of Pan can easily turn into *Pan's Labyrinth*. First because, as in Guillermo del Toro's film, pastoral idealization of national purity quickly transforms the imagined idyll into the scenario of ethnic and cultural cleansing and fratricidal violence. But the Land of Pan is also Pan's Labyrinth because it is a semiotic cage in which the "pastoral habitus" is trapped by its own reproduction: as we have seen in these pages, the "pastoral habitus" proves to be a persistent trope that conveys an imaginary of difference that is conceived as essential and immutable; but at the same time, its durability undermines such appeals to essentiality and immutability, since its allows for the analysis of its evolution, of how an apparently stable signifier is in fact constantly redeployed, resemanticized, and appropriated. By contrasting how pastoral has been used to convey different notions of collective purity in different historical periods, we can reveal the historicity of such notions of purity, their contingency on the ever-changing boundaries of ethnocentrism and the imagination of the cultural realm.

Pastoral and Ethnocentrism: Future Directions

This schematic overview reveals the necessity of developing a more thorough and comprehensive analysis of the historical evolution of the figure of the shepherd from the early modern period to the present; what happens from the late seventeenth century until the nineteenth century remains to be explored. Further, my analysis also issues a call to analyse the evolution of the racial "pastoral habitus" from blood purity to whiteness in contemporary Spain. My work here, due to limitations of space and disciplinary boundaries, could not give them the attention that they deserve.

By focusing on how Old Christians conceived of themselves as associated with an idealized idea of shepherding this book shows that such idealization aimed at further excluding Moriscos and *conversos* in early modern Spanish imaginary, beyond their physical expulsion and stigmatization. This strict focus provides a close reading of a delimited corpus of texts, and is not intended to convey the impression that pastoral imagery is exclusive to Christian identity. Far from that, if sheep herding was such a prestigious trope for collective identity in early modern Spain, it is natural that Moriscos and Sephardic Jews contested or negotiated such appropriation of the figure of the shepherd by Old Christians. The incorporation of this perspective is for another book, a project that would require a more sophisticated theoretical model and greater transcultural and linguistic competence.

There is some work done on early modern Sephardic pastoral, most notably about Samuel Usque's *Consolação às Tribulações de Israel* (1553), a pastoral dialogue that sums up the long history of persecution against Jews. When Constance Rose remarks that Usque "equates the classical Golden Age to the idyllic life in the land of Canaan" (145), she registers it as an exception, a diversion from the canonical conception of the pastoral, and Preto-Rodas offers a similar interpretation when situating Usque within the Iberian intellectual context (73). As this book has tried to demonstrate, there is nothing exceptional in Usque's association of pastoral idyll with ethnogenesis, and he may well be contesting previous appropriations of biblical pastoral; it also may be that, since his text predates the emergence of the pastoral romances in Spain with the publication of Montemayor's *La Diana* around 1559, pastoral ethnocentrism might have been influenced by Sephardic messianic pastorals like Usque's.

There is even less work done on the relation between Moriscos and the pastoral, even though there is evidence that Moriscos also participated in the cultivation of this literary and cultural fashion, such as the *Comedia pastoril de Torcato*, staged by the Moriscos in Aragon in 1573 (Yndurain). A cursory reading of *aljamiado* texts also seems to indicate that crypto-Islamic Moriscos were not alien to the language of sheep herding. If there is nothing written on the subject, it might be because the "pastoral habitus" that crystallized in early modern Spain prevents us from asking such questions of texts related to Islam.

Extending the argument of this study, we might ask if ethnocentrism is limited to dealing only with ethno-religious difference. The emphasis on how the pastoral is defined in opposition to Moriscos and *conversos* justifies the abstraction of internal differences within the Iberian Peninsula, to which much less attention has been devoted. However, because of the ability of pastoral romances to idealize local space, they are also privileged literary vehicles for conveying articulations of geopolitical distributions of power and for negotiating cultural homogenization and difference within the Iberian Peninsula. Here, I have not touched on the impressive production of Portuguese pastoral romances, most of them composed during the sixty-year period (1580–1640) in which Portugal was under the same monarchy as Castile, Aragon, and Navarre. A pastoral vision of early modern Basque identity was promoted by Joan Pérez de Lazarraga and Jerónimo de Arbolanche, as I have studied elsewhere.[19] Traditionally, the study of pastoral romances is centred in Castile, and this book has made only very limited attempts to counteract this vision. It would be interesting to explore whether pastoral romances composed in the kingdom of Aragon try to craft consciously a distinct literary identity in opposition to Castile. Works such as Bartolomé Ponce's *Clara Diana a lo divino* (1580), Antonio Lofrasso's *Los diez libros de fortuna de amor* (1573), Gaspar Mercader's *El prado de Valencia* (1600), and Jerónimo Jiménez de Urrea's now lost *La famosa Épila*, seem to formulate an alternative model of territorial idealization. For example, it is worth noting that Ponce and Lofrasso point to the Catalan Ausiàs March as their model and predecesor, instead of Garcilaso de la Vega. On the other hand, the Spanish empire extends far beyond the Iberian Peninsula. I have refrained from exploring how pastoral ethnocentrism is adapted and transformed in colonial Latin America or the Canary Islands, a much-needed project.

The emphasis on the association of ethnic identity and Arcadian imagery in Spanish pastoral romances should not be taken to imply that this cultural manifestation is unique to Spain and its empire. There is, as we have seen, a certain specificity in the case of early modern Spain due to the long history of cohabitation between Christianity, Judaism, and Islam. Yet, although there are methodological limitations to working with the national frame, a comparative study was beyond the scope of this book, and I have only gestured towards the possibilities, both in notes to the study and here. To begin with, the link of pastoral idealization of space with ethnocentrism can already be identified in the model for Spanish and European pastoral romances, since Sannazaro's *Arcadia* was a praise of Naples, its antiquities and its people, precisely at the moment in which the dominion of this Italian kingdom was disputed by France and Spain.[20] In the case of England, Jeffrey Knapp has analysed how John Lily's pastoral play *Galathea* undermines English ideas of national purity, and Françoise Lavocat points out how Honoré d'Urfé's pastoral romance *L'Astrée* combines nationalist myths with the Arcadian pastoral idealization (*Arcadies* 357). It is striking to realize that we as scholars tend to notice this link between pastoral and ethnocentrism only in the national, disciplinary, or religious tradition that is more immediate to us, always remarking on it as an exception to a pastoral model that is completely disconnected from such notions. This oblivion is particularly telling in the case of Lavocat, who is a comparatist extremely well versed in French, Italian, Spanish, and English pastoral literary traditions, but nonetheless perceives the link between pastoral and ethnocentrism only in the French tradition, neglecting to see that it is also at work in Spanish pastoral romances.

All these instances – Lavocat in the case of French pastoral romance, Constance Rose on Usque, and, I am sure, me many times writing this book – show how our disciplinary shrines prevent us from perceiving that what we see in one national tradition as an anomaly might be a general phenomenon, the shared common language of a larger debate of the past whose terms are not easily accessible to us. That is, it could be the case that it is not the precise appropriation of pastoral that is ethnocentric, but that there is something in the pastoral idealization of space that contains in itself the seed of ethnocentricity. No one scholar can be conversant, all at the same time, with the entire production of early modern European pastoral literature, the historiography of each European country, and the serendipitous proto-nationalist and ethnocentric discourses. As scholars trained in our own national traditions, it is somehow inevitable that we perceive these kinds of subtle ethnocentric connotations in those texts that are more immediate to us. Arguably, collaborative interdisciplinary work is required to explore how pastoral romances across the board are deeply engrained in the development of nationalist ideologies in Europe.[21]

This book aims to ask new questions about early modern pastoral romances in Spain, a genre that seems alien to modern readers, perhaps not so much because of the genre itself, but because of the kind of approaches that we have traditionally used to interpret it. In analysing its opposition to early modern ethno-religious diversity, this book has not even begun to mobilize the rich potentiality for debating pastoral romances (and the pastoral mode in general) as documents of ethnic identity. As well as opening pathways for my own future research on these issues, this book is meant to inspire others to follow or contest the paths thus marked out, to remedy its omissions and imperfections, and to point out that there is possibly not only one Pan's Labyrinth, but many.

Notes

Introduction: A Country of Shepherds

1 All translations are mine, unless otherwise indicated.

2 Among the extensive literature on the importance of sheep herding in medieval and early modern Spain, Carla Rahn Phillips and William D. Phillips's *Spain's Golden Fleece* continues to be the most comprehensive study in English. See also Marie-Claude Gerbet (*La ganadería medieval*), and the volumes edited by Joaquín Gómez-Pantoja (*Los rebaños de Gerión*) and Julian Pablo Díaz López and Antonio Muñoz Buendía (*Herbajes, trashumantes y estantes*). For a recent ecocritical reappraisal of herding in the Spanish empire, see Alves (72–88).

3 Spanish *raza* can be translated as race, stock, or bloodline, depending on the context; in this context, I am translating it as "race" to reflect the ambivalent meanings that reappear throughout similar texts under consideration here.

4 The role played by medieval and early modern Spanish contact with Islam in nineteenth-century colonialism is not the object of this book. On this, see, among others, Susan Martin-Márquez (12–63), and Eloy Martín Corrales (53–124).

5 Morisco studies is an ever-growing field across various languages and disciplines, recently intensified by the 400th anniversary of the expulsion in 2009. Some of the most recent books in English are L.P. Harvey, Mary Elizabeth Perry, Matthew Carr, and James B. Tueller. In Spanish, see Mercedes García-Arenal and Fernando Rodríguez Mediano, Rafael Carrasco, and the volume edited by Antonio Moliner Prada.

6 For the development of the statutes of blood purity, see Albert A. Sicroff, Juan Hernández-Franco, and Jaime Contreras Contreras. For their implantation in the New World, see María Elena Martínez and Karoline P. Cook. In

the case of the Moriscos, an indeterminate elite managed to obtain the status of "cristianos viejos de moros" (Childers, "Disappearing Moriscos").

7 Methodological caution is needed when dealing with these sources, especially with documents written by the defenders of the statutes of blood purity which, as propaganda, offer a monolithic view of ethnic difference (Márquez Villanueva, *El problema morisco* 166–71, and "Sobre el concepto"; Dadson 20). As Augustin Redondo shows, Old Christians were not necessarily more orthodox than New Christians ("La religion populaire").

8 Nonetheless, the existence of crypto-Jews and crypto-Muslims in early modern Spain after the forced conversions of the fifteenth and sixteenth centuries is undeniable. For crypto-Jews, see David Gitlitz (*Secrecy and Deceit*); for crypto-Muslims, see Harvey (102–21).

9 Yet phenotypic difference was not completely absent in medieval Iberian descriptions of Moors (as we will see in the conclusion). In the thirteenth-century *Primera crónica general*, some soldiers of the Almoravid army are described as black, while Christians are construed as white (Viguera Molins 66–7). In spite of migrations of Muslim Berbers and sub-Saharan people to al-Andalus, it seems that there was no physical difference between Moriscos and Old Christians in the early modern period (Vincent, "Qué aspecto físico" 338–9; López-Baralt 335–7; Fuchs "The Spanish Race" 94–5; Rafael Carrasco 119–20). Early modern critics of Moriscos and *conversos* never mention skin colour as part of their attacks. John Beusterien points out however that both systems of racial thinking were present in early modern Spain (13–32).

10 The position of Domínguez Ortiz is representative of the confusion on this matter, since he states in other parts of the same work that the doctrine of blood purity, at least since the middle of the sixteenth century, is not based on religious difference, which he describes several times as racial exclusion (*La clase social* 50). Henry Kamen ("Limpieza") also supports that racism is absent in early modern Spanish society by diminishing the real extent of the "statutes of blood purity." Although he is for the most part right in questioning the rigid vision of a homogeneous Christian society attached without exception to the doctrine of blood purity, Kamen falls too easily into the extreme view of negating its importance entirely. For a critique of Kamen, see Martínez (*Genealogical Fictions* 45).

11 On this, see also George Mariscal (11).

12 See also Martínez (*Genealogical Fictions* 59).

13 The risks of oversimplifying the search for historical continuities are exemplified by several mechanical accounts of cultural transmission. James Sweet, who mistakenly states that the concept of blood purity was based on skin colour (160), establishes the origins of North American racism from the

Islamic slave trade and the statutes of blood purity in Spain (162). Similarly, Jerome Friedman holds that the Spanish doctrine of blood purity is the immediate foundation of early modern anti-Semitism (27). The opposite and equally problematic view is represented by Ivan Hannaford and Audrey Smedley, for whom "racism" is a historical exception that takes place only in the Anglo-American colonies by the end of the seventeenth century.

14 See Matt Cartmill for a critique of the concept of "race" in physical anthropology. Livio Sansone also points out that "race is one of the many ways to express and experience ethnicity" (6).

15 See also Martínez (*Genealogical Fictions* 11 and 59).

16 Kamala Visweswaran (76) warns against the danger of merely replacing "race" with "culture" without previously describing the sociohistorical construction of "race." An alternative option would be that proposed by Faye Harrison, for whom "[the] quality of difference, whether constructed through a biodeterminist or a culturalist idiom, is what constitutes the social category and material phenomenon of 'race'" (613). Such a broad definition of "race" at the expense of "ethnicity" would be, in fact, more appropriate for early modern Spain. Nevertheless, Harrison simply diminishes the concept of "ethnicity" without integrating it into the concept of "race," leaving undefined the reach of both terms. A different negation of medieval European racism is that of Robert Bartlett's interpretation of medieval European thought: "The medieval situation was one where 'race' almost always means the same thing as 'ethnic group'" (53).

17 "As racism assumes more subtle and elusive forms in the contemporary world, it is being reconfigured without 'race' as a classificatory device for demarcating difference" (Harrison 610).

18 As Perceval shows in his analysis of the attitudes towards the Moriscos in early modern Spain, the differences between the defenders of the ethnocide (the acculturation and assimilation of the Moriscos) and the defenders of the genocide (the physical elimination or expulsion of the Moriscos) are not strictly defined, and their arguments tend to overlap (*Todos son uno* 91). For a technical description of ethnocide and genocide and their applicability to early modern Spanish ethnocentrism, see Stallaert (*Ni una gota* 70–92).

19 On the cultural cleansing of everything Moorish, see Perceval (*Todos son uno* 46) and Milhou ("Desemitización" 44–50). More recently, Barbara Fuchs has retaken Milhou's contribution, complicating the concept of cultural cleansing of everything Moorish. She shows that it was not a uniform social strategy and that in many instances recognized Moorish cultural practices were openly embraced as signs of aristocratic, local, or even national identity (*Exotic Nation* 11–30). Something similar could be said about the Jewish element in Spanish culture, with which this book deals only tangentially. The reason is

that, although the influence of Jewish culture is undeniable in Spanish culture, it is not as visible as Moorish influence, especially during the sixteenth and seventeenth centuries. Furthermore, it is the Islamic presence that conditioned the chronology of ethnogenesis articulated around the concept of the *reconquest* (Stallaert, *Etnogénesis* 58–9).

20 For a reappraisal of the intellectual legacy of Américo Castro, see the volume edited by Eduardo Subirats.

21 I use the term "*reconquest*" because it is still widely used in Spanish scholarship to refer to the territorial expansion of Iberian Christian kingdoms during the Middle Ages. It is important to note nonetheless that it is an anachronistic concept coined in the context of nineteenth-century Spanish nationalism (Ríos Saloma 25–39).

22 See Fuchs ("1492": 493–6) for a critique of the excesses of teleological narratives articulated around 1492.

23 Even less so in medieval Europe, where, as Bartlett states, there was "no direct connection between ethnic and political homogeneity" (53).

24 For a recent debate on early modern Spanish identity, see the volume edited by Alain Tallon, as well as that edited by Antonio Álvarez-Ossorio Alvariño and Bernardo J. García García. Tamar Herzog (94–118) suggests that the concept of the "Spanish people" emerged only within the context of the American colonization. In a pre-statistical era, it is hard to measure the extent that the feeling of national identity may have had among the Spanish population (Ballester Rodríguez 297–312).

25 See for example Antonio Domínguez Ortiz and Bernard Vincent (111) and Miguel Ángel Ladero Quesada (*Los mudéjares* 148), who reproduce the traditional image that the Moriscos were devoted mainly to agriculture, but seldom to sheep herding.

26 For scattered references to Moriscos working as shepherds, see Nicolás Cabrillana (55), Julio Caro Baroja (*Los moriscos* 115–16), José Enrique López de Coca Castañer (107–10), Margarita Birriel Salcedo (45), Ladero Quesada (*Los mudéjares* 79), José Antonio Fernández Otal (244–8), Javier Castillo Fernández (238), José Luis Castán Esteban (241–50), Lorenzo Cara Barrionuevo (207), Gregorio Colás Latorre ("Los moriscos en Aragón" 190–2 and 196), and Serafín de Tapia (35n51).

27 According to Ruth MacKay there is no substantial evidence to support the widespread assumption that there was some kind of causal link between professional occupation and "limpieza de sangre," since, beyond the scorn of certain jobs in literary texts, normative treatises, and guild ordinances, the fact is that lawsuits and memorials do not mention the issue of blood purity at least until the eighteenth century (183–6).

28 The most recent and comprehensive analysis of Pedro de Valencia's defence of the Moriscos is Grace Magnier.

29 This is not to suggest that the early modern period was different from previous periods, since the trope was fundamental and pervasive and simply reflects the scope of this particular project. José Antonio Maravall (187–8) traces the most remote allegorical uses of the trope of the good shepherd both in the Christian tradition and medieval Spanish literature. For a broader view of the classical, popular, and religious precedents of the figure of the shepherd, see the study by López Estrada (*Los libros de pastores*).

30 "Romance" here in the context of sixteenth-century pastoral and sentimental fictional forms generally has a prose framework but often includes other genres (poetry, drama), and is not to be confused with Spanish *romances* (ballads).

31 For the literary and mythological sources in the Spanish pastoral literature, see López Estrada (*Los libros de pastores* 57–151), Avalle-Arce (*La novela pastoril* 13–34), and Menéndez Pelayo (679–711). For the influence of the Spanish pastoral novel in Europe, see Barbara Matulka and Eugenia Fosalba (*La Diana*).

32 On Navagero's visit to Granada and his impressions, see Cammy Brothers.

33 Navarrete also interprets Boscán's comments about the origins of poetic forms within the social anxieties of genealogical purity: "By seeking a legitimate Greco-Roman ancestry for Spanish poetry, he betrays the ethnic preoccupations that lay behind Spanish alterity" (71). The "genealogical" suspicion about the traditional Castilian verses was explicitly made by Argote de Molina in his *Discurso sobre la poesía castellana* (1575), where he considered the possibility that the eight-verse old ballads (romances) "pudo ser que todos lo tomassen de la poesia Barbara de los Arabes" [they could have been taken from the barbaric poetry of Arabians], transcribing some Arabic verses about the fall of Granada to prove his point (209–11).

34 Roland Greene draws the intersections between Petrarchan amatory poetry and the empire in a wider European context. Even though Greene refers to the colonial experience in America, his analysis would be valid as well for the situation within Spain, in which the physical and cultural presence of Moriscos creates a domestic colonial system (Childers, *Transnational Cervantes* 171–7). For a similar analysis of Garcilaso de la Vega, see Beverley ("Humanism") and Mar Martínez-Góngora (517–18). On the contrary, Javier Lorenzo sees Petrarchan forms as a vehicle for political dissidence.

35 See Milhou ("Desemitización"), Fuchs (*Exotic Nation* 20–3), and Perceval (*Todos son uno* 46).

36 It is commonly assumed that the *princeps* of *La Diana* was printed in 1559. There is however no date on the title page.

37 Abraham Madroñal Durán has recently edited *El teatro pastoril en la ribera del Tajo* (190–214).

38 For a comprehensive overview of the editorial history of the Spanish pastoral novel, see the bibliographical guide by López Estrada, Huerta Calvo, and Infantes de Miguel.

39 Moll ("La narrativa" 38n22) shows that the belief in a 1662 Madrid edition of Montemayor's *La Diana* is based on a typographical error.

40 Constance Rose (156–7) also identifies the rise of Iberian pastoral romances by the late 1550s as a *converso* reaction against inquisitorial persecution.

41 As Francisco Márquez Villanueva remarks, there were authors like Lope de Vega who cultivated both themes simultaneously (*El problema morisco* 173–4). However, even if for some individual authors they were interchangeable motifs, each of them was overcharged with different cultural and political connotations.

42 The *zambras* were a Morisco dance (see chapter 4).

43 See Américo Castro ("Lo hispánico" 57–63), Constance Rose, and Marcel Bataillon, who believe that the literary prestige of the figure of the shepherd by the middle of the sixteenth century comes from a group of *converso* writers including Bernardim de Ribeiro, Feliciano de Silva, Alonso Núñez de Reinoso, and Jorge de Montemayor, and that it served as the vehicle for expressing the melancholy for lost medieval *convivencia*. They left unexplained, however, how such a strategy could pass unnoticed by their Old Christian contemporaries, such that Montemayor's *La Diana* turned out to be one of the greatest best-sellers in early modern Spain. To solve this paradox within the concept of "*converso* literature," Colbert I. Nepaulsingh suggests two levels of reading: one orthodox, aiming at Inquisitorial censorship, and a Judaizing reading that was encrypted so that only an ideal reader would be able to interpret it. One of the problems of such analysis is that, as Dayle Seidenspinner-Núñez (16) points out, the ideal *converso* reader proposed by Nepaulsingh seems highly improbable by the middle of the sixteenth century, when Hebrew culture had been progressively lost by the descendants of *conversos* who remained in the Iberian Peninsula, even for those who were crypto-Jewish. While I do not follow their line of inquiry here, it must be acknowledged that these scholars were pioneers in identifying the possible social values of the pastoral genre.

44 For an overview of the representation of women in the Spanish pastoral romance, see also Begoña Souviron López's *La mujer en la ficción arcádica*.

45 On how the pastoral space is constructed upon denial of the other, see also Jean-Pierre Van Elslande (394), although he refers primarily to class difference.

46 Hernández-Pecoraro alludes in passing to the focus of the present project when she lists the many alterities negated in pastoral literature: "Partially informed by comparisons to the fantasy of a past Golden Age, the ideal of a

homogeneous and harmonious Christian land set a standard for Spanish national identity during this period. Unfortunately, it could only be secured by repressive mechanisms of Inquisitorial trials, general expulsions, and the illusive standard of blood purity" (*Bucolic* 37–8).

47 See also Martínez (*Genealogical Fictions* 61) for an analysis of *limpieza de sangre* as an operative fiction. Other scholars have argued that, at least since the 1550s, the real aim of the discourses of blood purity was to dissolve traditional aristocratic circles of patronage, thus allowing for the social ascension of a new class of *letrados* into the body of the ecclesiastical and administrative positions (Poole 369; Samson 824–8).

48 This kind of analysis is also taken by Dominick Finello (*Pastoral*), who blends a reading of Cervantes's treatment of pastoral with analysis of other non-literary expressions of pastoral, such as sheep herding and public performances.

49 Italian has a very similar lexical bond between *pastore* (but also *pecoraro*) and *pastorale*; however, the latter lacks the derogative connotations of Spanish *pastoril*. Both *pastoril* and *pastoral* appear to be clearly differentiated in Castilian medieval texts, and it is not until the middle of the sixteenth century, arguably because of the influence of Italian literature, that *pastoral* began to be used as a synonym for *pastoril*, as it can be shown by searching the CORDE database.

50 Another Spanish word for "shepherd," and with the same semantic range as in English, was *ovejero*, but it was used very rarely.

1. Sheep Herding and Discourses on Race

1 One notable exception is Margaret Marek, who is currently exploring how pastoral romances reflect early modern sheep herding ("Itinerant Shepherds"; "Intertextual Textiles").

2 This is of course only one of the many approaches in which the relation between early modern Spaniards and their conception of animals can be considered. The field of animal studies is growing rapidly, as shown in the recent books by Abel Alves and John Beusterien (whose work on canines in Cervantes and Velázquez was released as this book is being sent to print), as well as Adrienne Martín's ongoing project on animals in Golden Age literature and art.

3 See also Thomas F. Glick (103–4).

4 All the early mentions of "Merino" in the fourteenth century are summarized by José María Sánchez Benito (257–65).

5 Eduardo Laguna Sanz provides further archeological and historical elements in support of the Iberian hypothesis.

6 For a similar critical approach to British historiography on sheep breeds, see Franklin (80–94).

7 There is no consensus about the etymological origins of the word "race," for which Arabic, Latin (Spitzer 65–6), and Germanic roots have been proposed, nor on how and when it took its current meaning in Spanish. For Joan Corominas (4:800–2), Castilian *raça* is a loan from Italian or Provençal, while L.P. Harvey suggests that it might have been in Castilian that the term evolved to have its current meaning (7n4). It might even be argued that the semantic slippage was made when Jews and Muslims were required to wear a distinctive sign; the coincidence of proverbial expressions about "en el mejor paño cae la raça" (Covarrubias 851) [the best cloth is not free from defect] with the measure that put a notorious mark in the cloth (*paño*), suggests that the mark could be read as the *raça* that was to be avoided, hence being conceived as the sign of the genealogy of its bearer.

8 For the use of the vocabulary of animal breeding to convey notions of race in medieval and early modern Spain, see Nirenberg ("Race" 77–9), Martínez (*Genealogical Fictions* 28), and Alves (87–8). On aims at rendering visible the genealogical *raça*, see Beusterien (127). There seems to be nothing exceptional in the Spanish fixation with sheep herding as the source for their nascent racial, pseudobiological vocabulary. As Roger J. Wood points out, it was the raising of the Merino sheep in eighteenth-century Europe, once Spain lost the monopoly of the breed (whose exportation had been prohibited until then), that contributed to emerging scientific terminology about heredity.

9 The image of the Good Shepherd was first outlined in Ezekiel 34, and later on applied to the figure of Christ. All my quotations from the Bible come from Casiodoro de la Reina's Spanish translation of 1569. Even though it is unlikely that early modern Spaniards read this version directly, due to the Catholic interdiction of translating the Bible into the vernacular, it offers a more accurate rendition on how they would have translated and interpreted it. All quotations from the Bible in English are mine, in order to reflect the Spanish translation, although in some instances I have considered as well the Douay-Rheims 1899 American edition.

10 For similar appropriations of John (10:16) in imperial terms, see Christopher Maurer and Benjamin Nelson (118–19).

11 I cite from the Spanish translation made by Luis A. Díaz y Díaz.

12 The Spanish original is ambiguous: when the manuscript R-M says that "en esto no tiene culpa," it can be translated either as "they [the people] are not to be blamed," or "he [Henry IV] is not to be blamed." I have chosen the former because it serves to emphasize religious freedom, but the latter would be equally valid.

13 For the debates over the 1547 statutes of blood purity and their repercussions, see Sicroff (132–72).

14 The play is no longer extant, but an account of the festivities written by Baltasar Porreño describes it and copies some fragments. This account is reproduced as an appendix by Jerónimo López de Ayala (169–75).

15 The revival of the Pauline reading of the pastoral metaphor had already been advanced by the University of Alcalá in a letter issued opposing the statute promoted by Siliceo, stating that baptism erases genealogical differences among Christians and therefore "no queda más que un solo rebaño y un solo pastor" (qtd. in Sicroff 134–5) [there is only one flock and one shepherd].

16 The *roña* was an animal mange typical of sheep (Covarrubias 915).

17 The play *Cortes de la muerte* was probably composed by Michael de Carvajal with some interpolations by Luis Hurtado de Toledo, who assumed the final responsibility of publishing a highly polemical text. Early modern Spanish "ley" can be translated as both "law" and "religion."

18 A treatise written in 1586 against the introduction of the statutes of blood purity in the Religious Order of Saint Jerome, attributed to Gaspar de Uceda, complains that, with their implementation, *conversos* "han de ser almagrados y notados" (97) [will be marked and defamed], implying that the statutes of blood purity play the same role that the marks of infamy did in the previous centuries.

19 For the relation between Fray Cipriano de la Huerga and Fray Luis de León, see Bataillon (133–43) and Francisco Javier Fuente Fernández. On inquisitorial persecution of Fray Luis, see Hermenegildo ("Fray Luis") and Francisco Javier Perea Siller (1118–20). In the Salamanca edition of 1587, Fray Luis added the name "Pastor," to expand on Fray Cipriano de la Huerga's Pauline interpretation of the Good Shepherd (224–38).

20 The adjective *roñoso* (mangy), derived from *roña*, had in early modern Spanish a meaning limited purely to sheep herding at this time: "Roñoso, el que tiene esta sarna" (Covarrubias, "roña" 915).

21 The criticism of the "shepherd" as potential introducer of heresy is always latent in the metaphor, as the shepherd Aminta states in Torquemada's *Coloquios satíricos* when contrasting the profession of the real shepherd to that of the pastor, since the latter "ynficionan sus ovejas con el mal enxemplo de su vida y excesos" (1:306) [infect the sheep with the bad example of their lives and excesses].

22 Similarly, Damián de Vegas's *Poesía cristiana, moral y divina* (1590) [*Christian, Moral, and Divine Poetry*] recommends priests to actively engage in the purification of the Christian community by expelling undesirable elements: "Cuando alguna en el rebaño / tan roñosa venga a hallarse, / que no dejando de curarse, / hiciese a las otras daño, / procurad echarla de él" (Sancha 496) [When a sheep in the herd is mangy and does not let itself be cured, try to expel it from the herd so that it does not harm other sheep].

23 Francisco Vázquez García also analyses the image of the Good Shepherd in conjunction with the emergence of biopower, although he articulates his study around the traditional opposition between menacing wolves and the herd that needs the shepherd's protection ("Métaphores" 200). This rhetoric of externalization was certainly used against the Morisco: Jaime Bleda considered that the Moriscos "no eran ouejas, sino lobos carniceros" (Corónica 882) [they were not sheep, but butchering wolves]; Aznar Cardona called them "lobos entre las ouejas" (2:63v) [wolves among sheep]; and Pedro González de Mendoza praised Philip III for expelling the Moriscos because: "A lupis agnos separa" (qtd. in Morón 492) [he separates wolves from sheep]. However, it was the conception of the Moriscos as belonging to the Christian fold that triggered a radically different discourse which was saturated with concepts of contagion and heredity and which therefore altered the symbolic ties between the monarch and his vassals.

24 See Stallaert (*Ni una gota* 284), and Lapeyre (28–36 for Valencia; 116–19 for Aragon; 150–8 for Castile and Granada; 180–3 for Andalusia and Murcia). The *Relaciones topográficas*, begun in 1574, were one of Philip II's most ambitious projects of gathering statistical information about his territories; in the list of questions addressed to each town, it included a specific question about the Moriscos living in each place (Campos y Fernández de Sevilla).

25 This new form of managing populations is manifested in the discourses that conceive of the ruler as a doctor who needs to purify and heal the nation state (Stallaert, *Ni una gota* 213–18; Vázquez García, *La invención* 55–137).

26 In 1638, during Philip IV's reign, Bartolomé Jiménez Patón writes that the expulsions of both Jews and Moriscos were necessary preventive measures "para que no inficionaran con su roña las ouejas de nuestra ley Euangelica" (1v) [so they would not infect the sheep of our Evangelic faith with their mange].

27 Foucault establishes a very vague chronology, arguably because this is not his priority when presenting the concept. Vázquez García is the most recent scholar to systematically apply Foucault's concept of biopolitics to early modern Spain; he does not address Foucault's chronology, merely stating that the early modern period is characterized by the use of "absolutist biopolitics," focused on the mercantilist problem of depopulation (*La invención* 16).

28 Foucault mentions in passing that one of the tasks of the Good Shepherd is to "arrange their mating in order to produce the most vigorous and fertile sheep that produce the best lambs" (*Security* 143). Unfortunately, he does not explore the socio-political implications of the language of breeding and does not integrate this insight in his discussion on biopower.

29 Examples for such uses in early modern Europe can be found in Ania Loomba and Jonathan Burton (100–1; 114–15; 236–7; 249), and Wendy

Doniger. Javier Moscoso studies the systematization of this theory in the eighteenth century.

30 See Javier González Rovira (28–9). A parallel moral and political reading of this biblical theme is found in seventeenth-century emblem books. Juan de Solórzano Pereira's *Emblemata Centum, regio politica* (1653) follows Francisco Vallés in emblem 29 ("Sceptrorum imitatio potentissima"), in his notion that the herd stands for the king's vassals and the rods symbolize the sceptre. Thus the biblical passage serves to convey the idea that the king should be a model for his subjects so they will imitate his ways. Most of Solórzano Pereira's emblems are translated to Spanish in Andrés Mendo's *Príncipe perfecto y ministros ajustados* (1662). Mendo's translation is not verbatim, and his emblem 9 is a summary of Solórzano Pereira's, excising all references to generation (42–6). By construing the king as having the ultimate responsibility for his flock's habits, the latent biopolitical reading of the shepherd is progressively neutralized. See Taggard (111–13) for both Solórzano Pereira and Andrés Mendo. For a comprehensive use of the figure of the shepherd in emblem books, see Alonso Rey.

31 Most of the literary and pictorial references on the life of Jacob are found in Edward Glaser and Taggard.

32 Pedro Orrente was an acclaimed painter who worked for both the royal court and local patrons, travelling constantly between his native Murcia, and Madrid, Toledo, Valencia, and Italy. As part of his cursus honorum, he applied to become a "familiar del Santo Oficio" in Murcia in 1625, although he would not be accepted until 1633, probably thanks to his contact with his influential patrons (Agüera Ros 88–90). On the diffusion of Orrente's pastoral compositions and his school of imitators, see especially Diego Angulo Íñiguez and Alfonso E. Pérez Sánchez (261–93), complemented by José Carlos Agüera Ros (107–8).

33 Taggard suggests that Murillo's series was commissioned around 1660–70 by the Sevillian Manuel Luis de Guzmán y Zúñiga, IV Marquis of Villamanrique, who would have intended to reconfigure himself as a pious Christian not interested in political matters, and that the paintings symbolize the triumph of the Catholic faith (106–15). However, as Taggard herself acknowledges, there is little evidence to prove why and who commissioned both Orrente's and Murillo's cycles on the life of Jacob (75–7), and therefore any interpretation of such agency remains a matter of opinion.

34 Some nuance is needed here: Orrente's sheep are not white, but lightly reddish. This may actually reflect a direct observation of Spanish sheep, which looked red because of the *almagre* [red ochre paste] that shepherds used to distinguish their flocks. Jacopo Bassano's *The Journey of Jacob*

(c. 1578), which might have been the model for Orrente and Murillo, also represents, like the latter, perfectly white sheep. For Bassano's painting of Jacob's livestock, see Paolo Berdini, who does not address the issue of their color (81–7).

35 For Ribera's patrons, see Brown.

36 On Olivares's attitude towards the statutes of blood purity, see Henry Kamen ("Una crisis de conciencia" 343–55), Henri Méchoulan (145–6), Juan Ignacio Gutiérrez Nieto (425–7), Antonio Domínguez Ortiz (*Los judeoconversos* 64–73), and Juan Hernández Franco (192–248).

37 Given the difficulties of political authors to craft a rhetorical opposition to Machiavelli, these biblical metaphors serve as a rhetorical instrument to reconcile biopolitics with Christian doctrine, a move within Catholic political thinking labelled "the Machiavellism of anti-Machiavellists" (Vázquez García, *La invención* 87–97).

2. Rustic Culture and the Invention of the Spanish People

1 In a 1973 article on "bad poets" in early modern narrative, Gonzalo Sobejano contextualizes this passage within the craving for novelties in the early modern stage, seemingly agreeing that "hacer una comedia de moros que fuesen pastores sin dejar de ser moros" [making a play about Moors who were shepherds without leaving off being Moors] reveals a "falta de propiedad" [lack of decorum] (317).

2 On pictorial representations of the Nativity, the most comprehensive study to date is that of Francisco Manuel Valiñas López (125–459). See also José María Palencia Cerezo (101–6) and Sánchez Cantón (45–71); Agüera Ros (108–10 and 115–16) offers a catalogue of the theme in the works of Pedro Orrente and his imitators.

3 As Paolo Berdini points out, after Trent artists also saw in the Adoration of the Shepherds a pretext to depict rural life (18–22). In other instances, the emphasis given to the Adoration of the shepherds may not be related solely to ethno-religious debates, since it also served to avoid the representation of Virgin's labour, which post-Tridentine morality perceived as indecorous (Valiñas López 128–9).

4 In Italy, the archbishop of Milan Federico Borromeo criticizes in *De pictura sacra* (1624) what he perceives as an excessive stylization of the shepherds in painting, in detriment to the attention that the figures of Christ and the Virgin should receive (47). See Berdini for Jacopo Bassano's *Adorations*, which greatly influenced Spanish artists.

5 For grotesque genealogies in the Spanish stage, see John Lihani. Stephen Gilman (20–1n3) and Alfredo Hermenegildo ("En torno a la burla" 52–62)

clearly link these genealogies of the dramatic shepherd to the doctrine of blood purity.

6 See Hermenegildo ("Sobre la dimensión"), Gilman, Márquez Villanueva ("La *Trivagia*" and "Sobre el concepto"), Gitlitz ("Lucas Fernández" 504), Yarbro-Bejarano, Wertheimer, Surtz ("Pastores judíos" 334–7), and Constance Rose (65–6).

7 For other cases in which patrons might have instructed artists to depict them as shepherds in their Adorations, see Valiñas López (329–31). This situation contrasts vividly with an anecdote recounted by Juseppe Martínez of a painter in Italy who, in revenge for not having received his due pay, depicted the portrayed commissioner wearing a shepherd's costume (Calvo Serraller 505).

8 For the identification of the monarchy with the Spanish people in the reign of Philip III, see Mateo Ballester Rodríguez (62–3).

9 It is true, as Ruth Fine notes, that Christian theologians often distinguished between biblical Hebrews that foreshadow the Christian people and the Jews who refused to accept Christ as the Messiah (437–9). However, the defenders of the statutes of blood purity usually conflated both when they claimed that the Christian faith of *conversos* could never be trusted.

10 Valiñas López collects several theological commentaries written in early modern Spain on the Adoration of the Shepherds (141–7). None of them mention the Jewishness of those first shepherds.

11 On literary representations of the Virgin of Guadalupe, see Françoise Crémoux; on the Virgin of Cabeza, see José Domínguez Cubero.

12 For the social class of the *caballeros villanos* and the privileges that the Military Orders obtained in wool trade, see Pedro García Martín (31), Marie-Claude Gerbet ("Ordres" 425–6), José María Mínguez Fernández (353), Enrique Rodríguez-Picavea Matilla (183–5), Carla Rahn Phillips and William D. Phillips (31–4), and Charles Julian Bishko (57–8).

13 In fact, livestock migrated through political and religious boundaries despite occasional military confrontations (Sánchez Belda 301; Glick 103–4; Rodríguez Molina 263–73; Ladero Quesada, "Almojarifazgo" 100).

14 For example, in the case of the Virgin of Guadalupe, the earlier document containing the legend of the shepherd dates from around 1440, which William Christian believes to be a fable (88–92).

15 For an analysis of this painting, see Rafael Frías Marín.

16 The devotion of Our Lady of Cabeza in Zújar (Granada) was the origin of the local "fiesta de moros y cristianos" that celebrates the conquest of the city by Christians (Muñoz Renedo 31–43).

17 A particular case is the cult of the Virgin of Aránzazu. Because the discovery happened in the intersection between Álava, Vizcaya, and Guipúzcoa, the shepherd served to bring together all Basques and to articulate a feeling of

Basque identity, especially in their migration to the New World (García Ayluardo).

18 Camps i Moranta and Seguí i Trobat transcribe only a selection of the 1642 investigation (57–63); for the complete text, see Gabriel Llompart.

19 See also Finello (*Evolution* 184–90) and Salomon (379–80).

20 See Jenaro Alenda y Mira and the eight volumes of poetic accounts of the game of canes edited by Antonio Pérez y Gómez.

21 In his study of European folklore, Peter Burke notes that the identification of the members of the upper classes with the "people" "seems to have gone furthest in Spain" (8).

22 As Otto and Pedersen suggest, people can become aware of the meaning of automatized rituals, and in this way, "intentional action can have an impact on unconscious routines" (21).

23 While direct confrontation between shepherds and Muslims is rare in pastoral romances, we can find a passage in Antonio Lofrasso's *Los diez libros de fortuna de amor* (1573) in which a shepherd fights Turkish and North African pirates with his crook (127v), which seems reminiscent of these plays.

24 There are certainly other cultural practices that might be considered as well, such as music (which is analysed in chapter 4), pastoral games (Castillo Martínez "Pancracios"), and diet.

25 Juan de Ávila was of *converso* origin and suffered in his own person the rigours of the statutes of blood purity (Roldán-Figueroa 1–4). Since he preached in Andalusia and Granada, his emphasis on the sartorial bond between sheep and shepherd may be intended for the cultural assimilation of the Moriscos.

26 On the aristocratic versions of rustic clothing, see Bernis (*Trajes* 134–5) and Ruth Anderson (115 and 248).

27 In the pastoral romance, extravagant shepherds sometimes chose different materials for their *pellicos*, like the one in Alonso Pérez's *Segunda Diana* wearing hyena's skins (150r), or in Bernardo de Balbuena's *Siglo de oro*, where one shepherd wears a *pellico* made out of lion's skins (312).

28 Covarrubias defines *xerga* as "tela gruessa, como de sayal" (1015) [coarse fabric, like that of rustic tunics], while the *brocado* is "la labor de las telas ricas de oro" (237) [the product of rich golden fabric]. In Bernardo de la Vega's *El pastor de Iberia*, the descriptions of lavish *pellicos* abound: one shepherdess is described wearing "una pellica, cuyo color no se pudo divisar, respecto de venir sembrada de joyas de inestimable valor" (152v) [a *pellica*, whose colour could not be discerned, because it was covered with invaluable jewelry]. In Francisco Bramón's *Los sirgueros de la Virgen* (1620), the shepherd Menandro wears "un pellico de terciopelo carmesi, cuyos aforros eran de armiños blancos" (81r) [a crimson velvet *pellico*, lined with white ermine].

29 See also Francisco Pacheco's *Arte de la pintura* (1649), which gives detailed instructions on how to depict the Nativity by adhering to the biblical text (602–8).

30 The sartorial heterogeneity in the Iberian Peninsula is reflected in many of the European costume books initiated with the influential *Trachtenbuch* (1529) by Christoph Weiditz, who traveled to Spain with the court of Charles V (Defert 26). As Díaz Viana remarks, cultural and institutional authorities render as uniform any expression considered as "traditional" in order to negate heterogeneity (104).

31 The quotation is found in "Discurso setenta y ocho, y último. Del vestido." The book resumes pagination after page 438v; therefore, the correct page number should be 494v.

32 For this rhetorical (but real) conception of blood purity, see Contreras Contreras (87–95).

33 For Moorish clothing in early modern Spain, see Bernis (*El traje* 461–82) and Rachel Arié.

34 Robert Ricard, using different sources, also arrives at the conclusion that "la *marlota* est bien sinon un *sayo vaquero*, tout au moins un vêtement rustique, et ... elle appartenait ... dans certains régions, au costume des gens de la campagne" (144) [even if the *marlota* may not have been exactly the same as a *sayo vaquero*, it was at least a rustic garment … that … belonged in certain regions to the attire of peasants]. Bernis argues that the *sayo morisco* (Moorish tunic) belonged to well-off Christian classes, suggesting that maybe the distinction between *sayo* and *marlota* comes from the use of different fabrics (*Trajes* 54).

35 "Serrano" is also a last name, quite possibly the origin of the "portal de los Serranos" in Valencia. The common term *serrano* derives from *sierra* (mountain), and was usually a synonym for rustic people.

36 On the penchant of Spanish and Valencian aristocracy for dressing up as Moors in the game of canes, see Irigoyen-García ("Poco os falta" 257–61).

37 The *sayagués* jargon is used only very rarely in pastoral romances. It appears in Bernardo de la Vega's *El pastor de Iberia* (1591), in which Artandro and Domenga, the captors of the shepherd Filardo, are derided because of their language (129v–30r).

38 The knowledge of "ancient speech" was indeed essential in the historiographical forgeries that infested early modern Spain. In 1595 the discovery of the *Lead Books* in Granada was made, which aimed at creating a syncretic vision of Christianity and Islam; they were written in a pseudo-archaic form of Arabic as part of their pretensions of authenticity (García-Arenal and Rodríguez Mediano 33–44).

39 I am not unaware that Hobsbawm assumes that the "invention of tradition" is a product of the Enlightenment, along with the emergence of modern nationalism. Yet researchers such as Noël Salomon have proved that there is ample evidence in early modern Spain for a construction of "popular culture" by elite classes. Ballester Rodríguez also identifies the invention of tradition as taking place in early modern Spanish society in the collection of proverbs as expressions of "popular wisdom," as well as in the premeditated cultivation of medieval poetic forms (348–50).

40 As Burke states, "the discovery of popular culture was part of a movement of cultural primitivism in which the ancient, the distant and the popular were all equated" (10).

3. In the Land of Pan: Pastoral Classicism and Historiography

1 The linguistic cleansing includes Charles V's 1526 prohibition of the Arabic language and the use of "nombre o sobrenombre que suene a moro" (qtd. in Gallego Burín and Gámir Sandoval 204–5) [name or last name that sounds Moorish], later enforced in 1566 by Philip II (Gallego Burín and Gámir Sandoval 274). For an analysis of early modern attitudes about the linguistic influence of Arabic in Castilian see Barbara Fuchs (*Exotic Nation* 24–30). Felipe Maíllo Salgado documents the disappearance of many Arabic words in Castilian in the transition between the Middle Ages and the early modern period, attributing it to the prohibitions of the sixteenth century (503). José María Perceval wonders whether this linguistic cleansing was part of a larger project of Spanish humanism (*Todos son uno* 199n827), and Serafín Fanjul takes for granted that the lexical cleansing reflects a genuine desire to revive a Latin identity (195).

2 For a concise account of the effect that humanistic geography had in the conception of ancient Iberian, see R.B. Tate. For a detailed study of the view of Iberian prehistory in medieval historiography, see Helena de Carlos Villamarín. For the case of Catalonia, where a parallel "recovery" of ancient geography was also made, see Mariàngela Villalonga.

3 For an overview of sixteenth-century uses of the term "restauración de España" (never *reconquista*, which is a neologism employed only after the nineteenth century), see Martín F. Ríos Saloma (41–94).

4 The search for the "original" place names is a pan-European phenomenon in early modern Europe. See Marian Rothstein (339) for the case of France, and Angus Vine for England. What is specific for Spain is that humanistic reconstruction of the geographical past was pitted against the Arabic legacy

– but also its Hebrew legacy, as has been studied by Dominique Reyre and Pauline Renoux-Caron.

5 In his work, Servet made a derogative comment, in passing, about Spanish intellectuals, who, according to him, "lingua Hispanica plus quam Latina in Academiis loqui gaudent, quin etiam Maurorum vocabula plurima usurpant" (Bullón y Fernández 207) [They rejoice to speak better Spanish than Latin in their schools, although they have taken over many words from the Moors] (trans. O'Malley 28).

6 See Sabine MacCormack, who states that the purpose of early modern Spanish humanists was "to link the Roman past to their own present" (98). See Santiago Quesada (187–90) for the recovery of ancient place names in local historiography.

7 Ortelius's *Theatrum orbis terrarum*, the predecessor of modern atlases, appeared in 1570 and included a fairly detailed map of early modern Spain. Three Spanish editions appeared between 1588 and 1612. On the relation of Ortelius with Spain, see Agustín Hernando.

8 As Hervé Guillorel states: "les toponymes peuvent être mobilisés pour prouver une thêse relative à des problèmes d'antériorité et de légitimité sur un territoire" (84) [place names can be mobilized to prove an argument related to issues of anteriority and legitimacy over a territory]. Séverine Delahaye-Grélois also explores how the "poetic" substitution of the name Betis for Guadalquivir is ideologically driven (258).

9 All my quotations of Montemayor's *La Diana* come from Asunción Rallo's edition.

10 Similarly, Benjamin Nelson states that "[i]n the opening lines of this first Spanish pastoral in prose, Montemayor firmly grounds his bucolic setting within the realms of Imperial Spain" (69).

11 A document about the reformation of the blood purity statutes complains that "sin más razón tienen ... al de León y Asturias por Christiano viejo, y al de Almagro por confeso" (qtd. in Domínguez Ortiz, *La clase social* 230) [without any reason, one person from León and Asturias is regarded as Old Christian, while another from Almagro is held to be a *confeso* (someone convicted by the Inquisition)].

12 An anonymous satirical poem that circulated in Lima against Pedro Mexía de Ovando's *Ovandina* (1621) criticizes the pretension of extolling newly arrived lineages. What is striking is that the poem denounces the illegitimate pretensions of social ascension as if these people were trying to pass as shepherds: "si toda ella es ynvencion /sacada de tu archibon / mezclando hidalgos de barro / cubriendolos con çamarro /y haciendolos del tuson"

(qtd. in Cook 236n699) [if it is all invention / taken from your archive / mixing hidalgos from clay / covering them with *çamarro* (shearlings) / and making them of the Toison]. This text shows that the passing as shepherds, even if figuratively, is one of the strategies in which those who boasted of blood purity pretended to show off their lineage.

13 Belisa states that the dwellers in her town "son de los que en la gran España llaman libres, por el antigüedad de sus casas y linajes" (Montemayor 232) [are those that in the great Spain are called free, because of the antiquity of their households and lineages]. Like her, Dórida, Cinthia, Polydora, and, probably, the wise Felicia are also Galician, since they come from the other side of the Galician mountains (Montemayor 192).

14 "Vandalia" is also found in Cervantes's *La Galatea* (1585), where the shepherd Lisandro says that he was born "[e]n las riberas del Betis, caudalosísimo río que la gran Vandalia enriquece" (188–9) [on the shores of the Betis, the plentiful river that enriches the great Vandalia].

15 However, Dávalos y Figueroa admitted that the attribution of the name Soldina to Écija was likely hearsay and he could not find any authority to confirm it (173v). Neither "Sevilla" nor "Écija" are Arabic names, although they have been influenced by Arabic pronunciation over time. If Montemayor was referring to either "Sevilla" or "Écija," one wonders why he did not use the well-known Latin names "Hispalis" or "Astigi," respectively.

16 See Medina (122), Guadix (227), and Covarrubias (71–2).

17 Cómpluto is used in González de Bobadilla's *Ninfas y Pastores de Henares* (1583) to refer to the shepherdess Alberia "en la famosa Compluto criada y nacida" (144v) [born and raised in the famous Compluto]. This pastoral romance is located "[e]n las umbrosas riberas que el apazible Henares con mansas y claras olas fertiliza" (11v) [on the shady shores that the calm river Henares fertilizes with its peaceful and crystalline waves]; in spite of the many mentions of the river Henares throughout the romance, the author is careful not to mention "Alcalá de Henares" even once.

18 Medina (121), Guadix (762), and Covarrubias (778).

19 Mariana still referred in 1601 to the wider geographical interpretation of the term: "El reino de Toledo es asimismo parte de Castilla, el cual hoy se llama Castilla la Nueva, y antiguamente la Carpetania" (1:5) [The kingdom of Toledo belongs as well to that part of Castile that is called nowadays New Castile, and Carpetania in old times].

20 "Mantua Carpetana" appears in Gálvez de Montalvo's *El pastor de Fílida* (1582) when Alfeo recounts his own story to Siralvo: "Caballero soy, natural desta vecina Mantua" (512) [I am a noble from the neighbouring Mantua], as well as in González de Bobadilla's *Ninfas y pastores de Henares* (1587), where

the shepherds also speak of "la famosa Mantua Carpetanea" (114r) [the celebrated Mantua Carpetanea].

21 This is acknowledged by almost every early modern historian, such as Nebrija's (*Muestra* 220–1), Medina (70), Aldrete (351), and Covarrubias (212).

22 The examples in pastoral romances are countless, I here mention only a few: "El Betis caudaloso" (Botello de Carvallo 2:80); "En las riberas del caudaloso Betis" (González de Bobadilla 90r); Bernardo de la Vega's *El pastor de Iberia* (1591) begins "en las riberas del sagrado Betis" (8v) [on the shores of the sacred Betis]. Paradoxically, one of the pastoral romances that uses indistinctly both "Betis" and "Guadalquivir" is Gonzalo de Saavedra's *Los pastores del Betis* (1633). Lope in his *Arcadia* (1598) names it "Betis," including an erudite note that seems unnecessary for the potential early modern reader: "Llámase Guadalquivir, nombre que, como a otros ríos, le pusieron los africanos cuando ganaron a España" (424n145) [It is called Guadalquivir, a name that, like other rivers, was imposed by the Africans when they won Spain].

23 Medina talks about "el río que se llama Turia; y los moros le dijeron Guetalabiar, que quiere decir agua llana y blanca" (196) [the river named Turia; and the Moors named it Guetalabiar, which means calm and crystalline water]. Ortelius's map of Spain included in his *Theatrum Orbis Terrarum* (1570) shows it as "Guadalauiar rio." In 1585, the Flemish traveller Hendrick Cock refers to the river as "Guadalaviar" (qtd. in Mérimée lxxxvi n1). By the 1590s, the lexicographer Diego de Guadix states that its name is Guadalaviar, acknowledging in passing the alleged ancient name (668). Covarrubias only gives an entry for "Guadalabiar": "Río que corre por la ciudad de Valencia ... Vale río blanco (su nombre antiguo era Turia)" (661) [River that runs through the city of Valencia ... It means white river (its ancient name was Turia)]; in the entry for "Valencia" he only mentions the name "Guadalaviar," which means that this was still the common name by the beginning of the seventeenth century (990). Escolano follows Beuter very closely, citing almost the same classical sources, and naming the river "Turia" (272).

24 While most of the pastoral romances keep a very tight control over place names, avoiding Arabic names, there are a few exceptions, such as Lope's *Arcadia* (1598), where both modern and classical place names alternate (Morby, "Two Notes" 121). In Jerónimo de Tejeda's *Tercera Diana* (1627), Madrid, Alcalá de Henares, and the river Guadalquivir are always mentioned by their common names; arguably the circumstances of Tejeda are equally exceptional, since his is a late pastoral romance written for a French audience who would not understand the use of ancient place names of the Iberian Peninsula.

25 See Fernando Wulff (20) and MacCormack.

26 The influence of Annio de Viterbo in early modern historiography has been amply studied. See Jon Juaristi (20–5), José Antonio Caballero López (113–20), and Miguel Ángel López Trujillo (53–6).
27 Miguel Caxa de Leruela's *Restauración de la abundancia de España* (1631) comments on the classical references to livestock in ancient Iberia, passing over the Islamic period in silence (34–7).
28 For Covarrubias, the proper meaning of *mayoral* is that related to sheep herding: "El que assiste al govierno del ganado con mando, governando los demás pastores" (780) [He who helps in the management of livestock with authority, ruling over the other shepherds]. Although almost two centuries stretch between Corral and Covarrubias, a search in the CORDE database shows that *mayoral* already had the meaning of herdsman at least by 1250.
29 On Miguel de Luna, an official translator of Arabic, see Francisco Márquez Villanueva (*El problema morisco* 45–97), Fuchs (*Mimesis* 111–17), Elizabeth Drayson (47–58), and García-Arenal and Rodríguez Mediano (165–96).
30 For a revision of the real importance of Las Navas de Tolosa, see the special issue of *Journal of Medieval Iberian Studies* recently edited by Simon Doubleday and Miguel Gómez.
31 For the medieval sources on the shepherd of Las Navas de Tolosa, see Manuela García Pardo (219–33), Francisco García Fitz (47–9), and Rosado Llamas and López Payer (251–9), who include an appendix with many of the historiographical texts that address this episode (322–71).
32 Following Gonzalo Fernández de Oviedo, Argote de Molina even names the shepherd as the ancestor of the prestigious lineage of Cabeza de Vaca (1:28v).
33 By the beginning of the seventeenth century, the shepherd of Las Navas de Tolosa is identified as Saint Isidro, the patron of Madrid, among others, in Jaime Bleda's *Vida y Milagros de San Isidro el Labrador* (1622) (229–35), and Lope de Vega's hagiographic poem *Isidro* (1599) (582).
34 Mármol Carvajal describes the quality of pastures of different Morisco locations about twenty times (189, 191, 193, 194, 196, 201, 201, 202, 204, 205, 205, 207, 210, 213, 216, 264, 342). His insistence can be explained by the interest of the monarchy in promoting the repopulation of the kingdom of Granada.
35 Echave is not original in relating the mythical arrival of Tubal to the Iberian Peninsula with the introduction of livestock. He is probably taking it from Florián de Ocampo's *Corónica general de España* (1543), for whom Tubal "repartio las compañas que le quedaron por ella, para que la paciessen con sus ganados" (17r) [distributed his companions so they grazed the land with their livestock].
36 However, Iberian Jews maintained in the fifteenth century that Tubal, whose biblical and thus Hebrew origin they emphasized, was also their ancestor, and

that his arrival in the Iberian Peninsula before the death of Christ ultimately proved that they were not the descendants of deicidal Israelites (Lida de Malkiel 21–31).

37 For the current ethno-archaeological debate on the relevance of ancient sheep herding in the Iberian Peninsula, see Joaquín Gómez-Pantoja ("*Pastio agrestis*").

38 For the complex set of ideological connotations embedded in Aldrete's work see Kathryn A. Woolard (460–71).

39 I am thankful to José Ignacio Hualde for providing me with this information.

40 Some scattered examples might be found in Miguel de Luna's *Historia verdadera del rey don Rodrigo*. In describing the arrival of the king Abenrahmin at the river "que llamaron los moros Guid Arroman," Luna adds a note to clarify this place name: "Guidarroman quiere dezir, el rio de los ganados" (173) [Guidarroman means "river of livestock"]. More significant seems the topoetymology of "Guadiana" that he proposes: "Este rio llamaron los nuestros Guit Daina, que quiere dezir, rio de la Oveja, por ser manso" (302) [This river was called by our people *Guit Daina*, that means "river of the Sheep," because it is gentle].

41 Other place names that Guadix relates to sheep herding are: Arahal, Aznalmara, Daymora, Gemerendula, Maro, Mayrena, Monreal, Morana, Morrán, Moratalla, Moratiel, Moratón, Morea, Morella, Morián, Moriana, Morón, Mura, Redondela, and Saldaña. I have not taken into account all of the common words in the semantic field of sheep herding that he affirms to be from Arabic origins.

42 In the Biblioteca Nacional de España (Res. 208), there is a manuscript translation of the Gospels into Arabic which contains the approval of Guadix, among others (Bajo Pérez and Maíllo Salgado 125). For the intellectual attempts to conciliate Arabic with Christianity in early modern Spain, see García-Arenal and Rodríguez Mediano.

43 Not intending to condone or condemn Guadix's etymologies, it should be noted that, although they are probably not accurate, they are not entirely far-fetched either. "Roa" is similar to the Andalusi Arabic *ru*^{c}*a* (from Arabic *ru*^{c}*at*), meaning "shepherds"; *mor*^{c}*á* is the Andalusi form of Arabic *mar*^{c}*á*, meaning "pasture"; and *fahs al-ra*^{c}*iya* does mean "field of pastures." I am grateful to Mohamed Saadan for providing this information.

44 For a general overview of classical mythology in the Spanish pastoral romances, see Vicente Cristóbal López ("Mitología") and José María de Cossío (202–19).

45 As I have suggested elsewhere, the goddess Diana was also appropriated as a symbol of Old Christian values in early modern Spanish literature (Irigoyen-García, "Diana").

46 For the influence of Virgil and Sannazaro in Spanish pastoral literature, see, respectively, Marcial José Bayo and Rogelio Reyes Cano.

47 Examples abound in the pastoral romances, such as Balbuena's *Siglo de Oro* (93), Pérez's *Segunda Diana* (38r), Gálvez de Montalvo's *El pastor de Fílida* (524), González de Bobadilla's *Ninfas y pastores de Henares* (34r), Arbolanche's *Las Abidas* (61r), Arce Solorzeno's *Tragedias de amor* (61r), and Covarrubias Herrera's *La enamorada Elisea* (16r).

48 For the debates about Trinity in relation to evangelization in early modern Spain, see Louis Cardaillac (207–33) and Stallaert (*Ni una gota* 310).

49 Spanish *adiós* means "goodbye," but in this context, the typographical separation "a Dios" [to God] evokes the religious origin of the expression.

50 In Alonso Pérez's *Segunda parte de la Diana* (1563) Cardenia complains in identical terms to Fausto when she says that "[e]l officio de pastor / Mal t'esta Fausto a la ce, / Pues tan mal guardas la fe" (170r) [the profession of shepherd does not fit you well, I swear, Fausto, since you keep your faith (love) so badly]. Constance Rose suggests that this ambiguity is typical of *converso* texts (66–8).

51 The most famous case is Furió Ceriol, who states in his *El Concejo i Consejeros del Principe* (1559) that "[n]o hai mas que dos tierras en todo el mundo: tierra de buenos, i tierra de malos. Todos los buenos, agora sean Judios, Moros, Gentiles, Christianos, o de otra secta, son todos de una misma tierra, de una misma casa, i sangre: i todos los malos de la misma manera" (140) [there are but two lands in the world: the land of the good and the land of the bad. All the good people, be it Jews, Muslims, Gentiles, or Christians, or any other sect, all of them belong to the same land; they are from the same house and blood; and all the good people are the same]. See Schwartz (70–87) for a general account of tolerant attitudes towards religious heterogeneity in early modern Spain.

52 For religious attacks on the sensuality of pastoral romances, see Márquez Villanueva ("Sobre el contexto"), Werner Krauss ("Algunas observaciones" 683–8), and Juan Bautista Avalle-Arce (*La novela pastoril* 265–74).

53 The astrology section was replaced by a literary commentary praising Lope's contemporaries; thus S. Edwin Morby concludes that, rather than a case of religious censorship, it is a literary imposture of difficult attribution ("Páginas olvidadas" 85–6).

54 Márquez Villanueva strikingly states that "[e]ran sólo las paranoicas suspicacias de la instancia oficial española quienes conferían un aura de transgresividad a lo que en todas partes (sobre todo en la no menos católica Italia) se aceptaba como mero convencionalismo de un género culto" ("Sobre el contexto" 193–4) [It was only the paranoid suspicions of Spanish official

institutions who conferred an aura of transgression to what anywhere else (especially in no less Catholic Italy) was accepted as a mere conventionalism of an elite genre], even though all of the examples that he offers of ecclesiastical censure of the paganism of pastoral romances come from Italy, not from Spain.

55 The inconsistency of having two parallel sets of religious thought, one devotional (Christianity) and one profane (classical mythology) is exploited by the Inca Garcilaso de la Vega in the first part of his *Comentarios reales* (1609), where he defends the concept of "fábulas historiales" (historic fables) for the preservation of Amerindian cosmogonies as allegorical tales that do not collide with Christianity: "no hay que espantarnos de que gente que no tuvo letras con que conservar la memoria de sus antiguallas, trate de aquellos principios tan confusamente; pues los de la gentilidad del Mundo Viejo, con tener letras y ser tan curiosos en ella, inventaron fábulas tan dignas de risa ... también se pueden cotejar las de una gentilidad con las de la otra" (147) [we should not be surprised to see that people who did not have letters with which they could preserve the memory of their antiquities deal with the beginnings in such a confusing way, because those Gentile people of the Old World, having letters and being so skilful with them, invented such laughable fables ... We can also contrast those [fables] of one Gentile community and the other].

56 Similarly, Jesús Maestro suggests that the pagan elements in Cervantes's *La Galatea* serve to promote secularization (635–9). On the conflicting attitudes on paganism in early modern Spanish literature, see Otis H. Green.

57 Frederick de Armas traces the classical sources and analyses the use of prophecy in several Spanish pastoral romances, which he reads as a metaphor for poetic inspiration ("Caves" 337).

58 The only pastoral romances not located in Spain are Jerónimo de Covarrubias Herrera's *La enamorada Elisea* (1594), which takes place in Egypt, and Lope de Vega's two pastoral romances: *Arcadia* (1598), located in ancient Greece, and *Pastores de Belén* (1612), located in Israel.

59 An anonymous memorial attempting to rebuke the accusation of deicide usually raised against Jews stated that the killing of Christ was also commited by heathens: "no solo los judios le cometieron sino los gentiles (de quienes se precian que descienden los christianos viejos)" (qtd. in Domínguez Ortiz, *La clase social* 238) [not only the Jews committed it, but also the Gentiles (from whom Old Christians take pride in descending)], thus turning the accusation against Old Christians.

60 Morales's etymology is reproduced by other historians, such as Dávalos y Figueroa's *Primera parte de la miscelánea austral* (1602) (117v–18r), and Mariana's *Historia general de España* (1601) (1:10); these two historians did not subscribe to Morales's etymology, but neither did they reject it.

61 Aldrete was referring here to the geographical encyclopedia *Ethnica*, by Stephanus of Byzantium, a scholar who lived in the sixth century CE. On the rediscovery of Stephanus during the Renaissance, see Aubrey Diller (337–8). Aldrete was right in correcting Stephanus, since Pannonia was a Roman territory in the Balkans (Ilyria), but his suggestion that it should read "Pania" instead was merely a useful speculation.

62 A very similar account is found in Juan de Piña's *Epítome de las fábulas de la antigüedad* (1635) (9r).

63 Both José M. de Osma and Barbara E. Kurtz analyse Calderón de la Barca's allegorical play *El verdadero Dios Pan* (1670). Calderón's play is probably the most well known Christological reading of Pan, and it is even cited by Françoise Lavocat (*La Syrinx* 211–16) and Patricia Merivale (31). The pun with *pan* and the identification with Christ is already prevalent in religious literary texts by the time Bernardo de Aldrete and Juan de la Puente were writing their respective treatises, such as Lope de Vega's hagiographical poem *Isidro* (1599): "No venga fauno ni dría, / ni el Pan del arcadio suelo, / sólo ayuden a mi celo / la cristífera María, / y el Pan que bajó del Cielo. / Para hablar de un labrador, / éste es Pan, Dios y Pastor" (168) [Do not come, neither faun nor driad, nor Pan from Arcadian land; only the Christ-bearing Mary and the Pan (bread) that came from heaven may help me in my aim. To talk about a peasant, this one is Pan (bread), God, and Shepherd]. In Felipe Godínez's *Pastores de Belén*, the angel calls the shepherds to worship Christ: "Llegad, pastores dichosos, / adorareis al Dios Pan, / que Pan es el Dios de los pastores" (139r) [Come, blessed shepherds, and you will worship the god Pan, because Pan is the god of shepherds]. Juan de Barrionuevo y Moya's *Soledad entretenida* (1638) refers to "no Pan el de las fabulas, sino el verdadero Dios de los pastores" (112v) [not the God Pan of fictions, but the true God of shepherds]. See also Abel Alves (78–80), who reads the importance of Pan as an "allegory of the dominion of noble Spanish lords over people and other animals alike" (80).

64 Benito de Peñalosa y Mondragón (3v–5v), reproduces almost verbatim Juan de la Puente's passage in his *Cinco excellencias del Español* (1629). Caxa de Leruela's *Restauración de la abundancia de España* (1631) uses the Christological reading of Pan in order to defend the privileges of livestock owners (20–1).

65 I borrow the term from Maria Coroucli's "l'ancêtre absolu," which she uses to analyse how the figure of the shepherd serves as the search for the pure national ancestor in the Balkans (28). Although she analyses a different geographical area and historical period, her research parallels the claims made in the present book.

66 The interpretation that the Renaissance means the restoration of the Graeco-Latin past is not necessarily true in the early modern period, not even in an

aspect that seems to be settled now: linguistic evolution. Dante Alighieri's *De vulgari eloquentia*, written by the beginning of the fourteenth century, is arguably the first treatise to state that Romance languages derive from Latin (Eco 40). In Spain, authors like Nebrija, Valdés, or Aldrete defended the idea that Castilian derives from Latin, but their opinion was not shared by all of their contemporaries. Gregorio López Madera's *Discursos de la certidumbre de las reliquias descubiertas en Granada* (1595) and Francisco Bermúdez de Pedraza's *Antigüedad y excelencia de Granada* (1608) held the opposite view, that Castilian predated Latin (Binotti 76–86), and they were followed by authors like Correas, Jiménez Patón and Quevedo (Binotti 109–23). The reluctance to admit the precedence of Latin over Romance languages is common throughout early modern Europe (Eco 88–93). Specific to early modern Spain are the anxieties about dealing with the Roman past, coupled with no less stringent debates over the Hebrew and Arabic influence in Spanish culture.

4. The Moor in Arcadia

1 There are other versions of the story of *El Abencerraje*, all of them published around the same time. The one in the *Crónica del ínclito Infante don Fernando, que ganó a Antequera* (1561) is dedicated to Hierónimo Ximénez Dembún, lord of Moriscos in Aragon. A third version was published in Antonio de Villegas's compilation *Inventario* (1565); Villegas states that he obtained the licence to print the work in 1551, although it is unknown whether this previous version included the text of the *Abencerraje* or not. Villegas's version is generally favoured by modern editors of *El Abencerraje*. For an analysis of the different versions, see Francisco López Estrada (*El Abencerraje* 15–78).

2 For the opinion that the tale of *El Abencerraje* was used to defend the Moriscos or even the *conversos*, see María Soledad Carrasco Urgoiti (*The Moorish Novel* 60–72), André Stoll (156–64), George A. Shipley, Colbert I. Nepaulsingh (83–101), and Luis F. Avilés (470–1). Barbara Fuchs reads the novella as a complicated literary memorialization of the Moor (*Exotic Nation* 36–45). For Walter Holzinger, the text may be used to represent an Erasmian position (232–6). More in line with the very persuasive Foucaultian reading of Israel Burshatin, Laura Bass suggests that the homosocial bonds woven by Narváez around the Moorish couple is a metaphor for the repression of the Other, while Mary Gaylord ("Spain's" 125–30) sees in it a trope of the *reconquest*.

3 On the identification between Felismena and the tale of *El Abencerraje*, see Rosilie Hernández-Pecoraro ("Jarifa's Choice" 442–5) and Julio Baena (*Discordancias* 10–16). Begoña Souviron López ultimately suggests that it is

Montemayor himself who hides behind the fictitious character of Felismena ("*El Abencerraje*" 13).

4 Anton Guilio Brignole Sale's *Istoria Spagnuola* (1640–1) gave the name Felismena to one of the Moorish women (Carrasco Urgoiti, *El moro* 101), which may indicate that early modern readers perceived some sort of continuity between Felismena and the characters in her tale.

5 See Constance Rose (66–8) for the ambigous religious and sentimental meaning of "fe" (faith) in Spanish literature.

6 Similarly, when Sylvano was in despair for Diana's disdain for him, he "cometía cualquiera imaginación que en daño de su fe le sobreviniese" (119) [committed any act of imagination that occurred to him to damage his faith].

7 See Paloma Díaz-Mas (35–71) for an overview of Sephardic emigrations and L.P. Harvey (48–9) for emigration of Moriscos to North Africa in the sixteenth century.

8 Even though Bruno Damiani suggests that the idealization of moral virtues of the shepherds may be related to the identification of the figure of the shepherd with Old Christians (107), there are scattered comments in *La Diana* that seem to indicate that it is not a defence of the statutes of blood purity. Some scholars have suggested that Montemayor was of *converso* origin, such as Américo Castro ("Lo hispánico" 57–63), Marcel Bataillon (39–40), and Nepaulsingh (103–21). While I do not dispute these biographical interpretations, the question remains open to debate.

9 For the perception of Spain as a hybrid country in Europe, see J.N. Hillgarth (160–240) and Fuchs (*Exotic Nation* 115–38).

10 When Nicolas Colin published his French translation of *La Diana* in 1578, he based it on one of the earlier Spanish editions that did not include *El Abencerraje*. However, when Colin's translation was again reprinted in 1592, *El Abencerraje* was added to it (Fosalba, *La Diana* 229–30).

11 One of the versions of *El Abencerraje* was preserved because it was bound by the owner with one copy of *La Diana* (Cuenca, 1561) that lacked the Moorish tale (López Estrada, *El Abencerraje* 18).

12 Trying to imitate the editorial success of the combined edition of *La Diana* and *El Abencerraje*, or perhaps as a result of the influence of the Moorish ballads which were so successful in the 1580s, a few pastoral romances included Moorish compositions as well. Jerónimo de Covarrubias Herrera's *Los cinco libros de la enamorada Elisea* (1594) includes four poetic compositions of Moorish theme in its fifth book (Castillo Martínez, "Huellas"). In the anonymous *La pastora de Mançanares* (c. 1620), the shepherd Pánfilo spies the Moor Amete Cegrí, who laments the love between his beloved Celinda and his competitor Muza (vv. 7609–800). In Bernardo de la Vega's *El pastor de Iberia*

(1591) there is a ballad describing the fights of Arias Gonzalo against the Moors that is reminiscent of the frontier ballads in which the prowess of the Muslim is recognized (105r). In Jerónimo de Tejeda's *Tercera Diana* the shepherd Disteo tells a story about the confrontation between the Zegris and the Abencerrajes that has no connection with the main plot (2:79–88). However, none of these compositions have the reach or the repercussion of the inclusion of *El Abencerraje* in *La Diana*, and they are always clearly separated by a difference which can be narrative, temporal, or both, without interfering with the actions of the shepherds. Furthermore, these moments of maurophilia are located only in marginal pastoral romances, which were never reprinted again and were not included in any canonical repertoire of Spanish literature.

13 Contrary to Menéndez Pelayo, Eugenia Fosalba (*El Abencerraje* 84–105) and José Navarro Gómez defend Montemayor's participation in the rewriting of this version of *El Abencerraje*, basing their analysis on the analogies of style and plot.

14 Enrique Moreno Báez (1955), Francisco López Estrada and María Teresa López García-Berdoy (1993), and Juan Montero (1996) have published the edition of Valencia which they considered to be the first, probably from 1559. Julian Arribas (1996) stands alone in trying to reconstruct a hypothetical archetype primarily on the basis of the 1560 Zaragoza edition.

15 The editors that follow the 1561 Barcelona edition are Miguel Teijeiro Fuentes (1991) and Asunción Rallo Grus (1991). López Estrada (1946) also edited this version, but included *El Abencerraje* as a footnote.

16 In the 1993 edition of López Estrada and López García-Berdoy, the authors include the *Abencerraje* in an appendix.

17 See Willard King (113–23 and 147–51) and Dominick Finello (*Evolution*) for the importance of the literary academies in the composition of several pastoral romances.

18 The attachment that the Moriscos felt toward Saint John was amply acknowledged by Christian authors, who construed it as a fake sign of Christianization. See Pedro Aznar Cardona's *Expulsión justificada de los Moriscos Españoles* (1612) (2:51r) and Blas Verdú's *Discurso de la expulsión de los moriscos de España* (1612) (134).

19 Even though Jews are scorned in the brief reference to them by being called "infames personas," it was widely known that Mercader descended from a family of *converso* origin (Mérimée xxiii–xxiv).

20 The only exception, besides *El prado de Valencia*, is in Gonzalo de Saavedra y Torreblanca's *Los pastores del Betis* (1633). In this pastoral romance, the shepherds appear playing the game of canes "cubriendo sus animosos, y sueltos cuerpos con niuelados corchos, que cubiertos de blancas pieles,

remedauan las Africanas adargas" (55) [covering their brave and agile bodies with smoother corks, which covered with white skins looked like African *adargas* (Moorish shields)]. Therefore, although the text acknowledges that they are playing the game of canes, it makes sheep herding the material basis for a pseudo-Moorish spectacle. On the complex ideological value of the literary representation of the games of canes as a way of underlining a shared material culture, see Fuchs (*Exotic* 88–114). For the presence of the game of canes in the pastoral romances, see also Cristina Castillo Martínez ("Pancracios" 58–60).

21 These were actually the colours that Mercader frequently wore in public celebrations. One account of the festivities of the marriage of Philip III describes him during the celebration of a joust: "saco en esta jornada por enpressa, en la simera de su ielmo, una media luna grande de plata blanca toda guarnescida alrededor della de hermossas plumas blancas y naranxadas, ciendo los paramentos del y del caballo de las mismas colores, con las guarnisciones de la espada y daga y correas del mismo color naranxado" (Gauna 1:392) [that day he took as insignia, on top of his helmet, one big half moon made of white silver, decorated with splendid white and orange feathers, being the robes of him and his horse of the same colours, and the sword was garnished with the same orange colour].

22 The scarcity of means in the fiction is even more ironic if we take into consideration that in the real world Mercader paid for the organization of two similar naumachias during the visit of Philip III to Valencia in 1599 (Gauna 1:108–9 and 1:271–6).

23 The attachment of Valencian nobility to Moorish cultural practices is not exclusive to Mercader. On the contrary, he is only reproducing a more general, local taste for Moorish garments and games of canes. In the homonymous play *El prado de Valencia* (c. 1590), by Francisco Agustín Tárrega, the characters explicitly reveal their fondness for the game of canes as opposed to other chivalric European jousts (140). One ballad that circulated in Valencia during the wedding of Philip III in 1599 praised Valencian women and knights by comparing them with the characters of *El Abencerraje* (Gauna 2:834). Also during the same event, Mercader's wife performed a Morisco dance for the king (Gauna 2:751).

24 See Mérimée (xv–xxi) for a history of the barony of Buñol.

25 See Carrasco Urgoiti for an overview of the solidarity between Moriscos and Christian lords (*El problema morisco* 25–33). According to Harvey, this solidarity was interrupted in the 1570s, at least in Aragon, where, during a series of peasant revolts, the Moriscos realized that their lords were no longer in a position to protect them (254–63).

26 See Damián Fonseca's *Relación de la expulsión de los moriscos del Reino de Valencia* (1612) (42), and Gaspar Aguilar's epic poem *Expulsión de los moros de España* (1610) (187–8). According to Jaime Bleda's *Defensio Fidei in causa neophytorum* (1610), Mercader participated as well in the repression of the Morisco uprising in Laguar (595).

27 James Casey (516–23) and Eugenio Ciscar Pallarés (169) have suggested an alternative historiographical account for the sudden collaboration of the Valencian nobility with the expulsion, arguing that they had achieved such an extent of debt that they applauded the measure as a dramatic economic shift that would save them against their creditors, many of them their own Morisco vassals. Francisco Márquez Villanueva rejects their hypothesis, arguing that Valencia had been militarily occupied and that they had no choice but to obey the royal decision (*El problema morisco* 253–6).

28 It is possible that Cervantes started writing this work during his stay in Italy. On the date of *La Galatea*'s composition, see Geoffrey Stagg.

29 The debate about how we should read the references to the Moriscos in Cervantes's oeuvre has generated abundant bibliography. Some scholars consider that Cervantes's apparent approval of the expulsion is marked by irony: Fuchs (*Passing* 9–20), Márquez Villanueva (*Moros*), Salvador J. Fajardo (321–2), Carrasco Urgoiti ("Musulmanes" 75), Baena ("Sintaxis" 517–22), William Childers (*Transnational Cervantes* 170–7), José Luis Abellán (301), and Richard Hitchcock. Other scholars believe that Cervantes's opinion on the subject was not different from that of his contemporaries who applauded the expulsion: René Quérillacq (97–8), José María Perceval (*Todos son uno* 200–4), Martín de Riquer (168–70), and Michel Moner (98–100).

30 As Alain Milhou points out, the myth of the loss of Spain is characterized by its flexibility to serve a wide array of political positions ("De Rodrigue" 15). Thus, while it was mostly used to justify the repression of Moriscos and *conversos*, texts such as Fray Luis de León's "La profecía del Tajo" rewrote it to criticize the ethnocentric policy of Philip II (Márquez Villanueva, "Trasfondos" 182–94). The story of King Rodrigo is used by Miguel de Luna to defend the inclusion of the Moriscos in the national identity in his *Historia verdadera del rey don Rodrigo* (first part 1592; second part 1600) (Márquez Villanueva, *El problema morisco* 48–54; Drayson 47–58; Fuchs, *Mimesis* 111–13; Redondo, "Les divers visages" 362–4). For an overview of the treatment of King Rodrigo in Spanish literature and historiography, see the recent works by Patricia Grieve and Elizabeth Drayson.

31 Juan Luis Vives's *Instrucción de la mujer cristiana* (1528) also uses the case of Rodrigo to illustrate the undesirable effects of love (149). In Cristóbal Suárez de Figueroa's pastoral romance *La constante Amarilis* (1609) the shepherd

Coriolano writes a sonnet about the inconveniences of love by choosing the same topic, which provokes the community of shepherds to lament over the case (238).

32 For an overview of Cervantes's use of pastoral in his entire oeuvre, see Juan Bautista Avalle-Arce (*La novela pastoril* 229–63), Finello (*Pastoral*), and López Estrada ("Pastores en el *Quijote*").

33 In a different but complementary line of inquiry, Gaylord reads the pastoral element in *Don Quixote* as a projection of the colonial enterprise in the New World ("La Arcadia" 68).

34 In *La Galatea*, the shepherds play their instruments, including the *albogues*, to celebrate the announcement of the wedding between Daranio and Silveria (328–9), and in *Persiles* (1617) there is no trace of this definition. The *albogues* had already been referenced in *Don Quixote* when the characters listen to the music that comes from Camacho's wedding party (II, 19, 724). Therefore, we should exculpate Cervantes of the misleading definition, or rather, we should conclude that Don Quixote's "mistake" is intentional. For a recent study of Cervantes's musical sources and motifs, see Juan José Pastor Comín.

35 The *albogues* are mentioned in Gaspar Gil Polo's *Diana enamorada* (1564) (223); Bernardo González de Bobadilla's *Ninfas y pastores de Henares* (1583) (81v); Luis Gálvez de Montalvo's *El pastor de Fílida* (1582) (580); Juan de Arce Solorzeno's *Tragedias de amor* (1607) (3v); Gabriel del Corral's *La Cintia de Aranjuez* (1629) (35); and Pedro de Castro y Añaya's *Auroras de Diana* (1632) (63). The first three of these pastoral romances are present in Don Quixote's library (I, 6). Charlotte Stern ("Genesis" 425) records several references to the *albogues* in the early Castilian drama.

36 Conversely, the *albogues* are never mentioned in early modern descriptions of Morisco culture. In Juan Rufo's epic poem *La Austríada* (1584) the Moriscos of the Alpujarras are described heading to battle "al son de gaitas y de tamborinos" (14) [to the sound of bagpipes and tambourines]. According to Aznar Cardona, the typical instruments of the Moriscos were "gaytas, sonajas, adufes" (2:34v) [bagpipes, jingles, and Moorish tambourines]. One anonymous Morisco describes the musical instruments allowed in weddings without mentioning the *albogues* (*Tratado de los dos caminos* 289). Thus, with the exception of Covarrubias, neither Muslim nor Christian authors attribute to the Moriscos the use of the *albogues*.

37 As Pastor Comín has kindly pointed out in personal communication. For the artificiality of a strict differentiation between "Moorish" and "Christian" music in early modern Spain, see also Irigoyen-García ("La música" 46–50).

38 Covarrubias defines *marrano* as "el rezién convertido al christianismo" (791) [the recent convert to Christianity]. Although the term *marrano* is usually

applied to *conversos* of Jewish origin, in early modern Spain the term could refer as well to the Moriscos.

39 When Dorotea recounts her life in the first part of *Don Quixote*, emphasizing the blood purity of her parents, she exemplifies how connected the musical and racial vocabularies are: "son labradores, gente llana, sin mezcla de alguna raza mal sonante, y, como suele decirse, cristianos viejos ranciosos" (I: 28, 302) [they're farmers, simple folk, of pure blood unmixed with that of any ill-sounding races and, as it's often put, dyed-in-the-wool old Christians] (trans. Rutherford 250).

40 Carroll B. Johnson, in his own study of Cide Hamete as author of romances of chivalry, seems to question Walter's opinion, observing that Cide Hamete Benengeli had never been mentioned as an author of pastoral romances (*Transliterating* 220). He was not mentioned as an author of romances of chivalry either, which seems to be the underlying assumption of his book project – it should be noted that Johnson's comment is part of his unedited notes published posthumously, and therefore, he unfortunately did not have the chance to develop his interpretation further.

41 This would not be the first time that Cervantes reworks the Moorish and the pastoral themes at the narrative level. As Baena points out, there is a structural "mistake" at Juan Palomeque's inn in the first part of *Don Quixote*, in which characters have dinner twice within the same day, and which parallels an identical double dinner in Montemayor's *La Diana*. The detail that further suggests that Cervantes's "mistake" is motivated as a rewriting of *La Diana*, as Baena further notes, is that in both cases the double dinner frames the recounting of a Moorish tale – the *Abencerraje* in *La Diana* and the story of the captive in the first part of *Don Quixote* (*Discordancias* 4–10).

42 Michael Armstrong-Roche also points to the connection between the Visigoth myth and the statutes of blood purity in this description of Toledo (295–9).

5. Imagining the Spanish Arcadia after 1609

1 The expulsion of the Jews in 1492 was certainly a precedent as well, although in this case those who converted to Christianity could stay. It should be noted, however, that the expulsión of the *conversos* from Jewish origin was proposed and considered by the monarchy between 1597 and 1633, as Juan Ignacio Pulido Serrano has amply documented.

2 Other versions "a lo divino" were Ponce's *Primera parte de la clara Diana a lo divino* (1580), Bramón's *Los sirgueros de la Virgen sin pecado original* (1620), and Ana Francisca Abarca de Bolea's *Vigilia y octavario de San Juan Baptista* (1679). For an overview of this sub-genre, see Juan Bautista

Avalle-Arce (*La novela pastoril* 265–74) and Francisco López Estrada ("Los libros de pastores" 193–9); for an analysis of Ponce's *La clara Diana*, see López Estrada ("Los libros de pastores" 196–9) and Juan Montero.

3 On the biblical sources of Lope's *Pastores de Belén*, see María del Mar Cortés (193–7) and Antonio Carreño ("Introducción" 24–45).

4 *Pastores de Belén* was censored in modern times, when the 1941 children's edition expunged precisely the same episodes as the owners of the 1612 edition (Raynié 208–9).

5 Similar passages abound: "ya, pastores míos, habéis de entender una Iglesia y agregación de fieles y justos, y Cristo un Redentor, no de la terrena cautividad, sino de la espiritual redención" (418) [my dear shepherds, you must understand now that there is one Church and gathering of believers and the righteous, and Christ is the Redeemer, not of the earthly captivity, but of the spiritual redemption]; "Si salvarse Israel y Judá tiene / tan alta, tan divina profecía, / ya no a David nuestro pastor conviene, / sino al Hijo divino de María" (315) [If Israel wants to be saved and Judea has such divine prophecy, our shepherd is not David any longer, but the divine son of Mary]. As Françoise Lavocat points out, Lope's biblical shepherds are the guarantors of orthodoxy in prophesying the coming of Christianity (*Arcadies* 358).

6 Lope had already raised this issue in a dedicatory poem to Mateo Alemán's *San Antonio de Padua* (1604), stating that the work of this well known descendant of *conversos* "[c]onfirma la hidalguía / de Christo, por la parte de María" (qtd. in Pamp 82–5) [proves the *hidalguía* (nobility) of Christ, from Mary's side].

7 It is not until the second half of the sixteenth century that blood purity becomes part of the criteria for the concession of *hidalguía*, but many individuals were successful in obtaining it even when there was evidence that they were descendants of *conversos* or Moriscos (Fayard and Gerbet 64–6; Childers, "Disappearing Moriscos").

8 On the link between blood purity and the Spanish fixation on the notion of the Immaculate Conception see Henri Méchoulan (124–7) and Estrella Ruiz-Gálvez Prieto (216–24). In Godínez's *Coloquio primero de los Pastores de Belén* one of the shepherds gives Christ a newborn lamb, replicating the connection between blood purity and the whiteness of sheep: "Y yo al Niño un recental, / que porque no tiene mancha, / bien sé que le agradará" (140r–v) [And I (give) the Child a newborn lamb, and because it has no spot, I know for sure that he will like it].

9 Even Elaine Wertheimer, who supports the interpretation that the rustic characters of the early Castilian drama are despicable characters created by

converso playwrights who aimed to make fun of them, acknowledges that in some cases the dramatic shepherd may represent the Jew (113 and 147–57).

10 Travel accounts to the Holy Land evoked in early modern Bethlehem the medieval *convivencia* that was just on the brink of disappearance in the Iberian Peninsula by their own times, a place where Christians, Muslims, and Jews shared the same space; see Juan del Encina's *Trivagia* (1521) (Márquez Villanueva, "La *Trivagia*" 485), and Fray Diego de Mérida's *Viaje a Oriente* (1512), which, after acknowledging the cohabitation of Muslims, Christians, and Jews, depicts an encounter with some of the Sephardic Jews that were expelled from Spain (138).

11 According to Francisco Núñez Muley's *Memorandum* (1567), written to complain about Philip II's decree banning Morisco cultural practices, the prohibition of using Morisco names aimed at an effacement of genealogical memory similar to Herod's intention in *Pastores de Belén*: "pues que lo que toca en el sobrenonbre morisco, como se an de conoçer la gente y tratar con los sobrenonbres castellanos, e no mas perderse an las personas y los linajes moriscos, y no sabran con quien tratan ni conpran ni casan no conoçiendo el linaxe de rrayz" (229) [with respect to Morisco surnames: how are we supposed to know one another if we only make use of Castilian surnames? The people will know nothing of the person with whom they are speaking, from whom they are purchasing, and with whom they are marrying, given that they have no knowledge of his or her lineage] (trans. Barletta 88).

12 However, recalling the cases of ethnic engineering in the Bible does not necessarily entail a critique of them. On the contrary, ancient cases can be instrumentalized to justify modern ones, as the Dominican Fray Blas Verdú does in his 1612 defence of the expulsion of the Moriscos: "Si Dauid por el poco respeto que le tuuo Nabal Carmelo, estaua y con razon, resuelto a passar a cuchillo todos los de su casa, porque no quedasse rastro de tan ruyn sangre y solar: por que no podra hazer otro tanto el Rey nuestro Señor, a lo menos desterrando los Moriscos de España?" (142r) [If David, because of the lack of respect that Nabal of Carmel showed to him, resolved with a good reason to behead all his lineage so there would be no trace of such wicked stock and place, why should not the King, our Lord, do the same, or at least expel the Moriscos from Spain?].

13 On the other hand, Lope reproduces the biblical racial classification taken from Genesis 9:25. As told in *Pastores de Belén*, after the deluge Noah distributed the world among his sons, assigning Asia to Sem, Africa to Cam, and Europe to Japhet (459). Cam was then condemned to serve his two brothers for seeing Noah's nudity (552). The equation of the three sons of

Noah with the racial classification of the people in the world was established in Europe and its colonies to justify the subjection of Black people when the slavery trade in sub-Saharan Africa was developed in the fifteenth century (Braude 127–9).

14 About the Ethiopians: "particularmente nos fue de sumo contento ver los Etíopes, sus bailes, sus canciones y sus fervorosos deseos" (536) [it especially caused us great happiness to see the Ethiopians, their dances, their songs, and their fervent wishes]. The Ethiopians stand for Black people, who constituted a very important community in early modern Spain, both as slaves and freemen (Martín Casares, Blumenthal). Because phenotype was secondary when compared to the discourse of blood purity and because phenotypic difference was very rarely related to discourses of sheep herding and the pastoral community, I have barely touched their case when discussing the Spanish racial system, but will return to this issue in the conclusion.

15 For example, Huarte de San Juan states in his *Examen de ingenios para las ciencias* (1575) that "con haber más de doscientos [años] que vinieron de Egipto a España los primeros gitanos, no han podido perder sus descendientes la delicadeza de ingenio y solercia que sacaron sus padres de Egipto" (523) [even though the first Gypsies came from Egypt to Spain more than two hundred years ago, their descendants could not lose the delicate inventiveness and wit that their parents brought from Egypt]. On Gypsies in early modern Spain, see María-Helena Sánchez Ortega, Richard Pym, Lou Charnon-Deutsch, and George Mariscal (11–13).

16 "Una rica danza ordenan / de ricas ropas vestida, / matizando aljófar y oro / por las labradas camisas. / Los tocados aderezan / de corales y amatistas, / de flores, de perlas y oro, / y cuentas de aguas marinas. / Con laúdes y salterios, / y con sonajas repican / adufes y cascabeles, / a cuyos sones relinchan" (522) [They organize an elegant dance, dressed in luxurious garments, pearls and gold shining in their embroidered shirts. They decorate their head dresses with corals and amethysts, golden and pearl flowers and aquamarine beads. They shriek with the sound of their lutes and psalteries, and the tambourines and rattles ring their jingle bells].

17 We should not idealize Lope's attitude towards Gypsies, since the image he offers of them in other works is rather negative (Charnon-Deutsch 21–2).

18 Neither were Gypsies, from our historical perspective; but as we have seen, the early modern historical vision conflated them with ancient Egyptians.

19 For other moments in Lope's oeuvre in which Muslims and Jews appear as interchangeable figures, see Diane Pamp (50). For a similar interpretation of Godínez's play *De buen moro buen cristiano*, see Carrasco Urgoiti ("*De buen moro*" 564–6).

20 For the propaganda campaign of the monarchy, see Márquez Villanueva (*El problema morisco* 98–195). As Trevor Dadson has demonstrated, the asserted effectiveness of the expulsion was itself an act of propaganda, since many Moriscos, like Cervantes's Ricote, managed to stay or return to Spain.

21 About the same date, Lope wrote the play *La villana de Getafe* in order to contest the rumour that he was a Morisco (Márquez Villanueva, "Lope" 172–82).

22 The formulaic style of this epic poem sheds little light on *El premio de la constancia*. While there are abundant references to the frontier wars of the ancestors of Íñigo Manrique de Lara against North African and Iberian Muslims, there is no mention of the expulsion of the Moriscos.

23 See Avalle-Arce (*La novela pastoril* 195). Strikingly, Avalle-Arce denies that *El premio de la constancia* is a pastoral romance based on the geographical origin of Arsindo: "disfraz parcial del autor, es un hidalgo nacido en Manilva y educado en Ronda, ajeno por lo tanto, al mundo pastoril" (*La novela pastoril* 195) [partial disguise of the author, he is a noble born in Manilva and raised in Ronda, alien therefore to the pastoral world]. Avalle-Arce does not explain why living between Manilva and Ronda disqualifies the main character (and the entire fiction) as properly pastoral. One may wonder whether he would have said the same if the plot had been located on the banks of the river Tajo or in the Mantua Carpetanea.

24 From the descriptions within the fiction it seems that, when Arsindo states that he comes from Munda, he is actually referring to Ronda (even though there is a village named Monda within the same area): "De la Gentilica Munda, / origen noble y antiguo, / arruynados no del tiempo, / mas sublimados del mismo" (90v) [From pagan Munda, noble and ancient origin, not yet wrecked by time, but rather sublimated by it]. This is a common confusion at that time, as Covarrubias attests: "Ronda. Lugar de la Andaluzía; está puesta en un rivaço, cinco leguas de Córdova; conserva casi el nombre antiguo, por averse llamado Arrunda, *alias* Munda" (914) [Ronda. Place in Andalusia which stands on a rift, five (Spanish) leagues from Cordoba; it almost preserves its ancient name, because it was called Arrunda, otherwise Munda]. Behind this identification, we find the same polemics and uncertainties about recuperation of ancient place names analysed above. Guadix attributes to Ronda an Arabic etymology, disqualifying the identification between Ronda and Monda: "Ronda. Es *randa*, que, en arábigo, significa laurel ... Dize Abrahamo Ortelio que su nombre antiguo fue *Arunda*, aunque otros muchos authores dizen qu'este nombre, *Arunda*, es el nombre antiguo que tuvo una villa d'este mesmo obispado de Málaga, a que agora llaman *Monda*" (926) [Ronda. It is *randa*, which in Arabic means laurel. Abraham Ortelius says

that its ancient name was *Arunda*, although many other authors say that this name *Arunda* was the ancient name of a village within the same bishopric of Malaga which is now called *Monda*]. This polemic lasted until 1857, when the Real Academia de la Historia announced a competition to establish the location of the ancient Pompeyan city of Munda (Gascó 17).

25 The death of Alonso de Aguilar was memorialized in the "Coplas sobre lo acaescido en la Sierra Bermeja" (Benítez Sánchez-Blanco), which Pérez de Hita used to conclude the first part of *Guerras civiles de Granada* (1595).

26 For the frontier wars and the Morisco presence in Sierra Bermeja, see Joaquín Gil Sanjuan. Henri Lapeyre claims that Ronda was an exception to the expulsion of 1571 (180 and 204). On the Morisco population in Monda, see Alonso Franco Silva.

27 For López Estrada, the Andalusian location of *El premio de la constancia* determines the inclusion of the Moorish element ("Los libros de pastores" 174).

28 "La Mina" does exist and is included in almost every account of Ronda, such as Pedro de Medina's *Libro de grandezas y cosas memorables de España* (1545) (186) and Espinel's *Vida de Marcos de Obregón* (1618) (1:248–51). For a thorough study of La Mina and its legends, see Francisco Garrido Domínguez. The descent to the enchanted cave is a recurrent motif in the pastoral romances, as has been pointed out by John Cull (168–72), Pilar Berrio, and Frederick De Armas ("Caves"). Castillo Martínez compares this passage of *El premio de la constancia* with the episode of the cave of Montesinos in *Don Quixote* ("Cuevas" 472–3).

29 In other Spanish texts, the *morabitos* were the object of scorn. For Covarrubias: "Cerca de los árabes vale lo mesmo que en castellano llamamos ermitaño; éstos eran grandes vellacos hipocritones, exercitados en diversos linages de pecados" (814) [For the Arabs it is what we call hermit in Castilian; they were great liar hypocrites, used to different kinds of sins]. In the epic poem *Liga deshecha por la expulsión de los Moriscos de España* (1612), Juan Méndez de Vasconcelos describes one of these *morabitos* as: "Ministro falso de su vil Mahoma" (52r) [False minister of his vile Mohammed]. For the conception of the Moriscos as sorcerers, see Aznar Cardona (2:47v–9v).

30 Castillo Martínez ("*El premio*" 283), Berrio (322–4) and Avalle-Arce (*La novela pastoril* 194–5). Strikingly, Hugo Rennert does not even mention the story of King Zelimo in his synopsis of this pastoral romance (181–5).

31 In the case of the Castilian Moriscos, Lapeyre estimates that the crown gained about 300,000 ducats (195). Domínguez Ortiz and Vincent show that the final figure was probably much greater (284–5). Henry Lea estimates that the properties confiscated from the Moriscos were distributed to the Duke of Lerma and his family (372–3). For a recent study of how the seized properties of the Moriscos of La Mancha were distributed after the expulsion, see

Francisco Javier Moreno Díaz del Campo (338–49). Early modern Spaniards were also concerned about the conflict between the alleged religious justification of the expulsion and the economic benefit that some would take from it. Pedro de Valencia, knowing the arguments in preparation for the expulsion, already anticipated that "[s]i se les quitan las haciendas, infamase todo el hecho, como procedido de aquesta codicia, aunque se le de otro color" (111) [if their properties are taken from them, the whole act is vilified as proceeding from that greed, even though another face is put on it].

32 See Perceval (*Todos son uno* 266). On the fiscal burden endured by Iberian Muslims since the Middle Ages, and even after they had been converted to Christianity, see Serafín de Tapia.

33 The order to pay for their deportation is conceptualized as a royal grace: "Pero bien permito que puedan llevar el dinero que huvieren menester assí para el tránsito que han de hazer por tierra como para sus embarcaciones por mar" (qtd. in Harvey 404) [But I allow that they take the money that they will need both for their transit by land as well as for the ships at sea]. The tactics adopted to defray the costs of the deportation are more openly acknowledged in the instructions made by the Marquis del Carpio for the case of the Moriscos departing from Seville in 1610 (Harvey 406; Lapeyre 184–9). Jorge Gil Herrera shows that the Crown, in spite of the initial prohibition to take currency out of Spain, ended up negotiating with the Moriscos.

34 As Perceval puts it more blatantly, "los moriscos pagan por existir o por dejar de existir ... Son una mina inagotable que reproduce la riqueza de su diferencia" (*Todos son uno* 268) [the Moriscos pay for existing and for ceasing to exist ... They are the inexhaustible trove that reproduces the wealth of their difference].

35 The legends of treasures buried by the Moors are widespread in Spain (Barrios Aguilera; Provansal; and Gómez Vozmediano 100–3). Delpech studies the pervasive myth of the hidden treasure of the Moors as a collective exorcism of the expulsion of the Moriscos, whose fantasmatic presence is kept nonetheless in the collective memory as a relationship between the alive and the dead (568). See Castillo Martínez, who also follows Delpech's interpretation of the hidden treasure of the Moor to explain the signification of King Zelimo ("*El premio*" 280–2).

36 Juan Andrés's anti-Islamic treatise *Confusión o confutación de la secta mahomética y del Alcorán* (1515) describes a very similar space in his critique of the Islamic paradise: "cómo estarán los moros echados y recolzados encima de las camas y catifas y estrados, los quales estarán reyendo y holgando sin cuydado y sin tristeza" (189) [how the Moors will be frolicking and lying down on top of their carpets and platforms, laughing and idling without any concern nor sorrow]. On the "estrado" and how it was adopted by Christians, see Fuchs (*Exotic Nation* 14–15).

37 In spite of the benevolent character of his evangelization, the Archbishop of Granada, Hernando de Talavera, aggressively tried to eradicate the Morisco custom of eating on the floor: "Luego que se convirtieron, convidava a los prinçipales para que aprendiesen las maneras de los xriptianos en el comer, i en las viandas, en todo lo al ... dioles mesas e manteles para en que comiesen porque no comiesen en el suelo" (*Breve suma* 151r, qtd. in Márquez Villanueva, *Investigaciones* 118n45) [After they converted, he invited the elite among them so they could learn the customs of Christians in their meals, in their foods, and the like ... he gave them tables and table cloths so they did not eat on the floor]. Bernardo Pérez de Chinchón's anti-Islamic treatise *Antialcorano* (1532) states that Christians show "en el comer más criança, assentámonos en alto y no en el suelo como bestias" (381) [more civility when eating, sitting high up and not on the floor like animals]. Aznar Cardona, in his *Expulsión justificada de los Moriscos Españoles* (1612) describes it in even harsher terms: "Eran brutos en sus comidas, comiendo siempre en tierra ... sin mesa, ni otro aparejo que oliesse a personas" (2:33r) [They were brutes in their meals, eating always on the floor ... without table or any other utensils appropriate for human beings].

38 Similar pastoral moments are also found in Méndez de Vasconcelos's epic poem *Liga deshecha por la expulsión de los Moriscos* (1612), where the Iberian rivers and their nymphs celebrate the expulsion of the Moriscos along with the exploits of Spaniards in North Africa (177r–9r and 186r–v).

39 For an analysis of the tensions between the Petrarchan imagery and the stereotypes about the Moriscos in *Expulsión de los moros de España*, see Mar Martínez-Góngora.

40 According to Aguilar, the alfaquíes of Valencia complied with the decree of expulsion stating "hagamos luego / que vean nuestros hijos naturales / de África los desiertos arenales" (179) [let's make our natural offspring see the sandy deserts of Africa].

41 For the apologists of the expulsion there was no doubt that the Moriscos were from a different stock than the rest of the Spaniards; see Aznar Cardona (2:17r) and Verdú (144r). Even for a defender of the Moriscos such as Pedro de Valencia, their Spanishness is based on the long period of their stay, and not on their genealogical origins: "son españoles como los demás que habitan en España, pues ha casi novecientos años que nacen y se crían en ella" (76–81) [they are Spaniards like the rest who inhabit Spain, because for almost nine hundred years, they have been born and raised in it]. The opposite view was however no less abundant. In his sermons from the beginning of the sixteenth century, Martín García defended the need to convert the Moriscos, because they "[s]unt hyspani, non armenii nec africani ... Sunt compatriotes hic nati et nutriti et conservati inter nos" (qtd. in Cardaillac 327) [they are Spaniards,

not Armenians or Africans. They are compatriots born here and raised and preserved among us]. Pérez de Chinchón's *Antialcorano* (1532) asked the Moriscos to convert to Christianity appealing to their common lineage: "porque somos nascidos y criados en una mesma tierra que es españa ... si bien mirássedes el linage de donde venís todos soys medio christianos" (297) [because we are all born and raised in the same land, which is Spain ... if you consider the lineage you come from, you will see that all of you are half Christian]. Luis de la Cueva's *Diálogos de las cosas notables de Granada* (1603) held that "[l]os moriscos del Alpuxarra eran tenidos por decendientes de Christianos" (66) [the Moriscos of the Alpujarra were regarded as descendants of Christians]. Juan de la Puente, even if he defended the expulsion, considered the Moriscos to be "Moros en lo secreto, en lo publico Christianos, y Españoles en la sangre" (22) [Moors in secret, Christian in public, and Spaniards in their bloodline]. In June 1611, the Council of State pushed for further expulsion of Moriscos who had been previously exempted, warning however that investigations should not go back more than two hundred years, because otherwise "there would be very few in Spain who would not have some Moorish parentage" (qtd. in Tueller 195).

42 There is also a second part published in Valencia in 1644, which will not be taken into consideration in this book because it does not have any pastoral elements. All my references to *Soledad entretenida* are to the first part.

43 The influences of maurophile literature are evident. The passage in which the Moor Abdalá is asked by his beloved Xarifa not to pass by her house ("que no pasase por su calle" 204v) is citing the well known Moorish ballad "Mira, Zaide, que te aviso." The games of canes in the Alhambra (216r, 220v) are reminiscent of Ginés Pérez de Hita's first part of the *Guerras civiles de Granada* (1595). The names of the Moorish women Xarifa and Daraxa, as well as the sentimental motifs, seem to have been borrowed, respectively, from *El Abencerraje* and from Alemán's "Historia de Ozmín y Daraja," included in the first part of *Guzmán de Alfarache* (1599).

44 The only biographical information appears in the title-page, which states that Barrionuevo y Moya is a priest from Villanueva, in the bishopric of Andújar. See also Castilla de la Vega (1–7).

45 See Avalle-Arce (*La novela pastoril* 227), King (147), and Castillo Martínez ("Introducción" xiv). Laura Zorrilla, who is currently preparing a critical edition, argues that *Soledad entretenida* should be regarded as a "novela de aventuras" (romance of adventures) and not as a pastoral romance.

46 Although the parallelisms with Góngora's *Soledades* and its typical poetic expressions are soon abandoned in favour of the peripeteia of the Byzantine romance, it is telling that Calisandro encounters Moors where the anonymous character of the *Soledades* meets American natives. It is this pastoral

encounter with the Others of the Spanish empire that Barrionuevo y Moya replicates in his work.

47 The Muslim shepherd is nearly absent in early modern Spanish literature, appearing only in eclogue nine of Encina's *Translación de las bucólicas de Virgilio*, first printed in 1496 (*Obra* 283–91). A conflation similar to that of Barrionuevo y Moya's *Soledad entretenida* is found a few years later in Philippe Quinault's play *La généreuse ingratitude* (1654), which represents the characters of *El Abencerraje* living as shepherds during their exile in Algiers (Carrasco Urgoiti, *El moro* 114). For analysis of Quinault's pastoralization of the Iberian Moors as part of the imperial confrontation between France and Spain, see Irigoyen-García ("Les bergers").

48 This symbolic hierarchy of herdsmen was not exclusive to Spain. In England, the eclogue *Hermes and Lycaon* (c. 1603) stages a theological debate between Lycaon, a Catholic goatherd, and Hermes, a Protestant shepherd, in which the latter prevails (Cooper 378). As Helen Cooper points out, the different perception of sheep and goats (and therefore, of shepherds and goatherds) probably comes from the biblical prophecy of Matthew 25:31–41, which states that on Judgment Day Christ will separate the goats from the sheep (366).

49 This legend circulated in popular culture, which degraded the Moor in the social hierarchy by transforming him into an outlaw (Carrasco Urgoiti, "Reflejos" 206).

50 The linguistic ideology that equates every kind of herdsman with his own flock causes the Moors to be characterized as goats themselves. During one skirmish, Calisandro describes his foes as being "acostumbrados a trepar como cabras por aquellas sierras" (197r) [used to climbing those mountains like goats]. The references to the Moriscos as *cabras* and *cabrones* to stress their deviant sexuality are frequent. Thus Bleda states that the Moors who flew from Valencia to North Africa did so in order to live with "los libidinosos cabrones Mahometanos" (*Corónica* 650) [the lustful Muslim billy goats]. In Ponce's *Primera parte de la clara Diana a lo divino* (1580), the shepherd Pluto (who is actually the devil), predicts the coming of his reign as a metaphorical transformation of the human livestock: "Ya todos los corderos son cabritos / Y cabras son tornadas las ouejas" (18v) [All the lambs have become little goats now, and sheep are goats].

51 On the use of Arabic by Old Christians, see Bernard Vincent ("Reflexión documentada" 115–17).

52 About the Moriscos who stayed or returned to Spain, see Dadson. On the presence of Muslims during the seventeenth century, see Vincent ("Musulmanes").

53 The containment of so-called "Moorish romances" (novela morisca) is not limited to its opposition to the pastoral romances. As Diane Sieber

demonstrates, works such as Ginés Pérez de Hita's *Guerras civiles de Granada* are classified as fictions even though they should properly belong to historiography (293).

Conclusion: Pan's Labyrinth

1 I am grateful to Abel Alves for suggesting the analogy with Guillermo del Toro's film.
2 For a broader study of anti-semitic fossils in Spanish, see Christiane Stallaert (*Ni una gota* 92–8).
3 Carmen de Burgos likely made use of the Spanish translation made by Juan Valera in 1880.
4 Indeed, both sides of the Spanish Civil War resorted to the *reconquest* trope to justify their cause. It is important to note that Muslims also fought for the Republic, although in smaller numbers (Sánchez Ruano 255–90).
5 Spanish *piara* is most often used to refer to a herd of pigs, a lexical choice that emphasizes again that Moors do not belong to the same stock (or flock) of Spaniards.
6 For a thourough analysis of pastoral tradition in Miguel Hernández work, see William Rose.
7 The article was first published in *La Gaceta Literaria* on 15 January 1932. In 1985, Giménez Caballero reproduced the text in his *Retratos españoles* (209–12), adding that he met Miguel Hernández again in Orihuela soon afterwards.
8 In Spanish, "estar como una cabra" (literally "to be like a goat") means to be completely nuts, hence Giménez Caballero's pun.
9 According to Miguel Núñez's memoirs, Giménez Caballero visited Miguel Hernández in prison in 1940, offering his liberation if he recanted his political position (164). Miguel Hernández refused and died in prison in 1942, right before Giménez Caballero published his fascist treatise. Therefore, it is possible that Giménez Caballero was thinking about Miguel Hernández when he wrote this piece. But how? Was he trying to counteract Miguel Hernández's appropriation of the figure of the shepherd for the Republican cause, or was he rather trying to rehabilitate his name posthumously? It is telling that Giménez Caballero does not mention this third encounter with Miguel Hernández in his own memoirs (*Retratos españoles* 209–12).
10 The title of the poem was changed to "Años triunfales" [Triumphant Years] in subsequent compilations.
11 The form "cagañeroz" used by Olalla is part of her debased linguistic characterization. It is a scatological distortion of *cabañero*, a synonym of shepherd derived from the *cabañas* (huts) that shepherds used as temporary

shelters when grazing their livestock far away from home, overlaid with *cagar* (to defecate), plus the derogative suffix *-oz*.

12 Lope de Rueda repeats the opposition in a ludicrous encounter between a Black woman slave and a shepherd in his "Paso de Ysacaro y la negra" (271–5).

13 For a study of this genre, see Enrique Martínez López, Baltasar Fra Molinero, and John Beusterien.

14 On this novel, see José Luis Aranguren Egozkue (141–5) and Edward Friedman.

15 On Lupercio Latrás's conflict with the Moriscos in Aragon, see L.P. Harvey (254–63).

16 A literal translation of "O belén da caixa" would be "The Nativity representation of the savings bank." The *cajas de ahorros* were, until recently, financial institutions that were required to invest a great part of their dividends in social and cultural activities. Os Resentidos (1984–94) was one of the first rock bands to sing in Galician.

17 Nge's social stigmatization is relative. He is not admitted to the church, and his own uncle, who lives in the same house, does not talk to him because he is black. But Nge can exercise his right to vote in the elections that are celebrated to designate a new mayor for the village. Thus, while he suffers exclusion from traditional institutions such as the family and the church, he is integrated in the democratic structure of post-Francoist Spain.

18 I am thankful to M. Luján Stasevicius for pointing at Bernardo Atxaga's text.

19 See my "Identidad colectiva" and "Jerónimo de Arbolanche." I have also found the exaltation of Galician identity in Juan Arce Solorzeno's *Tragedias de amor* (1607).

20 Furthermore, Sannazaro already provided the combination of pastoral and anti-Islamic propaganda in two of his farces celebrating the fall of Granada in 1492. For Sannazaro's farces, see Danielle Boillet.

21 I have myself devoted some pages elsewhere to explore how pastoral ethnocentrism circulates between France and the Iberian Peninsula both in the Middle Ages and the early modern period ("Les bergers").

Works Cited

Abarca de Bolea, Ana Francisca. *Vigilia y octavario de San Juan Baptista.* Zaragoza: Pascual Bueno, 1679.

Abellán, José Luis. "Cervantes y el problema morisco." Bernat Vistarini 297–301.

El Abencerraje (Novela y romancero). Ed. Francisco López Estrada. Madrid: Cátedra, 1997.

Acuña, Hernando de. *Varias poesías.* Ed. Luis F. Díaz Larios. Madrid: Cátedra, 1982.

Agüera Ros, José Carlos. *Pintores y pintura del Barroco en Murcia.* Murcia: Tabularium, 2002.

Aguilar, Gaspar. *Expulsión de los moros de España.* Ed. Manuel Ruiz Lagos. Alcalá de Guadaira: Guadalmena, 1999.

Aguirre de Cárcer Casarrubios, Luisa F. "El elemento árabe en *Marcos de Obregón.*" *Bulletin Hispanique* 98.2 (1996): 397–418.

Agustín, Miguel. *Libro de los secretos de agricultura, casa de campo y pastoril.* Madrid: Julián de Paredes, 1695.

Alcalá y Yáñez, Jerónimo. *El donado hablador Alonso, mozo de muchos amos. La novela picaresca española.* Ed. Ángel Valbuena Prat. Madrid: Aguilar, 1956. 1148–219.

Aldrete, Bernardo de. *Del origen y principio de la lengua castellana.* Hildesheim: Georg Olms Verlag, 1970.

Alemán, Mateo. *Guzmán de Alfarache.* Ed. José María Micó. 2 vols. Madrid: Cátedra, 1997.

Alenda y Mira, Jenaro. *Relaciones de solemnidades y fiestas públicas de España.* 2 vols. Madrid: Rivadeneyra, 1903.

Alonso Cortés, Narciso. "Sobre Montemayor y la Diana." *Artículos histórico-literarios.* Valladolid: Imprenta Castellana, 1935. 127–40.

Alonso de Herrera, Gabriel. *Obra de agricultura.* Ed. José Urbano Martínez Carreras. Biblioteca de Autores Españoles 235. Madrid: Atlas, 1970.

Alonso Rey, María Dolores. "Pastores en los libros de emblemas españoles." *Imago: Revista de Emblemática y Cultura Visual* 1.1 (2009): 27–36.
Althusser, Louis. *Lenin and Philosophy and Other Essays*. New York: Monthly Review Press, 2001.
Álvarez Martí-Aguilar, Manuel. "Modelos historiográficos e imágenes de la antigüedad: *El cerco de Numancia* de Miguel de Cervantes y la historiografía sobre la España antigua en el siglo XVI." *Hispania Antiqua* 21 (1997): 545–70.
Álvarez Ossorio-Alvariño, Antonio, and Bernardo J. García García, eds. *La monarquía de las naciones: Patria, nación y naturaleza en la Monarquía de España*. Madrid: Fundación Carlos de Amberes, 2004.
Alves, Abel. A. *The Animals of Spain: An Introduction to Imperial Perceptions and Human Interaction with Other Animals, 1492–1826*. Brill: Leiden, 2011.
Amanece que no es poco. Dir. José Luis Cuerda. Compañía de Aventuras Comerciales / RTVE, 1988.
Anderson, Benedict. *Imagined Communities: Reflections on the Origin and Spread of Nationalism*. London: Verso, 1991.
Anderson, Ruth Matilda. *Hispanic Costume 1480–1530*. New York: Hispanic Society of America, 1979.
Andrés, Juan. *Confusión o confutación de la secta mahomética y del Alcorán*. Ed. Elisa Ruiz García and María Isabel García-Monge. Mérida: Editora Regional de Extremadura, 2003.
Anglería, Pedro Mártir de [Pietro Martire d'Anghiera]. *Documentos inéditos para la historia de España*. Vol. 10, *Epistolario de Pedro Mártir de Anglería, II Libros XV–XXIV, Epístolas 232–472*. Ed. José López de Toro. Madrid: Góngora, 1955.
Angulo Íñiguez, Diego, and Alfonso E. Pérez Sánchez. *Historia de la pintura española: Escuela toledana de la primera mitad del siglo XVII*. Madrid: CSIC, 1972.
Aranguren Egozkue, José Luis. *Camón Aznar, literato*. Zaragoza: Museo e Instituto de Humanidades "Camón Aznar," 1989.
Arbolanche, Jerónimo. *Las Abidas*. Ed. F. González Ollé. 2 vols. Madrid: CSIC, 1972.
Arce Solorzeno, Juan de. *Tragedias de amor, de gustoso y apacible entretenimiento de historias, fabulas, enredadas marañas, cantares, bayles, ingeniosas moralidades del enamorado Acrisio, y su zagala Lucidora*. Madrid: Juan de la Cuesta, 1607.
Arco y Garay, Ricardo del. "La música y la danza en la obra de Cervantes." *Revista de ideas estéticas* 9 (1951): 253–70.
Argote de Molina, Gonzalo. *Discurso sobre la poesía castellana. La retórica en España*. Ed. Elena Casas. Madrid: Editora Nacional, 1980. 201–16.
– *Nobleza del Andaluzia*. Seville: Fernando Díaz, 1588.

Arié, Rachel. "Acerca del traje musulmán en España desde la caída de Granada hasta la expulsión de los moriscos." *Études sur la civilisation d'Espagne musulmane*. Leiden: Brill, 1990. 121–41.

Armstrong-Roche, Michael. *Cervantes' Epic Novel: Empire, Religion, and the Dream Life of Heroes in* Persiles. Toronto: U of Toronto P, 2009.

Arredondo, María Soledad. "Las críticas a los libros de pastores: De la ironía a la parodia." *Dicenda* 6 (1987): 349–58.

Atxaga, Bernardo. *Obabakoak*. Trans. Bernardo Atxaga. Madrid: Punto de Lectura, 2008.

Avalle-Arce, Juan Bautista. *La novela pastoril española*. Madrid: Istmo, 1974.

– *Las novelas y sus narradores*. Alcalá de Henares: Centro de Estudios Cervantinos, 2006.

Ávila, Juan de. *Obras completas*. Ed. Luis Sala Balust. 2 vols. Madrid: Editorial Católica, 1953.

Avilés, Luis F. "Los suspiros del *Abencerraje*." *Hispanic Review* 71.4 (2003): 453–72.

Aznar Cardona, Pedro. *Expulsión justificada de los Moriscos Españoles*. Huesca: Pedro Cabarte, 1612.

Baena, Julio. "De Belisa, su vaca y su becerro: Poesía como mercancía en Montemayor." *Calíope* 9.2 (2003): 75–91.

– *Discordancias cervantinas*. Newark: Juan de la Cuesta, 2003.

– "Sintaxis de la ética del texto: Ricote, en el *Quijote II*, la lengua de las mariposas." *Bulletin of Spanish Studies* 83.4 (2006): 507–24.

Bajo Pérez, Elena, and Felipe Maíllo Salgado. "Estudio introductorio." Guadix 13–131.

Balbuena, Bernardo de. *El Siglo de Oro en las selvas de Erífile*. Ed. José Carlos González Boixo. Xalapa: Universidad Veracruzana, 1989.

Balibar, Étienne, and Immanuel Wallerstein. *Race, Nation, Class: Ambiguous Identities*. Trans. Chris Turner. New York: Verso, 1991.

Ballester Rodríguez, Mateo. *La identidad española en la Edad Moderna (1556–1665): Discursos, símbolos y mitos*. Madrid: Tecnos, 2010.

Barrientos, Fray Lope de. *Vida y obras de Fr. Lope de Barrientos*. Ed. Luis G.A. Getino. Salamanca: Manuel Criado, 1927.

Barrionuevo y Moya, Juan de. *Primera parte de la soledad entretenida*. Écija: Luis Estupiñán, 1638.

Barrios Aguilera, Manuel. "Tesoros moriscos y picaresca." *Espacio, Tiempo y Forma* 9 (1996): 11–24.

Bartlett, Robert. "Medieval and Modern Concepts of Race and Ethnicity." *Journal of Medieval and Early Modern Studies* 31.1 (2001): 39–56.

Bass, Laura R. "Homosocial Bonds and Desire in the *Abencerraje*." *Revista Canadiense de Estudios Hispánicos* 24.3 (2000): 453–70.

Bataillon, Marcel. *Varia lección de clásicos españoles*. Madrid: Gredos, 1964.
Batlle Matabosch, Ramón. "La oveja merina." *La estética del vestir clásico: Antología del vestido, compuesta a la mayor gloria gremial de los tejidos españoles*. Ed. Pedro Roca Piñol. Tarrasa: Yuste, 1942. 41–4.
Bayo, Marcial José. *Virgilio y la pastoral española del Renacimiento (1480–1530)*. Madrid: Gredos, 1959.
Benítez Sánchez-Blanco, Rafael. "Lectura de las Coplas de Sierra Bermeja." *Revista de Literatura* 35.71–2 (1969): 73–90.
Bennassar, Bartolomé, and Lucile Bennassar. *Les chrétiens d'Allah*. Paris: Perrin, 1989.
Berco, Cristian. *Sexual Hierarchies, Public Status: Men, Sodomy, and Society in Spain's Golden Age*. Toronto: U of Toronto P, 2007.
Berdini, Paolo. *The Religious Art of Jacopo Bassano*. Cambridge: Cambridge UP, 1997.
Bernáldez, Andrés. *Memorias del reinado de los Reyes Católicos*. Ed. Manuel Gómez-Moreno and Juan de Mata Carriazo. Madrid: Real Academia de la Historia, 1962.
Bernat Vistarini, Antonio, ed. *Volver a Cervantes: Actas del IV Congreso Internacional de la Asociación de Cervantistas*. Palma: Universitat de les Illes Balears, 2001.
Bernis, Carmen. *Trajes y modas en la España de los Reyes Católicos*. Vol. 2, *Los hombres*. Madrid: CSIC, 1979.
– *El traje y los tipos sociales en* El Quijote. Madrid: El Viso, 2001.
Berrio, Pilar. "El viaje imaginario en la novela pastoril española." *Actas del IX Simposio de la Sociedad Española de Literatura General y Comparada*. Ed. Túa Blesa. 2 vols. Zaragoza: Universidad de Zaragoza, 1994. 1:315–25.
Beusterien, John. *An Eye on Race: Perspectives from Theater in Imperial Spain*. Lewisburg: Bucknell UP, 2006.
Beuter, Pere Antoni. *Crònica*. Ed. Enric Iborra. Valencia: Institució "Alfons el Magnànim," 1982.
Beverley, John. "Class or Caste: A Critique of the Castro Thesis." *Américo Castro: The Impact of His Thought*. Ed. Ronald E. Surtz, Daniel P. Testa, and Jaime Ferrán. Madison: Hispanic Seminary of Medieval Studies, 1988. 141–9.
– "Humanism, Colonialism and the Formation of the Ideology of the Literary: On Garcilaso's Sonnet 23: 'En tanto que de rosa y azucena.'" *New Hispanisms: Literature, Culture, Theory*. Ed. Mark I. Millington and Paul Julian Smith. Ottawa: Dovehouse, 1994. 53–68.
Bible, Douay-Rheims 1899 American edition. Bible Gateway, n.d. Web. <www.biblegateway.com>.

La Biblia, que es, los sacros libros del viejo y nuevo testamento. Trans. Casiodoro de Reina. Basilea, 1569. Biblioteca Online Classica Digitalia, Universidade de Coimbra, n.d. Web. <http://bdigital.sib.uc.pt>.

Binotti, Lucia. *La teoría del "Castellano Primitivo": Nacionalismo y reflexión lingüística en el Renacimiento español*. Münster: Nodus, 1995.

Birriel Salcedo, Margarita. "Datos sobre los oficios de los moriscos en la costa de Granada." *Actes du IV Symposium International d'Études Morisques sur: Métiers, vie religieuse et problematiques d'histoire morisque*. Ed. Abdeljelil Temimi. Zaghouan: Centre d'Études et de Recherches Ottomanes, Morisques, de Documentation et d'Information, 1990. 43–9.

Bishko, Charles Julian. "The Castilian as Plainsman: The Medieval Ranching Frontier in La Mancha and Extremadura." *The New World Looks at Its History*. Ed. Archibald R. Lewis and Thomas F. McGann. Austin: U of Texas P, 1963. 47–69.

Black, Jeremy. *Maps and Politics*. Chicago: U of Chicago P, 1997.

Bleda, Jaime. *Corónica de los moros de España*. Ed. facsimile Bernard Vincent and Rafael Benítez Sánchez-Blanco, Valencia: Universitat de València, 2001.

– *Defensio Fidei in causa neophytorum, siue Morischorum Regni Valentiae (la parte en castellano)*. Ed. facsimile. Valencia: Paris-Valencia, 1980.

– *Vida y milagros del glorioso San Isidro el Labrador*. Madrid: Tomás de Junta, 1622.

Blumenthal, Debra. "'La Casa dels Negres': Black African Solidarity in Late Medieval Valencia." Earle and Lowe 225–46.

Boillet, Danielle. "La participation d'un humaniste aux spectacles de la cour des aragonais de Naples: Les farces de J. Sannazaro." *La fête et l'Écriture: Théâtre de cour, cour-théâtre en Espagne et en Italie, 1450–1530*. Aix-en-Provence: Université de Provence, 1987. 233–56.

Boronat y Barrachina, Pascual. *Los moriscos españoles y su expulsión*. 2 vols. Valencia: Francisco Vives y Mora, 1901.

Borromeo, Federico. *Della pittura sacra*. Ed. Barbara Agosti. Pisa: Scuola Normale Superiore di Pisa, 1994.

Boscán, Juan. *Obra completa*. Ed. Carlos Clavería. Madrid: Cátedra, 1999.

Botelho de Carvalho, Miguel. *Prosas y versos del pastor de Clenarda*. Madrid: Viuda de Fernando Correa Montenegro, 1622.

Bourdieu, Pierre. *The Logic of Practice*. Trans. Richard Nice. Stanford: Stanford UP, 1990.

Bramón, Francisco. *Los sirgueros de la Virgen sin original peccado*. Mexico, 1620.

Braude, Benjamin. "The Sons of Noah and the Construction of Ethnic and Geographical Identities in the Medieval and Early Modern Periods." *William and Mary Quarterly* 54.1 (1997): 103–42.

Brothers, Cammy. "The Renaissance Reception of the Alhambra: The Letters of Andrea Navagero and the Palace of Charles V." *Muqarnas* 11 (1994): 79–102.

Brown, Jonathan. "Mecenas y coleccionistas españoles de Juseppe de Ribera." *Goya* 183 (1984): 140–50.

Bullón y Fernández, Eloy. *Miguel Servet y la geografía del renacimiento*. Madrid: CSIC, 1945.

Burgos, Carmen de. *La flor en la playa y otras novelas cortas*. Ed. Concepción Núñez Rey. Madrid: Castalia, 1989.

Burke, Peter. *Popular Culture in Early Modern Europe*. Burlington: Ashgate, 2006.

Burshatin, Israel. "Power, Discourse and Metaphor in the *Abencerraje*." *MLN* 99 (1984): 195–213.

Butler, Judith. *The Psychic Life of Power: Theories in Subjection*. Stanford: Stanford UP, 1997.

Caballero López, José Antonio. "Annio de Viterbo y la historiografía española del siglo XVI." *Humanismo y tradición clásica en España y América*. Ed. Jesús María Nieto Ibáñez. León: Universidad de León, 2002. 102–20.

Cabanelas Rodríguez, Darío. "Fray Diego de Guadix y su 'recopilación' de arabismos." *Antiqua et Nova Romania: Estudios lingüísticos y filológicos en honor de José Mondéjar*. Ed. Pilar Carrasco. 2 vols. Granada: Universidad de Granada, 1993. 1:234–46.

Cabranes-Grant, Leo. "Lope de Vega y los debates sobre el origen de la lengua castellana." *Bulletin of Hispanic Studies* 77 (2000): 147–68.

Cabrillana, Nicolás. "Esclavos moriscos en la Almería del siglo XVI." *Al-Andalus* 40 (1975): 53–128.

Calvo Serraller, Francisco. *Teoría de la pintura en el Siglo de Oro*. Madrid: Cátedra, 1981.

Camón Aznar, José. *El pastor Quijotiz*. Madrid: Espasa-Calpe, 1969.

Campos y Fernández de Sevilla, Francisco Javier. "Los moriscos en las 'Relaciones Topográficas' de Felipe II." *Anuario Jurídico y Económico Escurialense* 43 (2010): 413–30.

Camps i Moranta, Gabriel, and Gabriel Seguí i Trobat. *La llegenda hagiogràfica de la invenció de la Mare de Déu de Lluc*. Palma: Lleonard Muntaner, 2010.

Cara Barrionuevo, Lorenzo. "La ganadería de la ciudad de Granada y la Alpujarra en la primera mitad del siglo XVI." *La historia del Reino de Granada a debate: Viejos y nuevos temas; Perspectivas de Estudio*. Ed. Manuel Barrios Aguilera and Ángel Galán Sánchez. Málaga: Diputación Provincial de Málaga, 2004. 179–208.

Cardaillac, Louis. *Moriscos y cristianos: Un enfrentamiento polémico (1492–1640)*. Trans. Mercedes García-Arenal. Mexico: Fondo de Cultura Económica, 1979.

Carlos Villamarín, Helena de. *Las antigüedades de Hispania*. Spoleto: Centro Italiano di Studi sull'Alto Medioevo, 1999.
Caro Baroja, Julio. *Las falsificaciones de la Historia (en relación con la de España)*. Barcelona: Seix Barral, 1992.
– *Los moriscos del reino de Granada*. Madrid: Istmo, 1985.
Carr, Matthew. *Blood and Faith: The Purging of Muslim Spain*. New York: The New Press, 2009.
Carranza, Paul. "Empire Is a Small Space: Spanish Pastoral in Its Imperial Context." Diss. University of Pennsylvania, 2005.
Carrasco, Rafael. *Deportados en nombre de Dios: La expulsión de los moriscos; cuarto centenario de una ignominia*. Barcelona: Destino, 2009.
Carrasco Urgoiti, María Soledad. "*De buen moro, buen cristiano*: Notas sobre una comedia de Felipe Godínez." *Nueva Revista de Filología Hispánica* 30.2 (1981): 546–73.
– *The Moorish Novel*. Boston: Twayne, 1976.
– *El moro de Granada en la literatura*. Ed. facsimile Juan Martínez Ruiz. Granada: Universidad de Granada, 1989.
– "Musulmanes y moriscos en la obra de Cervantes: Beligerancia y empatía." *Fundamentos de Antropología* 6–7 (1997): 66–79.
– *El problema morisco en Aragón al comienzo del reinado de Felipe II*. Valencia: University of North Carolina, 1969.
– "Reflejos de la vida de los moriscos en la novela picaresca." *En la España Medieval 4: Estudios dedicados al profesor D. Ángel Ferrari Núñez*. Ed. Miguel Ángel Ladero Quesada. Madrid: Universidad Complutense, 1984. 1:183–223.
– "Vituperio y parodia del romance morisco en el romancero nuevo." *Culturas populares: Diferencias, divergencias, conflictos*. Ed. Yves-René Fonquerne and Alfonso Esteban. Madrid: Universidad Complutense, 1986. 115–38.
Carreño, Antonio. "Introducción." In Vega, Lope de. *Pastores de Belén*. 9–64.
– "La otra *Arcadia* de Lope de Vega: *Pastores de Belén*." *Homenaje al profesor Antonio Vilanova*. Ed. Marta Cristina Carbonell. 2 vols. Barcelona: Universidad de Barcelona, 1989. 1:137–55.
Cartagena, Alonso de. *Alonso de Cartagena y el* Defensorium Unitatis Christianae. Ed. and trans. Guillermo Verdín-Díaz. Oviedo: Universidad de Oviedo, 1992.
Cartmill, Matt. "The Status of the Race Concept in Physical Anthropology." *American Anthropologist* 100.3 (1999): 651–60.
Cascales, Francisco. *Tablas poéticas*. Ed. Benito Brancaforte. Madrid: Espasa-Calpe, 1975.
Casey, James. "La situación económica de la nobleza valenciana en vísperas de la expulsión de los moriscos." *Homenaje al Dr. D. Juan Reglà Campistol*. Ed. Miquel Tarradell. 2 vols. Valencia: Universidad de Valencia, 1975. 1:515–25.

Castán Esteban, José Luis. "El derecho de pastos de mudéjares y cristianos en los señoríos valencianos: El ducado de Segorbe." *VIII Simposio Internacional de Mudejarismo: De mudéjares a moriscos; Una conversión forzada*. 2 vols. Teruel: Centro de Estudios Mudéjares, 2002. 1:241–53.

Castilla de la Vega, Gabriel. "Introducción." *La comedia del santo monje cautivo*. By Juan de Barrionuevo y Moya. Villanueva de la Reina: Concejalía de Cultura, 1992. 1–7.

Castillo Fernández, Javier. "Luis Enríquez Xoaida, el primo hermano morisco del rey católico (análisis de un caso de falsificación histórica e integración social)." *Sharq al-Andalus* 12 (1995): 235–53.

Castillo Martínez, Cristina. "'Cuevas subterráneas,' 'maletas abandonadas' y otros paralelismos entre el *Quijote* y algunas novelas pastoriles del siglo XVII." Bernat Vistarini 471–8.

– "De las lágrimas a la risa: Análisis de la decadencia de los libros de pastores." *Memoria de la palabra: Actas del VI Congreso de la Asociación Internacional Siglo de Oro*. Ed. María Luisa Lobato and Francisco Domínguez Matito. Madrid: Iberoamericana, 2004. 497–509.

– "Huellas poéticas del Abencerraje en la novela pastoril *La enamorada Elisea*, de Jerónimo de Covarrubias." *Actas del Congreso Internacional de la Asociación Coreana de Hispanistas*. Ed. María Ángeles Álvarez Martínez and María Soledad Villarrubia Zúñiga. Alcalá de Henares: Universidad de Alcalá, 2004. 206–18.

– "Introducción." *Antología de libros de pastores*. Alcalá de Henares: Centro de Estudios Cervantinos, 2005. xi–xxv.

– "Pancracios, quinquercios, abecedarios ... y otros juegos pastoriles." *Nueva Revista de Filología Hispánica* 55.1 (2007): 51–75.

– "*El premio de la constancia y pastores de Sierra Bermeja* (Madrid, 1620), de Jacinto de Espinel Adorno: La experiencia del más allá." *La maravilla escrita, Antonio de Torquemada y el Siglo de Oro*. Ed. Juan Matas Caballero and José Manuel Trabado Cabado. León: Universidad de León, 2005. 273–86.

– "La violencia en los libros de pastores." *Revista de Literatura* 72.143 (2010): 55–68.

Castro, Américo. *España en su historia: Cristianos, moros y judíos*. Buenos Aires: Losada, 1948.

– "Lo hispánico y el erasmismo." *Revista de Filología Hispánica* 4.1 (1942): 1–66.

Castro y Añaya, Pedro de. *Auroras de Diana*. Ed. María Josefa Díez de Revenga. Murcia: Academia Alfonso X el Sabio, 1989.

Caxa de Leruela, Miguel. *Restauración de la abundancia de España*. Ed. Jean Paul Le Flem. Madrid: Instituto de Estudios Fiscales, 1975.

Cebrián, José. *Estudios sobre Juan de la Cueva*. Seville: Universidad de Sevilla, 1991.

Cervantes Saavedra, Miguel de. *Los baños de Argel*. Ed. Jean Canavaggio. Madrid: Taurus, 1983.

– *Don Quijote de la Mancha*. Ed. Martín de Riquer. Barcelona: Planeta, 1987.

– *Don Quixote*. Trans. John Rutherford. New York: Penguin, 2003.
– *La Galatea*. Ed. Francisco López Estrada and María Teresa López García-Berdoy. Madrid: Cátedra, 1999.
– *Novelas ejemplares*. Ed. Harry Sieber. 2 vols. Madrid: Cátedra, 1995.
– *Teatro completo*. Ed. Florencio Sevilla Arroyo and Antonio Rey Hazas. Barcelona: Planeta, 1987.
– *Los trabajos de Persiles y Sigismunda*. Ed. Juan Bautista Avalle-Arce. Madrid: Castalia, 1992.
– *The Trials of Persiles and Sigismunda: A Northern Story*. Trans. Celia Richmond Weller and Clark A. Colahan. Berkeley: U of California P, 1989.
Charnon-Deutsch, Lou. *The Spanish Gypsy: The History of a European Obsession*. Pennsylvania State UP: University Park, 2004.
Chevalier, Maxime. "*La Diana* de Montemayor y su público en la España del siglo XVI." *Creación y público en la literatura española*. Ed. Jean-François Botrel, Serge Salaün, and Andrés Amorós. Madrid: 1974. 40–55.
Childers, William. "Disappearing Moriscos." *Cross-Cultural History and the Domestication of Otherness*. Ed. Michal Jan Rozbicki and George O. Ndege. New York: Palgrave, 2012. 51–64.
– *Transnational Cervantes*. Toronto: U of Toronto P, 2006.
Christian, William A. *Apparitions in Late Medieval and Renaissance Spain*. Princeton: Princeton UP, 1981.
Ciscar Pallarés, Eugenio. "El régimen señorial en el reino de Valencia después de la expulsión de los moriscos: Los censos en especie." *Homenaje al Dr. D. Juan Reglà Campistol*. Ed. Miquel Tarradell. 2 vols. Valencia: Universidad de Valencia, 1975. 1:555–69.
Claramonte, Andrés de. *El valiente negro en Flandes*. Ed. Alfredo Rodríguez López-Vázquez. Alcalá de Henares: Universidad de Alcalá, 1997.
Clemencín, Diego, ed. Miguel de Cervantes. *El ingenioso hidalgo don Quijote de la Mancha*. 6 vols. Madrid: E. Aguado, 1833–9.
Los códigos españoles concordados y anotados. 12 vols. Madrid: Antonio de San Martín, 1873.
Colás Latorre, Gregorio. "Los moriscos aragoneses: Una definición más allá de la religión y la política." *Sharq al-Andalus* 12 (1995): 147–61.
– "Los moriscos en Aragón." Moliner Prada 179–209.
Collins, Marsha S. "The Shaping of Collective Memory in Lope's *Isidro* (1599)." *Bulletin of Spanish Studies* 86.3 (2009): 293–322.
Colmenares, Diego de. *Historia de la insigne ciudad de Segovia*. Segovia: Diego Díez, 1637.
Colonge, Chantal. "Reflets littéraires de la question morisque entre la guerre des Alpujarras et l'expulsion (1571–1610)." *Boletín de la Real Academia de Buenas Letras de Barcelona* 33 (1969–70): 137–243.

Contreras Contreras, Jaime. "Limpieza de sangre, cambio social y manipulación de la memoria." *Inquisición y conversos*. Ed. Ana María López Álvarez, María Luisa Menéndez Robles, and Ricardo Izquierdo Benito. Madrid: Asociación de Amigos del Museo Sefardí, 1994. 81–101.

Cook, Karoline P. "Forbidden Crossings: Morisco Emigration to Spanish America, 1492–1650." Diss. Princeton U, 2008.

Cooper, Helen. "The Goat and the Eclogue." *Philological Quarterly* 53.3 (1974): 363–79.

Las Coplas de Mingo Revulgo. Ed. Vivana Brodey. Madison: The Hispanic Seminary of Medieval Studies, 1986.

CORDE (Corpus diacrónico del español). *Real Academia Española*, n.d. Web. 20 Jul. 2012. <http://rae.es/rae.html>.

Corominas, Joan. *Diccionario crítico etimológico castellano e hispánico*. 6 vols. Madrid: Gredos, 1980.

Coroucli, Maria. "Bergers, héros et élites." *Nommer et classer dans les Balkans*. Ed. Gilles de Rapper and Pierre Sintês. Athens: École Française d'Athènes, 2008. 27–40.

Corral, Gabriel del. *La Cintia de Aranjuez*. Ed. Joaquín de Entrambasaguas. Madrid: CSIC, 1945.

Corral, Pedro del. *Crónica sarracina*. Ed. James Donald Fogelquist. 2 vols. Madrid: Castalia, 2001.

Correa, Pedro, ed. *Los romances fronterizos: Edición comentada*. 2 vols. Granada: Universidad de Granada, 1999.

Cortés, María del Mar. "La Biblia en Cervantes y la novela de la Edad de Oro." *La Biblia en la literatura española II: Siglo de Oro*. Ed. Rosa Navarro Durán. Madrid: Trotta, 2008. 181–232.

Cossío, José María de. *Fábulas mitológicas de España*. Madrid: Espasa-Calpe, 1952.

Covarrubias Herrera, Jerónimo de. *Los cinco libros intitulados de la enamorada Elisea*. Valladolid: Luis Delgado, 1594.

Covarrubias y Horozco, Sebastián de. *Tesoro de la lengua castellana o española*. Ed. Martín de Riquer. Barcelona: Alta Fulla, 2003.

Crémoux, Françoise. "Escenificación de un culto popular: La fortuna literaria de la Virgen de Guadalupe." *Actas del XIII Congreso de la Asociación Internacional de Hispanistas*. Ed. Florencia Sevilla Arroyo and Carlos Alvar Ezquerra. 4 vols. Madrid: Castalia, 2000. 1:476–84.

Cueva, Luis de la. *Diálogos de las cosas notables de Granada*. Ed. facsimile José Mondéjar. Granada: Universidad de Granada, 1993.

Cull, John T. "Los espacios de la maravilla en los libros de pastores españoles." *Loca ficta: Los espacios de la maravilla en la Edad Media y Siglo de Oro*. Ed. Ignacio Arellano Ayuso. Madrid: Iberoamericana, 2003. 165–88.

Dadson, Trevor J. "Official Rhetoric versus Local Reality: Propaganda and the Expulsion of the *Moriscos.*" *Rhetoric and Reality in Early Modern Spain.* Ed. Richard J. Pym. London: Tamesis, 2006. 1–24.

Damiani, Bruno. La Diana *of Montemayor as Social and Religious Teaching.* Lexington: UP of Kentucky, 1983.

Dangler, Jean. *Making Difference in Medieval and Early Modern Iberia.* Notre Dame: U of Notre Dame P, 2005.

Darst, David. "The Literariness of *El Abencerraje.*" *Estudios sobre el siglo de oro: Homenaje a Raymond M. MacCurdy.* Ed. Ángel González. Madrid: Cátedra, 1983. 265–73.

Dávalos y Figueroa, Diego. *Primera parte de la miscelánea austral.* Lima: Antonio Ricardo, 1602.

De Armas, Frederick A. "Caves of Fame and Wisdom in the Spanish Pastoral Novel." *Studies in Philology* 82 (1985): 332–58.

– *Don Quixote among the Saracens.* Toronto: U of Toronto P, 2011.

Defert, Daniel. "Un genre ethnographique profane au XVIe: Les livres d'habits (essai d'ethno-iconographie)." *Histoires de l'Anthropologie (XVIe–XIXe siècles).* Ed. Britta Rupp-Eisenreich. Paris: Klincksieck, 1984. 25–41.

Delage, Agnès. "*Mantua carpetana*: La construction d'une légende toponymique de la fondation de Madrid." Delpech, ed. 61–77.

Delahaye-Grélois, Séverine. "'Siendo al Betis un rayo de su trenza lo que es al Tajo su mayor tribute,' ou comment un poète du Siècle d'or utilisait les lieux communs." Delpech, ed. 255–62.

Delpech, François. "Un mito andaluz: El reino oculto de Boabdil y los moros encantados." *Las tomas: Antropología de la ocupación territorial del reino de Granada.* Ed. José Antonio González Alcantud and Manuel Barrios Aguilera. Granada: Diputación de Granada, 2002. 565–616.

Delpech, François, ed. *L'Imaginaire du territoire en Espagne et au Portugal (XVIe–XVIIe siècles).* Madrid: Casa de Velázquez, 2008.

Díaz del Castillo, Bernal. *Historia verdadera de la conquista de la Nueva España.* Ed. Carlos Pereyra. Madrid: Espasa-Calpe, 1992.

Díaz López, Julián Pablo, and Antonio Muñoz Buendía, eds. *Herbajes, trashumantes y estantes: La ganadería en la península Ibérica (Épocas medieval y moderna).* Almería: Instituto de Estudios Almerienses, 2002.

Díaz G. Viana, Luis. *Los guardianes de la tradición: Ensayos sobre la "invención" de la cultura popular.* Oiartzun, Gipuzkoa: Sendoa, 1999.

Díaz-Mas, Paloma. *Sephardim: The Jews from Spain.* Trans. George K. Zucker. Chicago: U of Chicago P, 1992.

Diego, Gerardo. "Cervantes y la música." *Anales Cervantinos* 1 (1951): 5–40.

Diller, Aubrey. "The Tradition of Stephanus Byzantius." *Transactions and Proceedings of the American Philological Association* 69 (1938): 333–48.

Domínguez Cubero, José. "La romería de la Virgen de la Cabeza a través de la pintura y literatura barrocas." *Giennium* 12 (2009): 305–42.
Domínguez Ortiz, Antonio. *La clase social de los conversos en Castilla en la edad moderna*. Granada: Universidad de Granada, 1991.
– *Los judeoconversos en España y América*. Madrid: Istmo, 1978.
Domínguez Ortiz, Antonio, and Bernard Vincent. *Historia de los moriscos: Vida y tragedia de una minoría*. Madrid: Revista de Occidente, 1978.
Doniger, Wendy. "The Symbolism of Black and White Babies in the Myth of Parental Impression." *Social Research* 70.1 (2003): 1–44.
Doubleday, Simon, and Miguel Gómez, eds. "On (De)Commemoration: Rethinking the Battle of Las Navas de Tolosa." Special issue of *Journal of Medieval Iberian Studies* 4.1 (2012): 1–3.
Drayson, Elizabeth. *The King and the Whore: King Roderick and La Cava*. New York: Palgrave, 2007.
Dudley, Edward. "Goddess on the Edge: The Galatea Agenda in Raphael, Garcilaso and Cervantes." *Calíope* 1.1–2 (1995): 27–45.
Dumora, Florence. "Toledo, Imperial Ciudad: Acercamiento a la historia y leyenda de un título." Delpech, ed. 35–59.
Earle, T.F., and K.J.P. Lowe, eds. *Black Africans in Renaissance Europe*. Cambridge: Cambridge UP, 2005.
Echave, Baltasar de. *Discursos de la antigüedad de la lengua cantabra bascongada*. 1607. Bilbao: Gran Enciclopedia Vasca, 1971.
Écija, Fray Diego de. *Libro de la invención de esta Santa Imagen de Guadalupe*. Ed. Arcángel Barrado Manzano. Cáceres: Seminarios de FET y de las JONS, 1953.
Eco, Umberto. *La búsqueda de la lengua perfecta*. Trans. Maria Pons. Barcelona: Grijalbo Mondadori, 1996.
Ehlers, Benjamin. *Between Christians and Moriscos: Juan de Ribera and Religious Reform in Valencia, 1568–1614*. Baltimore: Johns Hopkins UP, 2006.
Encina, Juan del. *Obra completa*. Ed. Miguel Ángel Pérez Priego. Madrid: Biblioteca Castro, 1996.
– *Teatro completo*. Ed. Miguel Ángel Pérez Priego. Madrid: Cátedra, 1991.
Encinar, Jesús. "La imagen de España." *El blog de Jesús Encinar*. 20 Aug. 2007. Web. 15 Oct. 2011. <http://www.jesusencinar.com/2007/08/la-imagen-de-es.html>.
Escolano, Gaspar. *Segunda parte de la década primera de la historia de la insigne y coronada ciudad y reino de Valencia*. Valencia: Patricio Mey, 1611.
Espinel, Vicente. *Vida de Marcos de Obregón*. Ed. Samuel Gili Gaya. 2 vols. Madrid: Espasa-Calpe, 1959.
Espinel Adorno, Jacinto de. *Descendencia ilustre de don Yñigo Manrique de Lara, alcayde de las fuerças y castillos de la ciudad de Malaga*. Málaga: Juan René, 1626.

– *El premio de la constancia, y pastores de Sierra Bermeja*. Madrid: Viuda de Alonso Martín, 1620.

Estévez Sola, Juan Antonio. "Algo más sobre los orígenes míticos de España." *Habis* 24 (1993): 207–17.

La expulsión de los moriscos del reino de Valencia. Valencia: Fundación Bancaja, 1997.

Fajardo, Salvador J. "Narrative and Agency: The Ricote Episode (*Don Quijote II*)." *Bulletin of Hispanic Studies* 78 (2001): 311–22.

Fanjul, Serafín. *Al-Andalus contra España: La forja del mito*. Madrid: Siglo XXI, 2000.

Faria e Sousa, Manuel. *The "Fortuna" of Manuel de Faria e Sousa: An Autobiography*. Ed. Edward Glaser. Münster Westfalen: Aschendorff, 1975.

Fayard, Janine, and Marie-Claude Gerbet. "Fermeture de la noblesse et pureté de sang en Castille à travers les procès de *hidalguía* au XVIème siècle." *Histoire, Économie et Société* 1.1 (1982): 51–75.

Feliciano Chaves, María Judith. "Mudejarismo in Its Colonial Context: Iberian Cultural Display, Viceregal Luxury Consumption, and the Negotiation of Identities in Sixteenth Century New Spain." Diss. University of Pennsylvania, 2004.

Fernández, Lucas. *Teatro selecto clásico de Lucas Fernández*. Ed. Alfredo Hermenegildo. Madrid: Escelicer, 1972.

Fernández Albaladejo, Pablo. "Entre 'godos' y 'montañeses': Avatares de una primera identidad española." Tallon 123–54.

Fernández-Cañadas de Greenwood, Pilar. *Pastoral Poetics: The Uses of Conventions in Renaissance Pastoral Romances:* Arcadia, La Diana, La Galatea, L'Astrée. Madrid: José Porrúa Turanzas, 1983.

Fernández de Navarrete, Pedro. *Conservación de monarquías y discursos políticos. Obras de don Diego de Saavedra Fajardo y del licenciado Pedro Fernández de Navarrete*. Biblioteca de Autores Españoles 25. Madrid: Atlas, 1947. 448–557.

Fernández Martín, Luis. *Comediantes, esclavos y moriscos en Valladolid (Siglos XVI–XVII)*. Valladolid: Universidad de Valladolid, 1988.

Fernández Otal, José Antonio. *La casa de ganaderos de Zaragoza: Derecho y trashumancia a fines del siglo XV*. Zaragoza: Institución Fernando el Católico, 1993.

Ferrer Valls, Teresa. "El duque de Lerma y la corte virreinal en Valencia: Fiestas, literatura y promoción social. *El prado de Valencia* de Gaspar Mercader." *Quaderns de Filologia: Estudis Literaris* 5 (2000): 257–71.

– *Nobleza y espectáculo teatral (1535–1622)*. Valencia: Universitat de València, 1993.

Ferris, José Luis. *Miguel Hernández: Pasiones, cárcel y muerte de un poeta*. Madrid: Planeta, 2010.

Fine, Ruth. "El entrecruzamiento de lo hebreo y lo converso en la obra de Cervantes: Un encuentro singular." *Cervantes y las religiones*. Ed. Ruth Fine and Santiago López Navia. Pamplona: Universidad de Navarra, 2008. 435–51.
Finello, Dominick. *The Evolution of the Pastoral Novel in Early Modern Spain*. Tempe: Arizona Center for Medieval and Renaissance Studies, 2008.
– *Pastoral Themes and Forms in Cervantes's Fiction*. Lewisburg: Bucknell UP, 1994.
Flecniakoska, Jean Louis. "Reflexions sur la parodie pastorale dans le *Quichotte*." *Anales Cervantinos* 8 (1959–60): 371–8.
Fonseca, Damián. *Relación de la expulsión de los moriscos del Reino de Valencia*. Ed. facsimile. Valencia: Sociedad Valenciana de Bibliófilos, 1878.
Forcione, Alban. "Cervantes en busca de una pastoral auténtica." *Nueva Revista de Filología Hispánica* 36.2 (1988): 1011–43.
Fosalba, Eugenia. *El Abencerraje pastoril*. Barcelona: Universitat Autònoma de Barcelona, 1990.
– *La* Diana *en Europa: Ediciones, traducciones e influencias*. Barcelona: Universitat Autònoma de Barcelona, 1994.
Foucault, Michel. "Of Other Spaces." Trans. Jay Miskowiec. *Diacritics* 16.1 (1986): 22–7.
– *Security, Territory, Population: Lectures at the Collège de France 1977–1978*. Ed. and trans. Graham Burchell. New York: Picador, 2007.
– *"Society Must Be Defended." Lectures at the Collège de France 1975–1976*. Trans. David Macey. New York: Picador, 2003.
Fra Molinero, Baltasar. *La imagen de los negros en el teatro del Siglo de Oro*. Madrid: Siglo XXI, 1995.
Franco Silva, Alonso. "Moriscos y cristianos en el obispado de Málaga: El caso de Monda." *VII Simposio Internacional de Mudejarismo: Teruel, 19–21 de septiembre de 1996*. Zaragoza: Centro de Estudios Mudéjares, 1999. 381–427.
Franklin, Sarah. *Dolly Mixtures: The Remaking of Genealogy*. Durham: Duke UP, 2007.
Frías Marín, Rafael. "Historia del cuadro y de los personajes retratados." *La romería de la Virgen de la Cabeza en una pintura del siglo XVII*. Córdoba: Obra Social y Cultural Cajasur, 1997. 47–68.
Friedman, Edward H. "Camón Aznar y la cuarta salida de don Quijote." *Ínsula* 60.697–8 (2005): 7–10.
Friedman, Jerome. "Jewish Conversion, the Spanish Pure Blood Laws and Reformation: A Revisionist View of Racial and Religious Antisemitism." *The Sixteenth Century Journal* 18.1 (1987): 3–30.
Fuchs, Barbara. *Exotic Nation: Maurophilia and the Construction of Early Modern Spain*. Philadelphia: U of Pennsylvania P, 2008.
– *Mimesis and Empire*. Cambridge: Cambridge UP, 2001.

– "1492 and the Cleaving of Hispanism." *Journal of Medieval and Early Modern Studies* 37.3 (2007): 493–510.

– *Passing for Spain: Cervantes and the Fictions of Identity*. Urbana: U of Illinois P, 2003.

– "The Spanish Race." Greer, Mignolo, and Quilligan 88–98.

Fuente Fernández, Francisco Javier. "La imagen del 'Buen Pastor' en fray Cipriano de la Huerga y fray Luis de León." *Fray Luis de León: Historia, humanismo y letras*. Ed. Víctor García de la Concha and Javier San José Lera. Salamanca: Universidad de Salamanca, 1996. 629–37.

Furió Ceriol, Fadrique. *El Concejo y Consejeros del Príncipe y otras obras*. Ed. Diego Sevilla Andrés. Valencia: Institución Alfonso el Magnánimo, 1952.

Gali Boadella, Montserrat. "Retórica del retablo de los Reyes: Catedral de Puebla de los Ángeles." *Identidad en el imaginario nacional*. Ed. Javier Pérez Siller and Verena Radkau García. Puebla: Instituto de Ciencias Sociales y Humanidades, 1998. 469–86.

Gállego, Julián. *Visión y símbolos de la pintura española del Siglo de Oro*. Madrid: Aguilar, 1972.

Gallego Burín, Antonio, and Alfonso Gámir Sandoval. *Los moriscos del reino de Granada según el sínodo de Guadix de 1554*. Ed. Darío Cabanelas Rodríguez. Granada: Universidad de Granada, 1968.

Gálvez de Montalvo, Luis. *El pastor de Fílida. Novelas de los siglos XV y XVI*. Ed. Marcelino Menéndez Pelayo. Nueva Biblioteca de Autores Españoles 7. Madrid: Rivadeneyra, 1907. 482–583.

García, Heliodoro. "El reformismo del *Pastor bonus* de Juan Maldonado." *Hispania Sacra* 35.72 (1983): 193–218.

García-Arenal, Mercedes. "Morisques et Gitans." *Mélanges de la Casa de Velázquez* 14 (1978): 503–10.

García-Arenal, Mercedes, and Fernando Rodríguez Mediano. *Un Oriente español: Los moriscos y el Sacromonte en tiempos de Contrarreforma*. Madrid: Marcial Pons, 2010.

García Ayluardo, Clara. "El milagro de la Virgen de Aránzazu: Los vascos como grupo de poder en la ciudad de México." *Manifestaciones religiosas en el mundo colonial americano*. Ed. Clara García Ayluardo and Manuel Ramos Medina. Mexico D.F.: Instituto Nacional de Antropología e Historia, 1997. 331–47.

García Fitz, Francisco. *Las Navas de Tolosa*. Barcelona: Ariel, 2005.

García Martín, Pedro. *La Mesta*. Madrid: Historia 16, 1990.

García Pardo, Manuela. "El pastor de las Navas de Tolosa: La realidad y la leyenda." *Historia, tradiciones y leyendas en la frontera: IV Estudios de frontera; Homenaje a Don Enrique Toral y Peñaranda*. Ed. Francisco Toro Ceballos and José Rodríguez Molina. Jaén: Diputación Provincial de Jaén, 2002. 215–28.

García Pedraza, Amalia. *Actitudes ante la muerte en la Granada del siglo XVII: Los moriscos que quisieron salvarse*. 2 vols. Granada: Universidad de Granada, 2002.
García Valdecasas, Amelia. "Decadencia y disolución del Romancero morisco." *Boletín de la Real Academia Española* 69 (1989): 131–58.
Garibay y Zamalloa, Esteban de. *Los quarenta libros del compendio historial de las chronicas y universal Historia de todos los Reynos de España*. Amberes: Christophoro Plantino, 1571.
Garrido Domínguez, Francisco. *La mina secreta del palacio del rey moro de Ronda: El esplendor de la Izna-Rand Onda musulmana*. Madrid: Albia, 1991.
Gascó, Fernando. "Historiadores, falsarios y estudiosos de las antigüedades andaluzas." *La antigüedad como argumento: Historiografía de arqueología e historia antigua de Andalucía*. Ed. José Beltrán and Fernando Gascó. Seville: Junta de Andalucía, 1993. 9–28.
Gauna, Felipe de. *Relación de las fiestas celebradas en Valencia con motivo del casamiento de Felipe III*. Ed. Salvador Carreres Zacarés. 2 vols. Valencia: Acción Bibliográfica Valenciana, 1926.
Gaylord, Mary. "La Arcadia nuevamente inventada del *Quijote* de 1605." *Cervantes y el* Quijote*: Actas del Coloquio Internacional. Oviedo, 27–30 de octubre de 2004*. Ed. Emilio Martínez Mata. Madrid: Arco, 2007. 55–68.
– "Spain's Renaissance Conquests and the Retroping of Identity." *Journal of Hispanic Philology* 16 (1992): 125–36.
Gerbet, Marie-Claude. *La ganadería medieval en la Península Ibérica*. Barcelona: Crítica, 2002.
– "Les Ordres Militaires et l'élevage dans l'Espagne médiévale." *En la España medieval* 5.1 (1986): 413–45.
Gil de Biedma, Jaime. *Moralidades*. Mexico D.F.: Joaquín Mortiz, 1966.
Gil Herrera, Jorge. "El botín de la expulsion: Proceso de recaudación de las 'mitades' y tasación de los bienes raíces dejados por los moriscos de Castilla." *Chronica Nova* 36 (2010): 43–65.
Gil Polo, Gaspar. *Diana enamorada*. Ed. Rafael Ferreres. Madrid: Espasa-Calpe, 1973.
Gil Sanjuan, Joaquín. "Disidentes y marginados de la Serranía de Ronda en el tránsito a los tiempos modernos." *Baetica* 13 (1991): 227–39.
Gilman, Stephen. "Relatos de conversos en la *Comedia Jacinta* de Torres Naharro." *Nueva Revista de Filología Hispánica* 17 (1966): 20–39.
Giménez Caballero, Ernesto. *El belén de Salzillo en Murcia (Origen de los Nacimientos en España)*. Madrid: Gaceta Literaria, 1934.
– *España nuestra: El libro de las juventudes falangistas*. Madrid: Vicesecretaría de Educación Popular, 1943.
– "Un nuevo poeta pastor." *La Gaceta Literaria* 121 (1932): 10–11.

– *Retratos españoles (bastante parecidos)*. Barcelona: Planeta, 1985.
Gitlitz, David. "Lucas Fernández y Pierre Menard: El auto de la pasión." *Encuentros and Desencuentros: Spanish Jewish Cultural Interaction throughout History*. Ed. Carlos Carrete Parrondo et al. Tel Aviv: University Publishing Projects, 2000. 503–15.
– *Secrecy and Deceit: The Religion of the Crypto-Jews*. Philadelphia: The Jewish Publication Society, 1996.
Glaser, Edward. "El patriarca Jacob, amante ejemplar del teatro del Siglo de Oro español." *Bulletin Hispanique* 68 (1956): 5–22.
Glick, Thomas F. *Islamic and Christian Spain in the Early Middle Ages: Comparative Perspectives on Social and Cultural Formation*. Princeton: Princeton UP, 1979.
Godínez, Felipe. *Coloquio primero y segundo de los Pastores de Belén. Autos sacramentales con quatro comedias nuevas y sus loas y entremeses: Primera parte*. Madrid: María de Quiñones, 1655. 134v–46r.
Gómez-Pantoja, Joaquín. "*Pastio agrestis*: Pastoralismo en Hispania romana." Gómez-Pantoja, ed. 177–213.
– ed. *Los rebaños de Gerión: Pastores y trashumancia en Iberia antigua y medieval*. Madrid: Casa de Velázquez, 2001.
Gómez Tejada, Cosme. *El filósofo: Ocupaciones de nobles y discretos contra la cortesana ociosidad*. Madrid: Domingo García y Morrás, 1650.
Gómez Vozmediano, Miguel F. "La expulsión de los moriscos granadinos de la Mancha a inicios del siglo XVII." *Chronica Nova* 36 (2010): 67–114.
Góngora, Luis de. *Soledades*. Ed. John Beverley. Madrid: Cátedra, 1991.
González Alcantud, José Antonio. *Lo moro: Las lógicas de la derrota y la formación del estereotipo islámico*. Barcelona: Anthropos, 2002.
González de Bobadilla, Bernardo. *Primera parte de las ninfas y pastores de Henares*. Ed. facsimile. Madrid: Ministerio de Cultura, 1978.
González Rovira, Javier. "Imaginativa y nacimientos prodigiosos en algunos textos del Barroco." *Criticón* 69 (1997): 21–31.
Graña, María Cecilia. "Lo pastoril entre el renacimiento y el barroco en dos antologías de hacia 1600." *Quaderni di lingue e letterature* 7 (1982): 53–69.
Granja, Fernando de la. "Fiestas cristianas en Al-Andalus (materiales para su estudio)." *Al-Andalus* 34.1 (1969): 1–53.
Green, Otis H. "'Fingen los poetas': Notes on the Spanish Attitude toward Pagan Mythology." *Estudios dedicados a Menéndez Pidal*. Madrid: CSIC, 1950. 1:275–88.
Greene, Roland. "Petrarchism among the Discourses of Imperialism." *America in European Consciousness, 1493–1750*. Ed. Karen Ordahl Kupperman. Chapel Hill: U of North Carolina P, 1995. 130–65.

Greer, Margaret R., Walter D. Mignolo, and Maureen Quilligan, eds. *Rereading the Black Legend: The Discourses of Religious and Racial Difference in the Renaissance Empires*. Chicago: U of Chicago P, 2007.

Grell, Chantal. "Annius de Viterbe et le roman des origines en France et en Espagne." Tallon 227–50.

Grieve, Patricia E. *The Eve of Spain: Myths of Origins in the History of Christian, Muslim, and Jewish Conflict*. Baltimore: Johns Hopkins UP, 2009.

Gruzinski, Serge. *The Mestizo Mind: The Intellectual Dynamics of Colonization and Globalization*. Trans. Deke Dusinberre. New York: Routledge, 2002.

Guadalajara y Xavier, Marcos de. *Memorable expulsión y justísimo destierro de los moriscos de España*. Pamplona: Nicolás de Assiayn, 1613.

Guadix, Diego de. *Recopilación de algunos nombres arábigos que los árabes pusieron a algunas ciudades y otras muchas cosas*. Ed. Elena Bajo Pérez and Felipe Maíllo Salgado. Gijón: Trea, 2005.

Guillén, Claudio. "Literature as Historical Contradiction: *El Abencerraje*, the Moorish Novel, and the Eclogue." *Literature as System: Essays toward the Theory of Literary History*. Princeton: Princeton UP, 1971. 159–217.

– *El primer Siglo de Oro: Estudios sobre géneros y modelos*. Barcelona: Crítica, 1988.

Guillorel, Hervé. "Toponymie et politique." *Noms et re-noms: La dénomination des personnes, des populations, des langues et des territoires*. Ed. Salih Akin. Rouen: Publications de l'Université de Rouen, 1999. 61–91.

Gutiérrez de los Ríos, Gaspar. *Noticia general para la estimación de las artes*. Madrid, 1600.

Gutiérrez Nieto, Juan Ignacio. "El reformismo social de Olivares: El problema de la limpieza de sangre y la creación de una nobleza de mérito." *La España del Conde Duque de Olivares*. Ed. John Elliott and Ángel García Sanz. Valladolid: Universidad de Valladolid, 1990. 417–41.

Haley, George. *Vicente Espinel and Marcos de Obregón: A Life and Its Literary Representation*. Providence: Brown UP, 1959.

Hannaford, Ivan H. *Race: The History of an Idea in the West*. Baltimore: Johns Hopkins UP, 1996.

Harris, Marvin. *Theories of Culture in Postmodern Times*. Walnut Creek, CA: AltaMira Press, 1999.

Harrison, Faye. "Expanding the Discourse on Race." *American Anthropologist* 100.3 (1998): 609–31.

Harvey, L.P. *Muslims in Spain 1500 to 1614*. Chicago: U of Chicago P, 2005.

Haywood, Charles. "Cervantes and Music." *Hispania* 31.2 (1948): 131–51.

Hermenegildo, Alfredo. "Fray Luis de León y su visión de la figura del Rey." *Letras de Deusto* 13.25 (1983): 169–78.

– "Sobre la dimensión social del teatro primitivo español." *Prohemio* 2 (1975): 25–50.

– "En torno a la burla de los linajes." *La palabra y el hombre* 23 (1977): 55–65.

Hernández, Miguel. *El pastor de la muerte. Obra completa.* Vol. 2, *Teatro*. Ed. Agustín Sánchez-Vidal and José Carlos Rovira. Madrid: Espasa-Calpe, 1992. 1815–2025.

Hernández del Portillo, Alonso. *Historia de Gibraltar*. Ed. Antonio Torremocha Silva. Algeciras: Centro Asociado de la UNED, 1994.

Hernández Franco, Juan. *Sangre limpia, sangre española: El debate de los estatutos de limpieza (siglos XV–XVII)*. Madrid: Cátedra, 2011.

Hernández-Pecoraro, Rosilie. *Bucolic Metaphors: History, Subjectivity, and Gender in the Early Modern Spanish Pastoral*. Chapel Hill: U of North Carolina P, 2006.

– "Jarifa's Choice: A Gendered Reading of *El Abencerraje* y *La hermosa Jarifa*." *Bulletin of Spanish Studies* 79 (2002): 429–46.

Hernando, Agustín. "The Contribution of Ortelius' *Theatrum* to the Geographical Knowledge of Spain." *Abraham Ortelius and the First Atlas: Essays Commemorating the Quadricentennial of his Death, 1598–1998*. Ed. Marcel van den Broecke, Peter van der Krogt, and Peter Meurer. Amsterdam: Hes, 1998. 239–62.

Herrera, Fernando de. *Poesía castellana original completa*. Ed. Cristóbal Cuevas. Madrid: Cátedra, 1985.

Herzog, Tamar. *Defining Nations: Immigrants and Citizens in Early Modern Spain and Spanish America*. New Haven: Yale UP, 2003.

Hillgarth, J.N. *The Mirror of Spain, 1500–1700: The Formation of a Myth*. Ann Arbor: U of Michigan P, 2003.

Hitchcock, Richard. "Cervantes, Ricote and the Expulsion of the Moriscos." *Bulletin of Spanish Studies* 81.2 (2004): 175–85.

Hobsbawm, Eric. "Introduction: Inventing Traditions." *The Invention of Tradition*. Ed. Eric Hobsbawm and Terence Ranger. Cambridge: Cambridge UP, 1983. 1–14.

Holzinger, Walter. "The Militia of Love, War, and Virtue in the *Abencerraje y la hermosa Jarifa*: A Structural and Sociological Reassessment." *Revista Canadiense de Estudios Hispánicos* 2.3 (1978): 227–38.

Huarte de San Juan, Juan. *Examen de ingenios para las ciencias*. Ed. Guillermo Serés. Madrid: Cátedra, 1989.

Huerga, Cipriano de la. "El sermón de los pendones." Ed. Javier Fuente Fernández. *Obras completas*. 10 vols. León: Universidad de León, 1991. 1:249–84.

Hurtado de Toledo, Luis. *Teatro pastoril en la ribera del Tajo. Las trescientas de Luis Hurtado, poeta castellano*. 1582. MS 107. Biblioteca Xeral, Universidade de Santiago de Compostela. 65r-90v.

Hurtado de Toledo, Luis, and Michael de Carvajal. *Cortes de casto amor y Cortes de la muerte*. Ed. facsimile Antonio Rodríguez-Moñino. Valencia: Bonaire, 1964.

Iglesias, Óscar. "El último cabrero vino de África." *El País*. 24 Sept. 2009. Web. 24 Sept. 2009. <http://elpais.com/diario/2009/09/24/galicia/1253787502_850215.html>.

Irigoyen-García, Javier. "'Les bergers qui conquirent l'Espagne': Pastoral, Transnational Ethnicity and Empire in Philippe Quinault's *La généreuse ingratitude*." *Transitions: Journal of Franco-Iberian Studies* 5 (2009): 38–56.

– "Diana and Wild Boar Hunting: Refiguring Gender and Ethno-Religious Conflict in the Pastoral Imaginary." *Bulletin of Hispanic Studies* 88.3 (2011): 273–87.

– "Identidad colectiva, genealogía y novela pastoril: El manuscrito en euskera de Juan Pérez de Lazarraga." *Hispanic Research Journal* 12.4 (2011): 291–305.

– "Jerónimo de Arbolanche's *Las Abidas* (1566) and the Mythical Origins of Spain." *Symposium* 67.2 (2013): 85–97 .

– "'La música ha sido hereje': Pastoral Performance and Cultural Hybridity in *Los Baños de Argel*." *Bulletin of the Comediantes* 62.2 (2010): 45–62.

– "'Poco os falta para moros, pues tanto lo parecéis': Impersonating the Moor in the Spanish Mediterranean." *Journal of Spanish Cultural Studies* 12.3 (2011): 355–69.

– "'¡Qué si destas diferencias de música resuena la de los albogues!': Lo pastoril y lo morisco en Cervantes." *Cervantes* 28.2 (2008): 119–46.

Jameson, Fredric. *The Political Unconscious: Narrative as a Socially Symbolic Act*. Ithaca: Cornell UP, 1981.

Jiménez Patón, Bartolomé. *Discurso en favor del santo y loable estatuto de la limpieza*. Granada: Andrés de Santiago Palomino, 1638.

Johnson, Carroll B. *Cervantes and the Material World*. Urbana: U of Illinois P, 2000.

– "Cervantes' *Galatea*: The Portuguese Connection." *Iberoromania* 23 (1986): 91–105.

– *Transliterating a Culture: Cervantes and the Moriscos*. Ed. Mark Groundland. Newark: Juan de la Cuesta, 2010.

Juaristi, Jon. *Vestigios de Babel: Para una arqueología de los nacionalismos españoles*. Madrid: Siglo XXI, 1992.

Kagan, Richard L. "Clío y la corona: Escribir historia en la España de los Austrias." *España, Europa y el mundo Atlántico: Homenaje a John H. Elliott*. Ed. Richard L. Kagan and Geoffrey Parker. Madrid: Marcial Pons, 2001. 113–47.

Kamen, Henry. "Una crisis de conciencia en la Edad de Oro en España: Inquisición contra 'limpieza de sangre.'" *Bulletin Hispanique* 88.3–4 (1986): 321–56.

– "Limpieza and the Ghost of Américo Castro: Racism as a Tool of Literary

Analysis." *Hispanic Review* 64.1 (1996): 19–29.
King, Willard F. *Prosa novelística y academias literarias en el siglo XVII*. Madrid: RAE, 1963.
Klein, Julius. *The Mesta: A Study in Spanish Economic History, 1273–1836*. Cambridge, MA: Harvard UP, 1920.
Knapp, Jeffrey. "Nations into Persons." *Rereading the Black Legend: The Discourses of Religious and Racial Difference in the Renaissance Empires*. Greer, Mignolo, and Quilligan 293–311.
Krauss, Werner. "Algunas observaciones sobre la novela pastoril española." *Eco* 23.138–9 (1971): 652–98.
– "Localización y desplazamientos en la novela pastoril española." *Actas del Segundo Congreso Internacional de Hispanistas*. Ed. Jaime Sánchez Romeralo and Norbert Polussen. Nijmegen: Asociación Internacional de Hispanistas / Instituto Español de la Universidad de Nimega, 1967. 363–9.
Kristeva, Julia. *Revolution in Poetic Language*. Trans. Margaret Wallet. New York: Columbia UP, 1984.
Krulls-Hepermann, Claudia. "Spanish Pastoral Novels of the Sixteenth Century: In Search of a Vanished Fascination." Trans. James Stack. *New Literary History* 19.3 (1988): 581–95.
Kurtz, Barbara E. *The Play of Allegory in the Autos Sacramentales of Pedro Calderón de la Barca*. Washington DC: Catholic U of America P, 1991.
Kusche, Maria. *Juan Pantoja de la Cruz y sus seguidores B. González de Villandrando y A. López Polanco*. Madrid: Fundación Arte Hispánico, 2007.
El laberinto del fauno [*Pan's Labyrinth*]. Dir. Guillermo del Toro. Estudios Picasso / Tequila Gang, 2006.
Ladero Quesada, Miguel Ángel. "Almojarifazgo sevillano y comercio exterior de Andalucía en el siglo XV." *Anuario de Historia Económica y Social* 2 (1969): 69–115.
– *Los mudéjares de Castilla y otros estudios de historia medieval andaluza*. Granada: Universidad de Granada, 1989.
Laguna Sanz, Eduardo. *Historia del merino*. Madrid: Ministerio de Agricultura, Pesca y Alimentación, 1986.
Lapeyre, Henri. *Geografía de la España morisca*. Trans. Luis C. Rodríguez García. Valencia: Diputación Provincial de Valencia, 1986.
Lasso de la Vega, Gabriel. *Manojuelo de romances*. Madrid: Saeta, 1942.
Lavocat, Françoise. *Arcadies malhereuses: Aux origines du roman moderne*. Paris: Honoré Champion, 1998.
– *La Syrinx au bûcher: Pan et les satyres à la Renaissance et à l'âge Baroque*. Geneva: Droz, 2005.
Lea, Henry Charles. *Los moriscos españoles: Su conversión y expulsión*. Trans. Jaime Lorenzo Miralles. Ed. Rafael Benítez Sánchez-Blanco. Alicante:

Diputación de Alicante, 1990.

León, Fray Luis de. *De los nombres de cristo*. Ed. Cristóbal Cuevas García. Madrid: Cátedra, 1982.

Lezra, Jacques. "'La mora encantada': Covarrubias and the Soul of Spain." *Journal of Spanish Cultural Studies* 1.1 (2000): 5–27.

Libro en que se qüentan los amores de Viraldo y Florindo, aunque en diverso estilo. Ed. Luis Gómez Canseco and Bernardo Perea Morillo. Salamanca: Universidad de Salamanca, 2003.

Lida de Malkiel, María Rosa. "Túbal, primer poblador de España." *Abaco* 3 (1970): 9–48.

Lihani, John. "Lucas Fernández and the Evolution of the Shepherd's Family Pride in Early Modern Drama." *Hispanic Review* 25.4 (1957): 252–63.

Linehan, Peter. *Past and Present in Medieval Spain*. Aldershot: Ashgate, 1992.

Lizarazu de Mesa, María Asunción. "Los pastores y la música." *Un camino de ida y vuelta: La trashumancia en España*. Ed. Luis Vicente Elías Pastor and Feliciano Novoa Portela. Barcelona: Lunwerg, 2003. 245–51.

Llompart, Gabriel. "Nostra Dona Santa Maria de Lluc." *Analecta Sacra Tarraconensia* 61–2 (1988): 189–268.

Lofrasso, Antonio. *Los diez libros de fortuna de amor*. Barcelona: Pedro Malo, 1573.

Loomba, Ania, and Jonathan Burton, eds. *Race in Early Modern England: A Documentary Companion*. New York: Palgrave, 2007.

Lopez, Robert Sabatino. "The Origin of the Merino Sheep." *The Joshua Starr Memorial Volume: Studies in History and Philology*. New York: Jewish Social Studies, 1953. 161–8.

López, Vicente Cristóbal. "Mitología clásica y novela pastoril." *Estudios sobre tradición clásica y mitología en el Siglo de Oro*. Ed. Jesús Ponce Cárdenas and Isabel Colón Calderón. Madrid: Clásicas, 2002. 109–22.

López-Baralt, Luce. "La estética del cuerpo entre los moriscos del siglo XVI, o de cómo la minoría perseguida pierde el rostro." *Le corps dans la société espagnole des XVIe et XVIIe siècles*. Ed. Augustin Redondo. Paris: Publication de la Sorbonne, 1990. 335–48.

López de Ayala, Jerónimo. *Toledo en el siglo XVI, después del vencimiento de las comunidades*. Madrid: Hijos de M.G. Hernández, 1901.

López de Coca Castañer, José Enrique. "El trabajo de mudéjares y moriscos en el reino de Granada." *Actas del VI simposio internacional de mudejarismo*. Zaragoza: Centro de Estudios Mudéjares, 1995. 97–136.

López de Enciso, Bartolomé. *Desengaño de celos*. Madrid: Francisco Sánchez, 1586.

López de Yanguas, Hernán. *Farsa turquesana (1529?)*. Ed. Julio F. Hernando and

Javier Espejo. Textos Lemir, 2002. Web. <http://parnaseo.uv.es/Lemir/Textos/index.htm>

– *Obras dramáticas*. Ed. Fernando González Ollé. Madrid: Espasa Calpe, 1967.

López de Zárate, Francisco. *Varias poesías*. Madrid: Viuda de Alonso Martín, 1619.

López Estrada, Francisco, ed. *El Abencerraje y la hermosa Jarifa: Cuatro textos y su estudio*. Madrid: Revista de Archivos, Bibliotecas y Museos, 1957.

– "Cotejo de las dos ediciones, Alcalá 1585 y París, 1611, de *La Galatea* de Cervantes." *Revista Bibliográfica y Documental* 2 (1948): 73–90.

– "Los libros de pastores." *La novela española en el siglo XVI*. By María Soledad Carrasco Urgoiti, Francisco López Estrada and Félix Carrasco. Madrid: Iberoamericana, 2001. 127–215.

– *Los libros de pastores en la literatura española: La órbita previa*. Madrid: Gredos, 1974.

– *Notas sobre la espiritualidad española de los siglos de oro: Estudio del* Tratado llamado el Deseoso. Seville: Universidad de Sevilla, 1972.

– "Pastores en el *Quijote*." *Anales Cervantinos* 37 (2005): 15–32.

– "La poesía pastoril en la aldea y corte de Segovia (1570)." *Estudios sobre el Siglo de Oro: Homenaje al profesor Francisco Ynduráin*. Madrid: Editora Nacional, 1984. 309–17.

– "Prólogo." Jorge de Montemayor. *Los siete libros de la Diana*. Madrid: Espasa-Calpe, 1970. i–ciii.

López Estrada, Francisco, and María Teresa López García-Berdoy. "Introducción." Cervantes, *La Galatea* 10–144.

López Estrada, Francisco, Javier Huerta Calvo, and Víctor Infantes de Miguel. *Bibliografía de los libros de pastores en la literatura española*. Madrid: Universidad Complutense de Madrid, 1984.

López Trujillo, Miguel Ángel. *Patrimonio: La lucha por los bienes culturales españoles (1500–1939)*. Gijón: Trea, 2006.

Lorenzo, Javier. "After Tunis: Petrarchism and Empire in the Poetry of Garcilaso de la Vega." *Hispanófila* 141 (2004): 17–30.

Luján de Sayavedra, Mateo. *Segunda Parte de la vida del pícaro Guzmán de Alfarache*. *Novelistas anteriores a Cervantes*. Ed. Buenaventura Carlos Aribau. Biblioteca de Autores Españoles 3. Madrid: Atlas, 1963. 365–430.

Luna, Miguel de. *Historia verdadera del rey don Rodrigo*. Ed. facsimile Luis F. Bernabé Pons. Granada: Universidad de Granada, 2001.

MacCormack, Sabine. "Visions of the Roman Past in Late Medieval and Early Modern Spain." *Genesis and Regeneration: Essays on the Conceptions of Origins*. Ed. Shaul Shaked. Jerusalem: The Israel Academy of Sciences and Humanities, 2005. 77–109.

MacKay, Angus. "Ritual and Propaganda in Fifteenth-Century Castile." *Past and*

Present 107 (1985): 3–43.

MacKay, Ruth. *"Lazy, Improvident People": Myth and Reality in the Writing of Spanish History*. Ithaca: Cornell UP, 2006.

Madroñal Durán, Abraham. "La importancia del medio toledano en la literatura del Siglo de Oro (el inédito *Teatro pastoril* de Luis Hurtado de Toledo)." *Jerusalén y Toledo: Historias de dos ciudades*. Ed. Manuel Casado Velarde, Ruth Fine, and Carlos Mata Induráin. Madrid: Iberoamericana, 2012. 175–214.

Maestro, Jesús G. "La fractura histórica de *La Galatea* de Miguel de Cervantes (problemas metodológicos de investigación histórico-literaria desde el materialismo filosófico)." *La fractura historiográfica: Las investigaciones de Edad Media y Renacimiento desde el tercer milenio*. Ed. Francisco Javier Burguillo and Laura Mier. Salamanca: Seminario de Estudios Medievales y Renacentistas, 2008. 629–41.

Magnier, Grace. *Pedro de Valencia and the Catholic Apologists of the Expulsion of the Moriscos: Visions of Christianity and Kingship*. Leiden: Brill, 2010.

Maíllo Salgado, Felipe. *Los arabismos del castellano en la baja edad media*. Salamanca: Universidad de Salamanca, 1991.

Malón de Chaide, Pedro. *La conversión de la Magdalena*. Ed. P. Félix García. 3 vols. Madrid: Espasa-Calpe, 1947.

Mann, Richard G. *El Greco y sus patrones*. Trans. Isabel Basilde. Madrid: Akal, 1994.

Maravall, José Antonio. *Utopía y contrautopía en el* Quijote. Santiago de Compostela: Pico Sacro, 1976.

Marchante-Aragón, Lucas A. "The King, the Nation, and the Moor: Imperial Spectacle and the Rejection of Hybridity in *The Masque of the Expulsion of the Moriscos*." *Journal for Early Modern Cultural Studies* 8.1 (2008): 98–133.

Marek, Margaret. "Intertextual Textiles and Wool Blends: Ovine Productions and Pastoral Arts." *Projections of the Early Modern Hispanic World in Contemporary Film and Literature*. Florida Atlantic University, Boca Raton. 26 Oct. 2012. Conference presentation.

– "Itinerant Shepherds and Arcadian Bounty: The Coincidental Flourishing of Transhumance and the *libros de pastores*." *55th Annual Meeting of the Renaissance Society of America*. Los Angeles. 21 March 2009. Conference presentation.

Mariana, Juan de. *Historia general de España. Obras del padre Juan de Mariana*. Ed. Francisco Pí y Margall. 2 vols. Madrid: Sucesores de Hernando, 1909–23.

Mariscal, George. "The Role of Spain in Contemporary Race Theory." *Arizona Journal of Hispanic Cultural Studies* 2 (1998): 7–22.

Mármol Carvajal, Luis del. *Historia del rebelión y castigo de los moriscos del Reino de Granada. Historiadores de sucesos particulares*. 2 vols. Biblioteca de

Autores Españoles 66. Madrid: Rivadeneyra, 1852. 1:123–365.
Márquez, Juan. *El governador christiano, deducido de las vidas de Moysen, y Josue, principes del pueblo de Dios*. Salamanca: Francisco de Cea Tesa, 1619.
Márquez Villanueva, Francisco. *Fuentes literarias cervantinas*. Madrid: Gredos, 1973.
– "Giovan Giorgio Trissino y el soneto de Hernando de Acuña a Carlos V." *Studia Hispanica in Honorem R. Lapesa*. 3 vols. Madrid: Gredos, 1972. 2:355–71.
– *Investigaciones sobre Juan Álvarez Gato*. Madrid: Real Academia Española, 1960.
– "Lope infamado de morisco: *La villana de Getafe*." *Anuario de Letras* 21 (1983): 147–82.
– *Moros, moriscos y turcos de Cervantes*. Barcelona: Bellaterra, 2010.
– *El problema morisco (desde otras laderas)*. Madrid: Libertarias, 1991.
– "Sobre el concepto de judaizante." *Encuentros and Desencuentros: Spanish Jewish Cultural Interaction Throughout History*. Ed. Carlos Carrete Parrondo et al. Tel Aviv: University Publishing Projects, 2000. 519–42.
– "Sobre el contexto religioso de *La Galatea*." *Actas del II Congreso Internacional de la Asociación de Cervantistas*. Ed. Giuseppe Grilli. Naples: Istituto Universitario Orientale, 1995. 181–96.
– "Transfondos de 'la profecía del Tajo': Goticismo y profetismo." *Varia lingüística y literaria: 50 años del CELL*. Ed. Martha Elena Venier. 3 vols. México D.F: Colegio de México, 1997. 2:177–201.
– "La *Trivagia* y el problema de conciencia religiosa de Juan del Encina." *La Torre* 1.3–4 (1987): 473–500.
Martín Casares, Aurelia. "Free and Freed Black Africans in Granada in the Time of the Spanish Renaissance." Earle and Lowe 247–61.
Martín Corrales, Eloy. *La imagen del magrebí en España: Una perspectiva histórica, siglos XVI–XX*. Barcelona: Bellaterra, 2002.
Martin-Márquez, Susan. *Disorientations: Spanish Colonialism in North Africa and the Performance of Identity*. New Haven: Yale UP, 2008.
Martín Romero, José Julio. "La temática pastoril en los libros de caballerías de la época de Felipe II." *Nueva Revista de Filología Hispánica* 57.2 (2009): 563–605.
Martínez, María Elena. *Genealogical Fictions: Limpieza de Sangre, Religion, and Gender in Colonial Mexico*. Stanford: Stanford UP, 2008.
Martínez-Góngora, Mar. "La problemática producción de la diferencia étnica: Imágenes de belleza petrarquista y génesis bíblica en la *Expulsión de los moros de España* de Gaspar Aguilar." *Revista de Estudios Hispánicos* 36.3 (2002): 501–21.
Martínez López, Enrique. *Tablero de ajedrez: Imágenes del negro heroico en la comedia española y en la literatura e iconografía sacra del brasil esclavista*. Paris: Centre Culturel Calouste Gulbenkian, 1998.

Marx, Leo. "Pastoralism in America." *Ideology and Classic American Literature*. Ed. Sacvan Bercovitch and Myra Jehlen. Cambridge: Cambridge UP, 1986. 36–69.

Matulka, Barbara. "On the European Diffusion of the 'Last of the Abencerrajes' Story in the Sixteenth Century." *Hispania* 16.4 (1933): 369–88.

Maurer, Christopher. "Un monarca, un imperio y una espada: Juan Latino y el soneto de Hernando de Acuña sobre Lepanto." *Hispanic Review* 61.1 (1993): 35–51.

McKendrick, Melveena. *Theatre in Spain, 1490–1700*. Cambridge: Cambridge UP, 1992.

Méchoulan, Henri. *El honor de Dios: Indios, judíos y moriscos en el Siglo de Oro*. Trans. Enrique Sordo. Barcelona: Argos Vergara, 1981.

Medina, Pedro de. *Libro de grandezas y cosas memorables de España. Obras de Pedro de Medina*. Ed. Ángel González Palencia. Madrid: CSIC, 1944. 1–257.

Méndez de Vasconcelos, Juan. *Liga deshecha por la expulsión de los Moriscos de los Reynos de España*. Madrid: Alonso Martín, 1612.

Mendo, Andrés. *Príncipe perfecto y ministros ajustados*. Lyon: Horacio Boissat, 1662.

Menéndez Onrubia, Carmen. "Felipe Godínez: 'Auto y coloquio de los pastores de Belén.'" *El teatro menor en España*. Madrid: CSIC, 1983. 173–86.

Menéndez Pelayo, Marcelino. *Orígenes de la novela*. Vol. 2, *Novelas sentimental, bizantina, histórica y pastoril*. Madrid: CSIC, 1961.

Menéndez Pidal, Ramón. *Romancero tradicional*. Vol. 1, *Romanceros del rey Rodrigo y de Bernardo del Carpio*. Madrid: Gredos, 1957.

Menocal, María Rosa. *The Ornament of the World: How Muslims, Jews, and Christians Created a Culture of Tolerance in Medieval Spain*. Boston: Little, Brown, 2002.

Mercader, Gaspar. *El prado de Valencia*. 1600. Ed. Henri Mérimée. Toulouse: Edouard Privat, 1907.

Mérida, Fray Diego de. "Viaje a Oriente." Ed. Antonio Rodríguez Moñino. *Analecta Sacra Tarraconensia* 18 (1945): 115–87.

Mérimée, Henri. "Introduction." Mercader xv–cx.

Mérique, Cyril. "Un mensaje rescrito: Los cuatro introitos del *Aucto de la oveja perdida* de Joan Timoneda." *Criticón* 102 (2008): 69–83.

Merivale, Patricia. *Pan the Goat-God: His Myth in Modern Times*. Cambridge. MA: Harvard UP, 1969.

Mesa, Cristóbal de. *Las Navas de Tolosa*. Madrid: Viuda de P. Madrigal, 1594.

Middlebrook, Leah. *Imperial Lyric: New Poetry and New Subjects in Early Modern Spain*. University Park: Pennsylvania State UP, 2009.

Milhou, Alain. "De Rodrigue le pécheur à Ferdinand le restaurateur." *Pouvoir Royale et absolutism dans l'Espagne du XVIe siècle*. Toulouse: Presses Universitaires du Mirail, 1999. 13–29.

– "Desemitización y europeización en la cultura española desde la época de los Reyes Católicos." *Cultura del Renaixement: Homenatge al pare Miquel Batllori.* Ed. Manuel Fernández Álvarez. Bellaterra: Universitat Autónoma de Barcelona, 1993. 35–60.
– "Propaganda mesiánica y opinión pública: Las reacciones de las ciudades del reino de Castilla frente al proyecto fernandino de cruzada (1510–11)." *Homenaje a Jose Antonio Maravall.* Ed. María del Carmen Iglesias, Carlos Moya, and Luis Rodríguez Zúñiga. 3 vols. Madrid: Centro de Investigaciones Sociológicas, 1985. 3:51–62.
Mínguez Fernández, José María. "Ganadería, aristocracia y reconquista en la Edad Media castellana." *Hispania* 42 (1982): 341–54.
Moliner Prada, Antonio, ed. *La expulsión de los moriscos.* Barcelona: Nabla, 2009.
Moll, Jaime. "Aproximaciones a la sociología de la edición literaria." *La edición literaria: Actas del I Congreso Internacional de Hispanistas del Siglo de Oro.* Ed. Pablo Jauralde Pou, Alfonso Rey, and Dolores Noguera. London: Tamesis, 1990. 61–8.
– "La narrativa castellana a comienzos del siglo XVII: Aspectos editoriales." *Anales Cervantinos* 40 (2008): 31–46.
Moner, Michel. "El problema morisco en los textos cervantinos." *Las dos grandes minorías étnico-religiosas en la literatura española del Siglo de Oro: Los judeoconversos y los moriscos.* Ed. Irene Andres-Suárez. Paris: Les Belles Lettres, 1995. 85–100.
Monroy y Silva, Cristóbal. *El pastor más perseguido y finezas de Raquel.* Valencia: Viuda de Joseph de Orga, 1764.
Montemayor, Jorge de. *La Diana.* Ed. Asunción Rallo. Madrid: Cátedra, 1999.
– *La Diana.* Ed. Juan Montero. Barcelona: Crítica, 1996.
– *Los siete libros de la Diana.* Ed. Francisco López Estrada. Madrid: Espasa-Calpe, 1946.
– *Los siete libros de la Diana.* Ed. Francisco López Estrada and María Teresa López García-Berdoy. Madrid: Espasa-Calpe, 1993.
– *Los siete libros de la Diana.* Ed. Enrique Moreno Báez. Madrid: Real Academia Española, 1955.
– *Los siete libros de la Diana.* Ed. Miguel Teijeiro Fuentes. Barcelona: Promociones y Publicaciones Universitarias, 1991.
– *Los siete libros de la Diana.* Ed. Julian Arribas. Woodbridge: Tamesis, 1996.
Montero, Juan. "La *Clara Diana* (Épila, 1580) de fray Bartolomé Ponce y el canon pastoril." *Criticón* 61 (1994): 69–80.
Montero Reguera, José. "Historia, política y literatura en *La Galatea* de Miguel de Cervantes." *Romeral: Estudios filológicos en homenaje a José Antonio Fernández Romero.* Ed. Inmaculada Báez and María Rosa Pérez. Vigo: Universidade de Vigo, 2002. 329–41.

Montoro, Adrian G. "'Libertad cristiana': Relectura de *Marcos de Obregón.*" *Modern Language Notes* 91 (1976): 213–30.

Morales, Ambrosio de. *Las antigüedades de las ciudades de España*. Alcalá de Henares: Juan Íñiguez de Lequerica, 1575.

Morby, S. Edwin. "Páginas olvidadas de la *Arcadia* de Lope." *Boletín de la Real Academia Española* 48 (1968): 77–86.

– "Two Notes on *La Arcadia.*" *Hispanic Review* 36.2 (1968): 110–23.

Moreno Díaz del Campo, Francisco Javier. "La venta de bienes raíces de los moriscos de La Mancha: Una desamortización en tiempos de Cervantes." *Cervantes y la economía*. Ed. Miguel Ángel Galindo Martín. Cuenca: Universidad de Castilla-La Mancha, 2007. 329–51.

Morínigo, Marcos A. "El teatro como sustituto de la novela en el Siglo de Oro." *Revista de la Universidad de Buenos Aires* 2.1 (1957): 41–61.

Morón, Ciriaco. "Una visión inédita de la expulsión de los moriscos." *Salmanticensis* 6.2 (1959): 483–502.

Moscoso, Javier. "Los efectos de la imaginación: Medicina, ciencia y sociedad en el siglo XVIII." *Asclepio* 53.1 (2001): 141–71.

Mujica, Barbara. *Iberian Pastoral Characters*. Washington DC: Scripta Humanistica, 1986.

Mulcahy, Rosemarie. "*The Meeting of Jacob and Rachel*: Murillo's 'Jacob' Cycle Complete." *Burlington Magazine* 135.1079 (1993): 73–80.

Muñoz Renedo, Carmen. *La representación de "Moros y cristianos" de Zújar: Cautiverio y rescate de Nuestra Señora de la Cabeza de Zújar*. Madrid: CSIC, 1972.

Navarrete, Ignacio. *Orphans of Petrarch: Poetry and Theory in the Spanish Renaissance*. Berkeley: U of California P, 1994.

Navarro Gómez, José. "El autor de la versión del *Abencerraje* contenido en la *Diana* ¿era Montemayor?" *Revista de Literatura* 39 (1978): 101–4.

Nebrija, Antonio de. *Muestra de la istoria de las antigüedades de España. Gramática de la lengua castellana*. Ed. González-Llubera. Oxford: Oxford UP, 1926. 203–28.

– *Vocabulario de romance en latín*. Ed. Gerald J. MacDonald. Philadelphia: Temple UP, 1973.

Nelson, Benjamin J. "Tending to Empire: The Spanish Pastoral Novel and Its Reflection upon Imperial Spain." Diss. University of Chicago, 2007.

Nepaulsingh, Colbert I. *Apples of Gold in Filigrees of Silver: Jewish Writing in the Eye of the Spanish Inquisition*. New York: Holmes and Meier, 1995.

Nirenberg, David. *Communities of Violence: Persecution of Minorities in the Middle Ages*. Princeton: Princeton UP, 1996.

– "Race and the Middle Ages: The Case of Spain and Its Jews." Greer, Mignolo, and Quilligan 71–87.

Núñez, Miguel. *La revolución y el deseo: Memorias*. Barcelona: Península, 2002.
Núñez de Reinoso, Alonso. *Historia de los amores de Clareo y Florisea, y de los trabajos de Isea*. 1552. *Novelistas anteriores a Cervantes*. Ed. Buenaventura Carlos Aribau. Madrid: Atlas, 1963. 431–68.
Núñez de Velasco, Francisco. *Diálogos de contención entre la milicia y la ciencia*. Valladolid: Miguel Sánchez, 1614.
Núñez Muley, Francisco. *A Memorandum for the President of the Royal Audiencia and Chancery Court of the City and Kingdom of Granada*. Ed. and trans. Vincent Barletta. Chicago: U of Chicago P, 2007.
– "Memorial de Francisco Núñez Muley." Ed. Raymond Foulché-Delbosc. *Revue Hispanique* 6 (1899): 205–39.
Ocampo, Florián de. *Los cinco primeros libros de la corónica general de España*. Alcalá de Henares: Juan Íñiguez de Lequerica, 1578.
O'Malley, Charles Donald, ed. and trans. *Michael Servetus: A Translation of his Geographical, Medical, and Astrological Writings*. Philadelphia: American Philosophical Society, 1953.
Oropesa, Alonso de. *Luz para conocimiento de los gentiles*. Ed. and trans. Luis A. Díaz y Díaz. Madrid: Universidad Pontificia de Salamanca, 1979.
Orozco Días, Emilio. *Temas del barroco de poesía y pintura*. Granada: Universidad de Granada, 1947.
Ortelius, Abraham. *Theatrum orbis terrarum*. Ed. facsimile R.A. Skelton. Amsterdam: N. Israel, 1964.
Os Resentidos. "O belén da Caixa." *Jei*. Gasa, 1990. LP.
Osma, José M. de. "Introducción." *El verdadero Dios Pan, auto sacramental alegórico*. By Pedro Calderón de la Barca. Lawrence: U of Kansas P, 1949. 1–52.
Otto, Ton, and Poul Pedersen. "Disentangling Traditions: Culture, Agency and Power." *Tradition and Agency: Tracing Cultural Continuity and Invention*. Ed. Ton Otto and Poul Pedersen. Aarhus: Aarhus UP, 2005. 11–49.
Pacheco, Francisco. *El arte de la pintura*. Ed. Bonaventura Bassegoda i Hugas. Madrid: Cátedra, 1990.
Palencia Cerezo, José María. "La adoración de los pastores como tema franciscano y su evolución en la pintura cordobesa: Una aproximación al eslabón perdido de los Sarabia." *El franciscanismo en Andalucía: San Francisco en la cultura y en la historia del arte español*. Ed. Manuel Peláez del Rosal. Córdoba: Cajasur, 2001. 101–22.
Pamp, Diane J. *Lope de Vega ante el problema de la limpieza de sangre*. Northampton, MA: Smith College, 1968.
La pastora de Mançanares y desdichas de Pánfilo. Ed. Cristina Castillo Martínez. Salamanca: Universidad de Salamanca, 2005.
Pastor Comín, Juan José. *Cervantes: Música y Poesía: El hecho musical en el pensamiento lírico cervantino*. Vigo: Academia del Hispanismo, 2007.

Peñalosa y Mondragón, Benito. *Libro de las cinco excellencias del Español, que despueblan a España para su mayor potencia y dilatación*. Pamplona: Carlos Labayen, 1629.

Perceval, José María. "Lope de Vega y la imagen del poder monárquico en las fiestas celebradas en honor de los matrimonios reales de 1615." *Anuario Lope de Vega* 10 (2004): 63–84.

– *Todos son uno. Arquetipos, xenofobia y racism: La imagen del morisco en la Monarquía Española durante los siglos XVI y XVII*. Almería: Instituto de Estudios Almerienses, 1997.

Perea Siller, Francisco Javier. "Genealogía y escritura en Fray Luis de León: Valoración del campo léxico de *linaje* en *De los nombres de Cristo*." *Revista Agustiniana* 42.129 (2001): 1115–37.

Pérez, Alonso. *Segunda parte de la Diana de George de Montemayor*. Amberes: Pedro Bellero, 1581.

Pérez, Joseph. *Historia de una tragedia: La expulsión de los judíos de España*. Barcelona: Crítica, 2004.

Pérez de Chinchón, Bernardo. *Antialcorano: Diálogos Christianos (conversión y evangelización de moriscos)*. Ed. Francisco Pons Fuster. Alicante: Universidad de Alicante, 2000.

Pérez de Hita, Ginés. *Guerras civiles de Granada. Novelistas anteriores a Cervantes*. Ed. Buenaventura Carlos Aribau. Madrid: Atlas, 1963. 513–686.

Pérez y Gómez, Antonio, ed. *Relaciones poéticas sobre las fiestas de toros y cañas*. 8 vols. Cieza: El Ayre de la almena, 1971–3.

Perry, Mary Elizabeth. *The Handless Maiden: Moriscos and the Politics of Religion in Early Modern Spain*. Princeton: Princeton UP, 2007.

Phillips, Carla Rahn, and William D. Phillips Jr. *Spain's Golden Fleece: Wool Production and the Wool Trade from the Middle Ages to the Nineteenth Century*. Baltimore: Johns Hopkins UP, 1997.

Piña, Juan de. *Epítome de la primera parte de las fábulas de la antigüedad*. Madrid: Imprenta del Reyno, 1635.

Plato. *Republic*. Trans. George Maximilian Anthony Grube. Indianapolis: Hackett, 1992.

Ponce, Bartolomé. *Primera parte de la clara Diana a lo divino*. Zaragoza: Lorenzo de Robles, 1599.

Poole, Stafford. "The Politics of *limpieza de sangre*: Juan de Ovando and His Circle in the Reign of Philip II." *The Americas* 55.3 (1999): 359–89.

Porcel, Baltasar. *Los Chuetas mallorquines: Quince siglos de racismo*. Palma de Mallorca: Miguel Font, 1986.

Preto-Rodas, Richard A. "Samuel Usque's *Consolação às Tribulações de Israel* as Pastoral Literature Engagée." *Hispania* 73.1 (1990): 72–6.

Provansal, D. "Tesoros y apariciones: La prohibición de la riqueza." *Demófilo: Revista de Cultura Tradicional de Andalucía* 15 (1995): 37–61.

Puente, Juan de la. *Conveniencia de las dos Monarquías Católicas, la de la Iglesia Romana y la del Imperio Español.* Madrid: Imprenta Real, 1612.

Pulido Serrano, Juan Ignacio. "La expulsión frustrada: Proyectos para la erradicación de la herejía judaica en la Monarquía Hispana." *La declinación de la monarquía hispánica en el siglo XVII.* Ed. Francisco José Aranda Pérez. Cuenca: Universidad de Castilla-La Mancha, 2004. 891–904.

Pym, Richard J. *The Gypsies of Early Modern Spain, 1425–1783.* New York: Palgrave, 2007.

Quérillacq, Réné. "Los moriscos de Cervantes." *Anales Cervantinos* 30 (1992): 77–98.

Querol Gavaldá, Miguel. *La música en las obras de Cervantes.* Barcelona: Comtalia, 1948.

Quesada, Santiago. *La idea de ciudad en la cultura hispana de la edad moderna.* Barcelona: Universitat de Barcelona, 1992.

Quinault, Philippe. *La généreuse ingratitude.* Paris: Toussainct Quinet, 1656.

Rallo Grus, Asunción. "Introducción." Montemayor *La Diana* 9–99.

– *Los libros de antigüedades en el Siglo de Oro.* Málaga: Universidad de Málaga, 2002.

Raynié, Florence. "*Pastores de Belén*, de Lope de Vega: ¿Una novela para niños?" *Didáctica (Lengua y Literatura)* 14 (2002): 195–200.

Redondo, Augustin. "Les divers visages du thème (wisi)gothique dans l'Espagne des XVIe et XVIIe siècles." *L'Europe Héritière de l'Espagne Wisigothique.* Ed. Jacques Fontaine and Christine Pellistrandi. Madrid: Casa de Velázquez, 1992. 353–64.

– "La religion populaire espagnole au XVIe siècle: Un terrain d'affrontement?" *Culturas populares: Diferencias, divergencias, conflictos.* Ed. Yves-René Fonquerne and Alfonso Esteban. Madrid: Universidad Complutense, 1986. 329–69.

Relaciones de los reinados de Carlos V y Felipe II. 2 vols. Madrid: Sociedad de Bibliófilos Españoles, 1950.

Relaciones históricas de los siglos XVI y XVII. Madrid: Sociedad de Bibliófilos Españoles, 1896.

Rennert, Hugo A. *The Spanish Pastoral Romances.* Philadelphia: U of Pennsylvania P, 1912.

Renoux-Caron, Pauline. "Une Terre Sainte espagnole? La région de Tolède dans *Historia de la muerte y glorioso martirio del Santo Inocente de la Guardia* (1583) de fray Rodrigo de Yepes." Delpech, ed. 171–91.

Reyes Cano, Rogelio. *La* Arcadia *de Sannazaro en España.* Seville: Universidad de Sevilla, 1973.

Reyre, Dominique. "Topónimos hebreos y memoria de la España judía en el Siglo de Oro." *Criticón* 65 (1995): 31–53.

Ricard, Robert. "Espagnol et portugais 'marlota': Recherches sur le vocabulaire du vêtement hispano-mauresque." *Bulletin Hispanique* 53.2 (1951): 131–56.

Ríos Saloma, Martín F. *La Reconquista: Una construcción historiográfica (siglos XVI–XIX)*. Madrid: Marcial Pons, 2011.

Riquer, Martín de. "El siglo del Quijote (1580–1680)." *Historia de España* 26.2. Ed. Ramón Menéndez Pidal. Madrid: Espasa Calpe, 1986. 160–70.

Robert, Ulysse. *Les signes d'infamie au Moyen Age*. Paris: Honoré Champion, 1891.

Rodríguez Becerra, Salvador. "El santuario de la Virgen de la Cabeza de Andújar según el manuscrito de Juan de Ledesma (1633)." *Boletín del Instituto de Estudios Giennenses* 202 (2010): 57–77.

Rodríguez Molina, José. *La vida de moros y cristianos en la frontera*. Alcalá la Real: Alcalá Grupo Editorial, 2007.

Rodríguez-Picavea Matilla, Enrique. "La ganadería en la economía de frontera: Una aproximación al caso de la meseta meridional castellana en los siglos XI–XIV." *Identidad y representación de la frontera en la España medieval (siglos XI–XIV)*. Ed. Carlos de Ayala Martínez, Pascal Buresi, and Philippe Josserand. Madrid: Casa de Velázquez, 2001. 181–203.

Rojas Villandrando, Agustín de. *El viaje entretenido*. Ed. Jean Pierre Ressot. Madrid: Castalia, 1972.

Roldán-Figueroa, Rady. *The Ascetic Spirituality of Juan de Ávila (1499–1569)*. Leiden: Brill, 2010.

Romano de Thuesen, Evalia Ana. "Transcripción y edición del *Catálogo Real de Castilla*, autógrafo inédito de Gonzalo Fernández de Oviedo y Valdés." 3 vols. Diss. U of California-Santa Barbara, 1992.

Rosado Llamas, María Dolores, and Manuel Gabriel López Payer. *La batalla de las Navas de Tolosa. Historia y mito*. Andújar: Caja Rural de Jaén, 2001.

Rosal, Francisco del. *Diccionario etimológico: Alfabeto primero de Origen y Etimología de todos los vocablos originales de la Lengua Castellana*. Ed. facsimile Enríquez Gómez Aguado. Madrid: CSIC, 1992.

Rose, Constance Hubbard. *Alonso Núñez de Reinoso: The Lament of a Sixteenth-Century Exile*. Rutherford, NJ: Fairleigh Dickinson UP, 1971.

Rose, William. *El pastor de la muerte: La dialéctica pastoril en la obra de Miguel Hernández*. Barcelona: Puvill, 1983.

Rosselló i Verger, Vicenç M. "La designació dels grans rius valencians." *Toponímia, geografia i cartografia*. Valencia: Universitat de València, 2004. 35–50.

Rothstein, Marian. "Etymology, Genealogy, and the Immutability of Origins." *Renaissance Quarterly* 43.2 (1990): 332–47.

Rouanet, Léo, ed. *Colección de autos, farsas y coloquios del siglo XVI*. 4 vols. Barcelona: L'Avenç, 1901.

Rueda, Lope de. *Pasos*. Ed. José Luis Canet Vallés. Madrid: Castalia, 1992.

Rufo, Juan. *La Austriada. Poemas épicos* 2. Ed. Cayetano Rosell. Biblioteca de Autores Españoles 29. Madrid: Rivadeneyra, 1854. 1–136.

Ruiz, Juan, Arcipreste de Hita. *Libro de buen amor*. Ed. Alberto Blecua. Madrid: Cátedra, 2008.

Ruiz-Gálvez Prieto, Estrella. "*Sine Labe*: El inmaculismo en la España de los siglos XV a XVII: La proyección social de un imaginario religioso." *Revista de Dialectología y Tradiciones Populares* 63.2 (2008): 197–241.

Ruiz Lagos, Manuel, ed. *Moriscos: De los romances del gozo al exilio*. Seville: Guadalmena, 2001.

Saavedra y Torreblanca, Gonzalo de. *Los pastores de Betis*. Ed. Martín de Saavedra y Guzmán. Trani: Lorenzo Valery, 1633.

Sáez, Ricardo. "*Hidalguía*: Essai de definition: Des principes identificateurs aux variations historiques." *Hidalgos et hidalguía dans l'Espagne des XVI*e*–XVIII*e *siècles*. Ed. Martine Lambert-Gorges and Bernard Lavallé. Paris: CNRS, 1989. 23–45.

Salazar, Adolfo. "Música, instrumentos y danzas en las obras de Cervantes." *Nueva Revista de Filología Hispánica* 2 (1948): 21–56.

Salcedo Olid, Manuel de. *Panegírico historial de Nuestra Señora de la Cabeza de Sierra Morena*. Madrid: Julián de Paredes, 1677.

Salomon, Noël. *Lo villano en el teatro del Siglo de Oro*. Trans. Beatriz Chenot. Madrid: Castalia, 1985.

Salucio, Fray Agustín. *Discurso sobre los estatutos de limpieza de sangre*. Ed. facsimile Antonio Pérez y Gómez. Cieza: "… la fonte que mana y corre …", 1975.

Samson, Alexander. "The *adelantamiento* of Cazorla, *converso* Culture and Toledo Cathedral Chapter's 1547 *estatuto de limpieza de sangre*." *Bulletin of Spanish Studies* 84.7 (2007): 819–36.

Sancha, Justo de, ed. *Romancero y cancionero sagrados: Colección de poesías cristianas, morales y divinas*. Madrid: Rivadeneyra, 1855.

Sánchez Albornoz, Claudio. *España, un enigma histórico*. Buenos Aires: Sudamericana, 1956.

Sánchez Belda, Antonio, and María C. Sánchez Trujillano. *Razas ovinas españolas*. Madrid: Ministerio de Agricultura, 1979.

Sánchez Benito, José María. "Consolidación y práctica de la trashumancia en la Baja Edad Media castellana." *Itinerarios Medievales e identidad hispánica*. Ed. Elisa María Ferreira Priegue. Pamplona: Gobierno de Navarra, 2001. 257–92.

Sánchez Cantón, Francisco Javier. *Los grandes temas del arte cristiano en España* . Vol. 1, *Nacimiento e infancia de Cristo*. Madrid: Biblioteca de Autores Cristianos, 1948.

Sánchez de Badajoz, Diego. *Farsa de la Yglesia. Recopilación en metro*. Ed. Frida Weber de Kurlat. Buenos Aires: Universidad de Buenos Aires, 1968. 463–8.

Sánchez Ortega, María-Helena. "La oleada anti-gitana del siglo XVII." *Espacio, Tiempo y Forma: Serie IV: Historia Moderna* 4 (1991): 71–124.

Sánchez Ruano, Francisco. *Islam y Guerra Civil Española: Moros con Franco y con la República.* Madrid: La Esfera de los Libros, 2004.

Sannazaro, Iacobo. *Arcadia. Opere volgari.* Ed. Alfredo Mauro. Bari: Laterza & Figli, 1961. 1–295.

Sansone, Livio. *Blackness without Ethnicity: Constructing Race in Brazil.* New York: Palgrave, 2003.

Sarmiento, Fray Martín. *Cartas al Duque de Medina-Sidonia (1747–1770).* Ed. José Santos Puerto. Ponferrada: Instituto de Estudios Bercianos, 1995.

Schwartz, Stuart B. *All Can Be Saved: Religious Tolerance and Salvation in the Iberian Atlantic World.* New Haven: Yale UP, 2008.

Seidenspinner-Núñez, Dayle. "Inflecting the *Converso* Voice: A Commentary on Recent Theories." *La Corónica* 24.2 (1996): 6–18.

Shipley, George A. "La obra literaria como monumento histórico: El caso de *El Abencerraje*." *Journal of Hispanic Philology* 2.2 (1978): 103–20.

Sicroff, Albert A. *Los estatutos de limpieza de sangre: Controversias entre los siglos XV y XVII.* Trans. Mauro Armiño. Madrid: Taurus, 1985.

Sieber, Diane. "The Frontier Ballad and Spanish Golden Age Historiography: Recontextualizing the *Guerras civiles de Granada*." *Hispanic Review* 65.3 (1997): 291–306.

Smedley, Audrey. "'Race' and the Construction of Human Identity." *American Anthropologist* 100.3 (1999): 690–702.

Smieja, Florian. "La señora no es para la hoguera: El caso de la *Segunda parte de la Diana* de Alonso Pérez." *Actas del Sexto Congreso Internacional de Hispanistas.* Ed. Alan M. Gordon and Evelyn Rugg. Toronto: Asociación Internacional de Hispanistas, 1980. 715–18.

Smith, Paul Julian. "*La Celestina*, Castro, and the *Conversos*." *Representing the Other: "Race," Text and Gender in Spanish and Spanish American Narrative.* Oxford: Clarendon P, 1992. 27–58.

Sobejano, Gonzalo. "El mal poeta de comedias en la narrativa del siglo XVII." *Hispanic Review* 41 (1973): 313–30.

Solé-Leris, Amadeu. *The Spanish Pastoral Novel.* Boston: Twayne, 1980.

Solórzano Pereira, Juan de. *Emblemata Centum, regio politica.* Madrid: García Morras, 1653.

Souviron López, Begoña. "El *Abencerraje* en el templo de Diana." *Revista de Literatura* 61.121 (1999): 5–17.

– *La mujer en la ficción arcádica: Aproximación a la novela pastoril española.* Frankfurt: Vervuert, 1997.

Spitzer, Leo. "Raza del sol." *Hispanic Review* 10.1 (1942): 64–6.

Stagg, Geoffrey. "The Composition and Revision of *La Galatea.*" *Cervantes* 14 (1994): 9–25.

Stallaert, Christiane. *Etnogénesis y etnicidad: Una aproximación histórico-antropológica al casticismo*. Barcelona: Proyecto A, 1998.

– *Ni una gota de sangre impura: La España inquisitorial y la Alemania nazi cara a cara*. Barcelona: Galaxia Gutenberg, 2006.

Stern, Charlotte. "The Genesis of the Spanish Pastoral: From Lyric to Drama." *Kentucky Romance Quarterly* 25 (1978): 413–34.

– "Sayago and Sayagués in Spanish History and Literature." *Hispanic Review* 29.3 (1961): 217–37.

Stoll, André. "Abindarráez y Narváez: El último de los Abencerrajes, un cristiano noble y la persecución de los judíos conversos (un cuento del Renacimiento español)." *Averroes dialogado*. Ed. André Stoll. Kassel: Reichenberger, 1998. 141–84.

Suárez de Figueroa, Cristóbal. *La constante Amarilis*. Ed. María Asunción Satorre Grau. Textos Lemir, 2002. Web. 7 Oct. 2007. <http://parnaseo.uv.es/Lemir/Textos/Amarilis/index.htm>.

Subirats, Eduardo, ed. *Américo Castro y la revisión de la memoria* (*El Islam en España*). Madrid: Libertarias, 2003.

Surtz, Ronald E. "Cardinal Juan Martínez Silíceo in an allegorical *entremés* of 1556." *Essays on Hispanic Literature in Honor of Edmund L.King*. Ed. Silvia Molloy and Luis Fernández Cifuentes. London: Tamesis, 1983. 225–32.

– "Pastores judíos y reyes magos gentiles: Teatro franciscano y milenarismo en Nueva España." *Nueva Revista de Filología Española* 36.1 (1988): 333–44.

– "Sobre hidalguía y limpieza de sangre de la Virgen María, en el siglo XVII." *Cuadernos Hispanoamericanos* 372 (1981): 605–11.

Swaim, Kathleen M. "'Mighty Pan': Tradition and Image in Milton's Nativity *Hymn.*" *Studies in Philology* 68.4 (1971): 484–95.

Sweet, James H. "The Iberian Roots of American Racist Thought." *William and Mary Quarterly* 54.1 (1997): 143–66.

Taggard, Mindy Nancarrow. *Murillo's Allegories of Salvation and Triumph:* The Parable of the Prodigal Son *and* The Life of Jacob. Columbia: U of Missouri P, 1992.

Talavera, Hernando de. *Tratado de la misa*: *Escritores místicos españoles*. Vol. 1, *Fernando de Talavera, Alejo Venegas, Francisco de Osuna, Alonso de Madrid*. Ed. Miguel Mir. Nueva Biblioteca de Autores Españoles 16. Madrid: Bailly/Baillière, 1911. 79–93.

Tallon, Alain, ed. *Le Sentiment national dans l'Europe méridionale aux XVI*[e] *XVII*[e] *siècles*. Madrid: Casa de Velázquez, 2007.

Tapia, Serafín de. "La opresión fiscal de la minoría morisca en las ciudades castellanas." *Studia Historica: Historia Moderna* 4.3 (1986): 17–49.

Tárrega, Francisco Agustín. *El prado de Valencia*. Ed. José Luis Canet Vallés. London: Tamesis, 1985.

Tate, R.B. "La geografía humanística y los historiadores españoles del siglo quince." *Belfast Spanish and Portuguese Papers*. Ed. P.S.N. Russell-Gebbett, Arthur Terry, and Nicholas Grenville Round. Belfast: Queen's University of Belfast, 1979. 237–42.

Tejeda, Jerónimo de. *Tercera Parte de la Diana de Montemayor*. Paris: A costa del Autor, 1627.

Tellechea Idígoras, J. Ignacio. *El obispo ideal en el siglo de la Reforma*. Madrid: Instituto Español de Historia Eclesiástica, 1963.

Torquemada, Antonio de. *Obras completas*. 2 vols. Ed. Lina Rodríguez Cacho. Madrid: Turner, 1994.

Trambaioli, Marcella. "Notas sobre el papel de Lenio en *La Galatea*: ¿Gracioso o 'pastor fino'?" *Romance Notes* 35.1 (1998): 45–51.

Tratado de los dos caminos, por un morisco refugiado en Túnez. Ed. Álvaro Galmés de Fuentes and Juan Carlos Villaverde Amieva. Madrid: Instituto Universitario Seminario Menéndez Pidal, 2005.

Tueller, James B. *Good and Faithful Christians: Moriscos and Catholicism in Early Modern Spain*. New Orleans: UP of the South, 2002.

Twiddy, Iain. "Seamus Heaney's Versions of Pastoral." *Essays in Criticism* 56.1 (2006): 50–71.

Uceda, Gaspar de. *El tratado de Uceda contra los estatutos de limpieza de sangre*. Ed. Elvira Pérez Ferreiro. Madrid: Aben Ezra, 2000.

Unamuno, Miguel de. *En torno al casticismo*. Ed. Jean-Claude Rabaté. Madrid: Cátedra, 2005.

Valencia, Pedro de. *Tratado acerca de los moriscos de España*. Ed. Joaquín Gil Sanjuán. Málaga: Algazara, 1997.

Valiñas López, Francisco Manuel. *La navidad en las artes plásticas del Barroco español*. Diss. Universidad de Granada, 2005.

Vallés, Francisco. *De Sacra Philosophia*. Turin: Heirs of Niccolò Bevilacqua, 1587.

Van Elslande, Jean-Pierre. "L'altérité arcadienne." *L'Autre au XVII*[ème] *siècle*. Ed. Ralph Heyndels and Barbara Woshinsky. Tübingen: Gunter Narr Verlag, 1999. 393–401.

Vázquez García, Francisco. *La invención del racismo: Nacimiento de la biopolítica en España, 1600–1940*. Madrid: Akal, 2009.

– "Les métaphores naturalistes et la naissance de la biopolitique en Espagne (XVIIe–XXe siècles)." *Cahiers Internationaux de Symbolisme* 116–18 (2007): 193–208.

Vega, Bernardo de la. *El pastor de Iberia*. Seville: Juan de León, 1591.

Vega, Garcilaso de la. *Poesías castellanas completas*. Ed. Elias L. Rivers. Madrid: Castalia, 1986.

Vega, Inca Garcilaso de la. *Comentarios reales*. Ed. Enrique Pupo-Walker. Madrid: Cátedra, 1999.

Vega, Lope de. *La Arcadia*. Ed. Edwin S. Morby. Madrid: Castalia, 1975.

– *La Dorotea*. Ed. José Manuel Blecua. Madrid: Cátedra, 1996.

– *La hermosura de Angélica*. *Obras completas: Poesía I*. Ed. Antonio Carreño. Madrid: Fundación José Antonio Castro, 2002. 609–970.

– *Isidro*. Ed. Antonio Sánchez Jiménez. Madrid: Cátedra, 2010.

– *Novelas a Marcia Leonarda*. Ed. Antonio Carreño. Madrid: Cátedra, 2002.

– *Los pastores de Belén*. Ed. Antonio Carreño. Barcelona: Promociones y Publicaciones Universitarias, 1991.

– *El prodigio de Etiopía*. Ed. John Beusterien. Vigo: Mirabel, 2005.

Vega Ramos, María José. "Las Indias interiores: Lope y la invención de *Las Batuecas del Duque de Alba*." *Anuario Lope de Vega* 2 (1996): 171–94.

Vélez de Guevara, Luis. *El Diablo Cojuelo*. Ed. Enrique Rodríguez Cepeda. Madrid: Cátedra, 1989.

– *La hermosura de Raquel: Primera parte*. *Flor de las comedias de España de diferentes autores: Quinta parte*. Barcelona: Sebastián de Cormellas, 1616. 133r–60v. Web. 28 Jan. 2010. <www.cervantesvirtual.com>.

Verdú, Fray Blas. *Engaños y desengaños del tiempo, con un discurso de la expulsión de los moriscos de España*. Barcelona: Sebastián Matheuad, 1612.

Viaje de Turquía. Ed. Fernando García Salinero. Madrid: Cátedra, 1995.

Vicente, Gil. *Compilaçam de todalas obras de Gil Vicente*. 2 vols. Ed. Maria Leonor Carvalhão Buescu. Porto: Imprensa Nacional, 1984.

Viguera Molins, María Jesús. "Al-Andalus como interferencia." *Comunidades islámicas en Europa*. Ed. Montserrat Abumalham Mas. Madrid: Trotta, 1995. 61–70.

Villalonga, Mariàngela. "La tradició dels tractats geogràfics humanístics a Catalunya." *Annals de l'Institut d'Estudis Gironins* 31 (1990–1): 99–112.

Villegas, Alonso de. *Flos Sanctorum: Segunda parte y Historia general en que se escriue la vida de la Virgen*. Toledo: Juan Rodríguez, 1583.

– *Fructus Sanctorum y quinta parte del Flos Sanctorum*. Cuenca: Juan Masselin, 1594.

Vincent, Bernard. "Carlos V en Granada." *Carlos V: Europeísmo y universalidad*. Ed. Juan Luis Castellano and Francisco Sánchez-Montes González. 5 vols. Madrid: Sociedad Estatal para la Conmemoración de los Centenarios de Felipe II y Carlos V, 2001. 1:283–90.

– "Les morisques et l'élevage." *Revue d'Histoire Maghrébine* 61–1 (1991): 155–62.

– "Musulmanes y conversión en España en el siglo XVII." *El río morisco.* Trans. Antonio Luis Cortés Peña. Valencia: Universitat de València, 2006. 75–87.

– "¿Qué aspecto físico tenían los moriscos?" *Andalucía moderna: Actas II Coloquio Historia de Andalucía.* 2 vols. Córdoba: Monte de Piedad y Caja de Ahorros de Córdoba, 1983. 2:335–40.

– "Reflexión documentada sobre el uso del árabe y de las lenguas románicas en la España de los moriscos (ss. XVI–XVII)." *El río morisco.* Trans. Antonio Luis Cortés Peña. Valencia: Universitat de València, 2006. 105–18.

Vine, Angus. "Etymology, Names and the Search for Origins: Deriving the Past in Early Modern England." *The Seventeenth Century* 21.1 (2006): 1–21.

Visweswaran, Kamala. "Race and the Culture of Anthropology." *American Anthropologist* 100.1 (1998): 70–83.

Vives, Juan Luis. *Instrucción de la mujer cristiana.* Buenos Aires: Espasa-Calpe, 1940.

Wade, Peter. *Race, Nature and Culture: An Anthropological Perspective.* London: Pluto, 2002.

Walter, Monika. "La imaginación del moro historiador y morisco traductor: Algunos aspectos de la ficticia autoría en *Don Quijote.*" *¿"¡Bon compaño, jura Di!"? El encuentro de moros, judíos y cristianos en la obra cervantina.* Ed. Caroline Schmauser and Monika Walter. Frankfurt am Main / Madrid: Verveurt / Iberoamericana, 1998. 35–48.

Weiditz, Christoph. *Authentic Everyday Dress of the Renaissance: All 154 Plates from the* Trachtenbuch. Ed. Theodor Hampe. New York: Dover, 1994.

Wertheimer, Elaine C. *Honor, Love, and Religion in the Theater before Lope de Vega.* Newark: Juan de la Cuesta, 2003.

Wood. Roger J. "The Sheep Breeders' View of Heredity before and after 1800." *Heredity Produced: At the Crossroads of Biology, Politics, and Culture, 1500–1870.* Ed. Staffan Müller-Wille and Hans-Jörg Rheinberger. Cambridge, MA: MIT Press, 2007. 229–50.

Woolard, Kathryn A. "Bernardo de Aldrete and the Morisco Problem: A Study in Early Modern Spanish Language Ideology." *Comparative Studies in Society and History* 44.3 (2002): 446–80.

Wulff, Fernando. *Las esencias patrias: Historiografía e historia antigua en la construcción de la identidad española (siglos XVI–XX).* Barcelona: Crítica, 2003.

Xenophon. *Memorabilia.* Trans. Amy L. Bonnette. Ithaca: Cornell UP, 2001.

Yarbro-Bejarano, Yvonne. "Juan del Encina and Lucas Fernández: Conflicting Attitudes towards the Passion." *Bulletin of the Comediantes* 36.1 (1984): 5–21.

Yndurain, Francisco. *Los moriscos y el teatro en Aragón: Auto de la destruyción de Troya y Comedia pastoril de Torcato.* Zaragoza: Institución Fernando el Católico, 1986.

Zorrilla Ortiz de Urbina, Laura. "Nuevos datos sobre la mezcla de géneros en el Barroco: El caso de la *Soledad entretenida*, de Juan de Barrionuevo." *AISO 2011: IX Congreso de la Asociación Internacional Siglo de Oro.* Université de Poitiers, Poitiers. 12 July 2011. Conference presentation.

Index

1492 (year): expulsion of non-converted Jews, 5, 44, 168, 189, 196, 215, 269n1; fall of Nasrid Granada, 5, 10, 16–17, 114, 125, 280n20; forced conversion of Jews, 5, 48, 54; origins of early Spanish drama, 79, 94; re-emergence of immemorial Spain, 127; setting of Lope de Vega, *Batuecas*, 109

Abarca de Bolea, Ana Francisca: *Vigilia y octavario de San Juan Baptista*, 18, 20. *See also* pastoral romances: Spanish
Abdera, 115
Abencerraje, El, 21, 103–4, 151, 152–61, 181, 184, 263nn1–2, 264n12, 269n41, 277n43; authorship, 265n13; and Mercader, *Prado*, 167; and Montemayor, *Diana*, 152–61; and Quinault, 278n47; and Spanish national identity, 29–30; textual history, 152, 263n1. *See also* maurophilia
Abidas, Las (Arbolanche), 18, 137, 260n47
Academia de los Nocturnos, 161
Acuña, Hernando de, 44
adargas, 164–5, 265n20
Adoration of the Shepherds, 14, 28–9, 78–85, 99–100, 250n3, 251n7; favoured by *conversos*, 192; paintings of, 56, 79–82, 99–100 (*see also individual artists*); and patron portraits, 83–5; representation of rustics, 98–100, 250n4; and royalty, 55–7, 78, 99
Africa: 8, 157, 198, 225; assigned to Cam, 271n13; in Espinel Adorno, *Premio*, 199, 200, 202–3, 207; as homeland of Moriscos, 207–8, 212, 276n41; in Luján de Sayavedra, 77, 105, 107; and Merino sheep, 36–8, 41; in Mira de Amescua, 103; Muslims, 103, 221–2, 257n22, 265n20; Spanish imperial projection in, 4, 221, 276n38
Aguilar, Alonso de, 198, 274n25
Aguilar, Gaspar: *Expulsión de los moros de España*, 30, 168, 204–6, 276n40
Agustín, Miguel: *Libro de los secretos de agricultura, casa de campo y pastoril*, 42
Alba, Duke of, 79, 109, 229
albogues (musical instrument), 176–9, 268nn35–6; in Cervantes, 174–80, 183–4, 268n34
Alcalá de Henares (Cómpluto, Complutum), 118–19, 169, 256n17, 257n24
Alcalá y Yáñez, Jerónimo de: *Alonso, mozo de muchos amos*, 24
Alcazarquivir, battle of, 10, 140
Aldana (Aragon), 132
Aldrete, Bernardo de: *Del origen y principio de la lengua castellana*, 116, 129–31, 144–5, 262n61, 262n63, 262n66
Alemán, Mateo: *Guzmán de Alfarache*, 77, 277n43; *San Antonio de Padua*, 270n6
Alfarache, Guzmán de. *See under* Alemán, Mateo; Luján de Sayavedra, Mateo
Alfonso VIII, 125
Alfonso X of Castile, 114, 121; *Primera crónica general*, 240n9; *Siete partidas*, 40, 108

Alfonso XI of Castile, 85
Alhambra, 15–16, 277n43
aljamiado, 103, 235
almagre, 39, 46–7, 50, 52, 249n34; distinguishing cultural practices, 40–1; ethno-religious communities, 39–43, 247n18
Almagro, 255n11
Almansur, 155
Almanzora (river), 127
Almería (Abdera), 115
Alonso Cortés, Narciso, 118
Alonso de Herrera, Gabriel: *Obra de agricultura*, 41–2
Alonso, mozo de muchos amos (Alcalá y Yáñez), 24
Alpujarras, Morisco uprising of the, 5, 127, 198, 268n36. *See also* Pérez de Hita, Ginés
Althusser, Louis, 25–6
Álvarez de Toledo, Fadrique, Duke of Alba, 79, 229; character in Claramonte, *Valiente negro*, 226
Alves, Abel, 245n2, 262n63
Amanece que no es poco (Cuerda), 230–3, 280n17
Andalusia, 10, 125, 118–19, 130, 169, 200, 206, 256n14; and Felismena, in Montemayor, *Diana*, 152, 155–6
Anderson, Benedict, 10, 75
Andrés, Juan: *Confusión o confutación de la secta mahomética y del Alcorán*, 275n36
ʿansara, 162–3
Antigüedad y excelencia de Granada (Bermúdez de Pedraza), 262n66
antiquity, 16, 112–47; and rustic clothing, 100, 128. *See also* archaeology; time: immemorial
anxiety, 60, 94, 138, 169, 230; around blood purity, 43; in Cervantes, *Galatea*, 119; cultural, 16, 21, 29, 94, 122, 262n66; about European perception, 158; about genealogical origins, 88, 243n33; about identity, 4, 30–1, 123
Arabic (language): *albogue*, 175–9; in Barrionuevo y Moya, 210, 211–13; etymologies from, 104, 132–4, 246n7, 259n41, 273n24; Guadix, Diego de, Arabist and translator, 132–4; place names, 113–20, 130–1, 162, 169, 172, 256n15, 257n24, 259n41, 259n43, 273n24; suppression of, 5, 254n1
Aragon, 10–11, 219, 227–8, 235–6; Moriscos of, 5, 48, 235
Arbolanche, Jerónimo de, 235; *Las Abidas*, 18, 137, 260n47
Arcadia (text). *See under* Sannazaro, Jacopo; Sidney, Philip; Vega, Lope de
Arcadia, Spanish (figure), 29, 90, 97, 134, 143, 147, 168, 173, 188, 204, 217; in Cervantes, *Don Quixote*, 173–4, 187; contested, 151–85, 189, 197; court as realization in Pantoja de la Cruz, *Birth*, 57; and cultural cleansing, 30, 169, 172, 179–80, 208, 217; as ethnocentric, 30, 154, 158, 173, 179; and history, 111, 113, 146; in Mercader, *Prado*, 168, 184; Montemayor, *Diana*, as model, 159, 161, 167; Moors in, 30, 159, 161–2, 206, 208, 217; and place names, 117, 121, 164
Arce Solorzeno, Juan de: *Tragedias de amor*, 18, 20–1, 140, 260n47, 268n35
archaeology: early modern, 8, 111, 116, 129; rustic speech as archaeological proof, 109
Argote de Molina, Gonzalo: *Discurso sobre la poesía castellana*, 243n33; introductory eulogy to Morales, *Antigüedades*, 114–15; *Nobleza del Andalucía*, 126, 258n32
Arias Montano, Benito, 116
aristocrats, 90, 163, 166, 167, 262n63, 273n23; and blood purity, 163, 245n47; in Mercader, *Prado*, 168; Moorish, 152–3, 213. *See also* class; festivals and festivities; *hidalguía*; identity: aristocratic
Armstrong-Roche, Michael, 269n42
Arribas, Julian, 161, 265n14
Astrée, L' (Urfé), 17, 236
Asturiano, Bernardo: *Pilgrimage to the Virgin of Cabeza*, 87
Asturias, 117–18, 255n11
Atxaga, Bernardo: *Obabakoak*, 232–3
authenticity, 90–2, 103, 109–10, 142, 253n38

Avalle Arce, Juan Bautista, 20, 24, 273n23
Ávila, Juan de, 95, 101, 179–80, 252n25
Aznar Cardona, Pedro: *Expulsión justificada de los Moriscos Españoles*, 12, 265n18, 268n36, 276n37

Baena, Julio, 71, 153, 173, 176, 269n41
Balbuena, Bernardo de: *Siglo de Oro en las selvas de Erífile*, 18, 137–8, 252n27, 260n47
Balibar, Étienne, 10, 29, 76, 147
ballads: 21–4, 57, 86, 212, 227; Juan de la Cueva, *Coro Febeo*, 108–9; medieval frontier, 21–2, 86, 264n12; Moorish, 21–4, 244n41, 264n12, 277n43
– titles: "Descúbranse ya esas caras," 23; "Después que el rey don Rodrigo," 123–4; "La merienda del moro," 86; "Mira, Zaide, que te aviso," 277n43; "Oídme, señor Belardo," 22–3; "Por Dios, señores Poetas," 23; "Por qué señores poetas," 23–4; "Señor Moro vagabundo," 23; "Tanta Zaida y Adalifa," 22
Ballester Rodríguez, Mateo, 11, 254n39
baños de Argel, Los (Cervantes), 93–4
Barrientos, Fray Lope de, 6, 27
Barrionuevo y Moya, Juan de, 277n44; *Primera parte de la soledad entretenida*, 18, 20, 30, 187, 208–16, 217, 227, 262n63, 277n45, 278n47
Bartlett, Robert, 241n16, 242n23
Basque (language), 18, 101, 131. *See also* Echave, Baltasar de
Basques, 101, 128–9, 131, 232–3, 235; Virgin of Aránzazu, 251n17
Bass, Laura, 263n2
Bassano, Jacopo: *The Journey of Jacob*, 249n34
Batlle Matabosch, Ramón, 37–8, 71–2
Batuecas del Duque de Alba, Las (Vega), 109
belén. See Nativity
belén de Salzillo en Murcia, El (Giménez Caballero), 223–4
Benengeli, Cide Hamete, 180–1, 187, 269n40. *See also* Cervantes, Miguel de: *Don Quixote*
Bermúdez de Pedraza, Francisco: *Antigüedad y excelencia de Granada*, 262n66
Bernis, Carmen, 96, 104, 253n34
Bethlehem: heterotopia, 193–4; in Lope de Vega, *Pastores*, 187–97; in Lope de Vega, *El prodigio de Etiopía*, 226; travel accounts, early modern, 271n10
Betis. *See* Guadalquivir; Seville
Beuter, Pere Antoni: *Primera parte de la història de València*, 120, 257n23
Beverley, John, 25
Bible: Gospels in Arabic translation, 133, 259n42; Gospels in Castilian translation, 246n9. *See also* Ezekiel; Genesis; John; Luke; Matthew
biopolitics: and Christian doctrine, 250n37; fascist, 38; Spain, early modern, 248n27
biopower, 58–9, 67–71, 75, 248n23, 248n28 and traditional iconography, 69–70; ruler as doctor, 248n25. *See also* engineering, ethnic; shepherd, figure of; Shepherd, Good: culling undesirable sheep
Blacks, 108, 110, 225–34; in Atxaga, 232–3; in Barrionuevo y Moya, 216; contemporary immigrants from Africa, 229–33; as shepherds, 195–6, 230–2, 272n14
Bleda, Jaime, 278n50; *Corónica de los Moros de España*, 12, 248n23; *Defensio Fidei in causa neophytorum*, 267n26; *Vida y milagros de San Isidro el Labrador*, 258n33
blood purity. *See* purity, blood
bohordos, 164
Borgoña, Juan de: *Nativity*, 79, 80
Borromeo, Federico: *De pictura sacra*, 250n4
Boscán, Juan, 15–17, 152, 158, 243n33; in Vélez de Guevara, *Diablo Cojuelo*, 117
Botelho de Carvalho, Miguel: *El pastor de Clenarda*, 18
Bourdieu, Pierre, 13, 76–7
Bramón, Francisco: *Los sirgueros de la Virgen sin original pecado*, 18, 252n28, 269n2. *See also* pastoral romances: Spanish
breeding, sheep, 28, 37–8, 41, 58, 69, 129, 248n28. *See also* Jacob: livestock; sheep: Merino
bucolic, 26–7; in *Abencerraje*, 153–4; in

Amanece que no es poco, 230–2; in Carmen de Burgos, 221; in Cervantes, *Don Quixote*, 180; in Cervantes, *Trabajos*, 182–3; and denial of history, 172; in Espinel Adorno, *Premio*, 205; in Lope de Vega, *Galatea*, 170; Menéndez y Pelayo on, 17–18; vs. pastoral, 27; in public celebrations, 90
Burgos, Carmen de: "En la guerra (Episodios de Melilla)," 220–1, 279n3
Burke, Peter, 252n21, 254n40
Burshatin, Israel, 153, 263n2
Busquets, Rafel: *Llibre de la invenció y miracles de Nostre Señora de Lluc*, 89

caballeros villanos, 86. *See also* aristocrats; class
Cabeza, Nuestra Señora de la, 85–6
Cabranes-Grant, Leo, 109
cabrero (goatherd), 27, 210–11, 278nn48–9; in *Amanece que no es poco*, 230–3; in Barrionuevo y Moya, 209–13, 216; in Gil de Biedma, 225; in Iglesias, "Último cabrero," 233; Moors and Moriscos as debased shepherds, 210–11, 278n47. See also *ovejero*; *pastor*
Calderón de la Barca, Pedro: *El verdadero Dios Pan*, 262n63
Calpe, Monte, 113, 198–9
çamarrón/çamarros. See under *zamarras*
Camón Aznar, José: *El pastor Quijotiz*, 227–8
canes, game of (*juego de cañas*), 90, 107, 163–6, 265n20, 266n23. See also *capellar*; cultural practices, Moorish; *marlota*
capellar (Moorish hood), 103, 165
Cardona, Sancho de, Admiral of Aragon, 167
Carducho, Vicente: *Diálogos de la pintura*, 98–9
Carpio, Marquis del, 275n33
Carranza, Paul, 20
Carrasco Urgoiti, María Soledad, 22
Carreño, Antonio, 188
Cartagena, Alonso de: *Defensorium Unitatis Christiane*, 45, 192
cartography, 29, 113–21, 254n2. *See also* place names
Carvajal, Michael de: *Cortes de la muerte*, 247n17
Cascales, Francisco: *Tablas poéticas*, 23
Casey, James, 267n27
castas, 9, 221–2. *See also* communities: ethno-religious
Castile, 10–11, 79, 85, 117, 235–6, 256n19; Basques in, 128; and Cervantes, *Galatea*, 169; and Giménez Caballero: *España nuestra*, 224; Jews, 215; kings, 40, 46–7, 52, 101, 114; Moriscos, 5, 48, 54; sheep herding and wool production, 3, 35, 122; and Spanish national identity, 4. *See also under* identity: Spanish
Castillo Martínez, Cristina, 203, 274n28
Castro, Américo, 9–10, 25, 82–3, 244n43, 264n8
Castro y Añaya, Pedro de: *Auroras de Diana*, 268n35
Catalonia, 10, 254n2
Catholic monarchs, 5, 10, 46, 51, 52, 94, 114, 126, 198
Caxa de Leruela, Miguel: *Restauración de la abundancia de España*, 258n27, 262n64
censorship, 138–9, 188, 244n43, 260n53; modern, 270n4
censuses. *See* Moriscos: censuses
Cervantes, Miguel de, 151–2, 168–84, 184–5, 245n48, 267n29, 268n34; *Los baños de Argel*, 93–4; *La casa de los celos*, 173; "El coloquio de los perros," 173; *Don Quixote*, 20, 30, 173–81, 187, 268n35, 269nn40–1, 273n20; *La Galatea*, 18, 20, 118–19, 138, 168–73, 256n14, 261n56, 267n28; *La Numancia*, 23; *Ocho comedias y ocho entremeses*, 173; *Los trabajos de Persiles y Sigismunda*, 173, 182–4
Charles V, 5, 15–16, 49, 176, 253n30; as Good Shepherd, 57; in López de Enciso, *Desengaño de celos*, 141; and prophecies of Good Shepherd, 44–5; suppression of Morisco cultural practices, 5, 16, 40, 176, 254n1
Chaves, Feliciano, 105
Chevalier, Maxime, 90, 160, 166

Childers, William, 92
chivalry, books of, 18, 173, 181, 269n40
Christ: birth, 14, 187–8; death of, 258n36, 261n59; Godínez, 270n8; Good Shepherd, 44–5, 47, 49, 194, 246n9, 270n5; *hidalguía* of, 190–2, 270n6; Jewish lineage, 142; Lope de Vega, *Pastores*, 187–97, 226; and paganism, 137, 141–2, 145–6, 233, 262n63
Christian, William, 251n14
Christians, New. See *conversos*
Christians, Old, 4, 5–6, 29, 42, 51, 67, 83, 88, 93, 124–5, 128, 137, 193, 240n7, 240n9, 259n45, 264n8; community, 25, 29, 76, 184; figured by the shepherd, 13–14, 26, 28–9, 79, 83, 85, 94, 192, 293. *See also* cultural practices: Old Christian; identity: Old Christian; rustics
Cid, El, 157, 222
cinco libros intitulados de la enamorada Elisea, Los. See under Covarrubias Herrera, Jerónimo de
Cintia de Aranjuez, La (Corral), 18, 268n35
Ciscar Pallarés, Eugenio, 267n2
Civil War, Spanish. *See* Spanish Civil War
Claramonte, Andrés de: *El valiente negro en Flandes*, 226
class, 4, 11, 25–6, 78–9, 84, 86, 90–2, 100, 110, 147, 213, 225, 229, 244n45; appropriation of popular culture, 84, 92, 252n21, 254n39; and Arcadia in Cervantes, *Don Quixote*, 173–4; interests, 168, 184; and *pellicos*, 96–100, 256n20, 273n23; struggle, 221–2. *See also* aristocrats; culture: popular
classicism, 17, 112–47, 206; cultural supersession, 29, 158; Gentile vs. Latin legacy, 147; and maurophilia, 162; negation of Moorish and Jewish legacy, 17, 29, 118, 147; and pastoral, 29, 112–15, 152. *See also* cleansing: cultural
cleansing: cultural, 16–17, 26, 29–30, 113–14, 168–84, 176, 234, 241n19, 254n1 (*see also* culling; purity, blood: cultural); ethnic, 59, 62, 66, 67, 94, 109, 114, 152, 184, 189, 200, 205, 216, 234; — and biopower, 68–9 (*see also* segregation)
clothing: aristocratic, 97, 107, 124–5; Gypsy, 195, 272n16; Jewish, 101; Moorish, 5, 78, 103–4, 105, 166, 202, 221, 266n23; pastoral, 76, 77–8, 94–107, 124–5, 128, 221, 252nn27–8. *See also* canes, game of; culture: popular; identity: Spanish; *pellicos*; rustics; sartorial practices; *sayo*; Toledo; *zamarras*
Cock, Hendrik, 257n23
Colmenares, Diego de: *Historia de la insigne ciudad de Segovia*, 115
Colonge, Chantal, 22
Columbus, Christopher, 26
Comedia pastoril de Torcato, 235
commoners, 84, 99, 110, 174. *See also* rustics; shepherds
communities: Black, 272n14; Christian, 38, 45, 247n22; ethno-religious, 7–8, 13, 52; imagined, 10, 75–6, 85, 174, 233; Jewish, 5; Morisco, 52; Old Christian, 13, 25, 28–9, 66–7, 94; pastoral, 28–9, 75–6, 101, 118–19, 121, 172; removal, 57, 187. *See also* sheep; Shepherd, Good; shepherds, community of
Cómpluto (Complutum). *See* Alcalá de Henares
Conde Lucanor (Juan Manuel), 108
conflict: ethno-religious, 5–6, 9, 10, 25, 29, 88, 93, 104, 147, 252n2. *See also* class
Confusión o confutación de la secta mahomética y del Alcorán (Andrés), 275n36
Conservación de monarquías (Fernández de Navarrete), 196
constante Amarilis, La (Suárez de Figueroa), 18, 267n31
contamination, 48, 146–7, 206. *See also* purity, blood; *raza*
Contreras Contreras, Jaime, 25
Conveniencia de las dos Monarquías Católicas (Puente), 145, 262nn63–4, 276n41
conversión de la Magdalena, La (Malón de Chaide), 138
conversos, 5–7, 28, 45, 48, 51, 55, 76, 82–3, 85, 89, 93, 110, 179, 234–5, 240n9, 251n9, 267n30, 268n38, 269n1, 270n7; and Adoration of the Shepherds, 192; defence of in *El Abencerraje*, 263n2;

shepherd as insult from *conversos*, 192–3, 215–16, 270n9; and stigmatization, 53, 190, 217, 247n18; writers, 24, 82–3, 170n6, 244n40, 244n43, 252n25, 264n8, 265n19. *See also* Muslims: conversion; purity, blood: statutes
convivencia, 5, 9, 137, 153, 163, 167, 180, 244n43, 260n51, 271n10. *See also* pluralism, ethno-religious; religion
Cooper, Helen, 278n48
Coplas de Mingo Revulgo. See *Mingo Revulgo, Coplas de*
Córdoba, Gonzalo de, 198
Coro Febeo de romances historiales (Cueva), 108–9. *See also* ballads; *fabla antigua*
Corónica de los Moros de España (Bleda), 12, 248n23
Coroucli, Maria, 262n65
Corral, Gabriel de: *La Cintia de Aranjuez*, 18, 268n35
Corral, Pedro del: *Crónica sarracina*, 123
Cortes de la muerte (Hurtado de Toledo), 50–1, 93, 247n17
Council of State: closure of expulsion of Moriscos, 216, 276n41; on expulsion of Gypsies, 195–6
Covarrubias Herrera, Jerónimo de: *Los cinco libros intitulados de la enamorada Elisea*, 18, 136–7, 260n47, 261n58, 264n12; *Segunda parte de la enamorada Elisea*, 19. *See also* pastoral romances
Covarrubias, Sebastián de: *Tesoro de la lengua castellana o española*, 39–40, 42, 43, 96, 104, 116, 163–4, 177–8, 178–9, 246n7, 247n16, 247n20, 252n28, 257n23, 258n28, 273n24, 274n29
cristianos viejos. *See* Christians, Old
cristianos nuevos. See *conversos*; Moriscos
Crónica sarracina (Corral), 123
Cuerda, José Luis, 230–3, 280n17
Cueva, Juan de la: *Coro Febeo de romances historiales*, 108–9
Cueva, Luis de la: *Diálogos de las cosas notables de Granada*, 276n41
culling, 37, 41–3; ecocide and ethnocide, 70–1, 241n18; model of cleansing and managing populations, 28, 48–50, 57–9, 70–1. See also *raza*; sheep; Shepherd, Good
cultural practices: Gypsy, 195; Jewish, 40–1, 241n19; Moorish, 8–9, 23–5, 30, 40–1, 78, 151, 162–6, 241n19, 275n36, 276n37; – and aristocratic identity, 184, 241n19, 266n23; – in Espinel Adorno, *Premio*, 201–4; – prohibition of, 5, 16, 54, 175–6 (*see also* aristocrats; canes, game of; clothing: aristocratic; cultural practices Morisco; culture: material; identity: aristocratic); Morisco, 23–4, 40–1, 102–7, 167, 181; – prohibition of, 5, 16, 40; Old Christian, 167, 179, 252n24. *See also* culture: popular; purity, blood: cultural; rustics
culture: fictitious agency of the people, 92–3, 252n22; fictitious political bond, 84, 92, 252n21, 254n39; material, 9, 122, 128, 167; popular, 29, 92, 110, 112. *See also* cultural practices; folklorism; tradition

Damiani, Bruno, 264n8
Dante Alighieri: *De vulgari eloquentia*, 262n66
Darst, David, 153
Dávalos y Figueroa, Diego: *Primera parte de la miscelánea austral*, 118, 256n15, 261n60
De Armas, Frederick, 181, 261n57, 274n28
death-function, 59. *See also* biopower
Defensorium Unitatis Christiane (Cartagena), 45, 192
Delahaye-Grélois, Séverine, 255n8
Del origen y principio de la lengua castellana. *See under* Aldrete, Bernardo de
Delpech, François, 275n35
Denia, Marquis of. *See* Sandoval y Rojas, Francisco de
de-Orientalization, 10; in and of Montemayor, *Diana*, 159, 161. *See also* re-Latinization
De pictura sacra (Borromeo), 250n4
Descendencia ilustre de don Yñigo Manrique de Lara (Espinel Adorno), 197, 273n22
Desengaño de celos (López de Enciso), 18,

140–2
determinism: biological, 241n16; ethno-religious, 6, 9–10
Diálogos de contención entre la milicia y la ciencia (Núñez de Velasco), 114
Diálogos de la pintura (Carducho), 98–9
Diana (mythology), 204, 259n45
Diana, La. See under Fernández de Córdoba, Gabriel; Gil Polo, Gaspar; Hernández, Gabriel; Montemayor, Jorge de; Pérez, Alonso; Tejeda, Jerónimo de
Díaz del Castillo, Bernal: *Historia verdadera de la conquista de la Nueva España*, 93
Díaz Viana, Luis, 92, 253n30
Diccionario de la Real Academia, 175
diez libros de la fortuna de amor, Los (Lofrasso), 18
difference: 52, 75, 78, 109; biological, 7, 38, 76, 221, 228–9, 232–3 (*see also* race); cultural, 15, 17, 75–6, 181; ethno-religious, 9, 13, 22, 29, 54–5, 75–6, 137, 190, 194, 240; – and clothing, 78, 105; – and racial difference, 5–6, 29, 227, 235 (*see also* diversity, ethno-religious); geopolitical, 225, 235 (*see also* Spain: regions); religious, 45–7, 52, 240n10; and sheep herding terminology, 38, 41, 43, 50, 76
Discurso de la expulsión de los moriscos de España (Verdú), 265n18, 271n12
Discursos de la certidumbre de las reliquias descubiertas en Granada (López Madera), 262n66
diversity, ethno-religious, 25, 51, 237; hierarchy within, 209, 215–16; in Lope de Vega, *Pastores*, 193–7; and sheep herding, 69–71. *See also* difference: ethno-religious
dogma, Catholic, 135–6
Domínguez Ortiz, Antonio, 6, 240n10, 274n31
Don Quixote. See under Cervantes, Miguel de
drama: and *conversos*, 82–3, 215, 270n9; Moriscos in, 103–4; pastoral, 14; vs. pastoral romances, 113; and rusticity, 113; Spain, 79–83, 93–4; unreliable ethnographic documents, 100. *See also* festivals and festivities; *officium pastorum*
Dudley, Edward, 169
Duero (river), 118, 204

Ebro (river), 169, 204, 228
Echave, Baltasar de: *Discursos de la antigüedad de la lengua cantabra bascongada*, 101, 128–9, 131, 258n35
Écija (Soldina; Astigi), 118, 256n15
Écija, Fray Diego de: *Libro de la invención de esta Santa Imagen de Guadalupe*, 85
Emblemata Centum, regio politica (Solórzano Pereira), 249n30
enamorada Elisea, La. See under Covarrubias Herrera, Jerónimo de
Encina, Juan del, 14, 79, 110; *Égloga de Mingo, Gil y Pascuala*, 101; *Translación de las bucólicas de Virgilio*, 278n47; *Trivagia*, 271n10
Encinar, Jesús, 225
engineering, ethnic, 28, 75, 189, 193–4, 271n12. *See also* biopower
En torno al casticismo. See under Unamuno, Miguel de
epic: of the expulsion, 30, 168, 204, 273n29, 276n38. *See also individual authors*
Escolano, Gaspar de: *Segunda parte de la década primera de la historia de la insigne y coronada ciudad y reino de Valencia*, 116, 257n23
España en su historia. See Castro, Américo
Espejo de religiosos, 95
Espinel Adorno, Jacinto de, 197–8; *Descendencia ilustre de don Yñigo Manrique de Lara, alcayde de las fuerças y castillos de la ciudad de Malaga*, 197, 273n22; *El premio de la constancia y pastores de Sierra Bermeja*, 18, 30, 187, 197–208, 217, 273n23, 274n27
Espinel, Vicente, 197–8; *Vida de Marcos de Obregón*, 198, 210–11
Ethiopians, 60, 195–6, 226, 272n14. *See also* Blacks
ethnic cleansing. *See* cleansing: ethnic
ethnicity, 7–8, 75–6, 241n14, 241n16. *See*

also communities: ethno-religious; identity: ethno-religious; shepherd, figure of
ethnocentrism, 6–8, 28, 51, 73, 195, 233–7; contested, 151–84; ethnocide and genocide, 241n18; exacerbated by expulsion of Moriscos, 186–7; exposed in Cervantes, *Don Quixote*, 173–4; and pastoral, 205, 220, 228; and zootechnical discourses, 39
ethnocide, 186–7, 241n18. *See also* cleansing; culling
ethnogenesis, 100–1, 128, 134, 235, 241n19
etymology: of *albogue*, in Cervantes, *Don Quixote*, 174–80; from Arabic, 132–4; early modern, 129–34, 142–6, 254n1
Eusebius of Cesarea, 145
evangelic message, 48, 190, 192. *See also* evangelization
evangelization, 44, 139; of Moriscos, 51, 167, 276n37; universal, 78
exclusions, 4, 6–8, 25, 28, 30, 39, 76, 217, 220, 227; and discourse of economics (biopower), 69; in pastoral romances, 146–7, 174, 216; racial, 240n10, 280n17. *See also* inclusions and exclusions; Jews: expulsion; Moriscos: expulsion
expulsion. *See under* Jews; Moriscos
Expulsión de los moros de España. *See under* Aguilar, Gaspar
Expulsión justificada de los Moriscos Españoles. *See under* Aznar Cardona, Pedro

fabla antigua, 108–9, 253n38. *See also* Arabic; *sayagués*
faith (*fe*), 136–7, 155–8, 184, 260n50, 264n6; religious, 44, 47, 51, 54, 79, 89, 176–7, 248n26, 249n33, 251n9
famosa Épila, La, 19, 236
Fanjul, Serafín, 9, 254n1
Faria e Sousa, Manuel de, 19
Farsa del Sacramento llamada de los lenguajes, 93
Farsa nuevamente compuesta sobre la felice nueva de la concordia y paz (López de Yanguas), 176
Farsa turquesana (López de Yanguas), 93
Felismena (*La Diana*), 152, 154–7, 181, 263n3, 264n4
Ferdinand of Aragon, 10, 51, 94. *See also* Catholic monarchs
Fernández Albaladejo, Pablo, 121
Fernández de Castro, Pedro, 21
Fernández de Córdoba, Gabriel: *Tercera y cuarta parte de la Diana*, 19
Fernández de Navarrete, Pedro: *Conservación de monarquías*, 196
Fernández de Oviedo, Gonzalo, 258n32
Fernández, Lucas, 14, 79, 83
Fernández-Cañadas de Greenwood, Pilar, 24
festivals and festivities, 29, 89–94, 162–3, 188
fictionality, 90–1, 110
filósofo, El (Gómez Tejada), 60
Finello, Dominick, 90–1, 245n48
Flecniakoska, Jean Louis, 173
Flos Sanctorum, Quinta parte del (Villegas), 100–1, 145
folklorism, 92, 194. *See also* culture: popular
Forcione, Alban, 180
Fosalba, Eugenia, 265n13
Foucault, Michel, 58–9, 67–9, 193, 248nn27–8
Franco, Francisco, 30, 37–8, 219, 223, 225, 227
frontier: in literature: 21–2, 86, 152–4, 209–13, 217, 264n12; Marian apparitions, 125; socio-economic system, 86; wars, 36, 273n22. *See also* ballads
Fuchs, Barbara, 9, 166–7, 177, 241n19, 263n2
Furió Ceriol, Fadrique: *El Concejo i Consejeros del Príncipe*, 260n51

gaita zamorana, 178–9
Galatea, La. *See under* Cervantes, Miguel de
Galicia, 117–18, 228–9, 233, 256n13, 280n16, 280n19
Gállego, Julián, 14, 79
Gálvez de Montalvo, Luis: *El pastor de Fílida*, 18, 20–1, 97, 256n20, 260n47, 268n35
ganaderos. *See* livestock owners
García Ferrer, Pedro, 83

García Martín, Pedro, 37
Garcilaso. *See* Vega, Garcilaso de la
Gaylord, Mary, 263n2, 268n33
genealogy, 5–10, 16, 21, 84, 88, 105, 246n7–8, 276n41; and blood purity, 117, 126, 189, 250n5; of Castilian poetic forms, 15–16, 243n33; difference, 28–9, 50; erasure, 45, 52, 247n15; Hebrew books of lineages, 193–4; institutional records of Morisco lineages, 54–5; invisible difference, 41, 94; manipulation of, 66; mixed (Herod), 196; Moriscos and Old Christians, in Espinel Adorno, *Premio*, 206–8; pastoral, 117–18; purity, 31, 66, 103, 105, 180; rustic, 82; sheep herding, 28, 37, 43, 71, 75; shepherds, 142, 146. *See also* purity, blood; rustics; shepherds
Genesis, 36, 59–60, 63, 121, 193, 271n13
genocide, 193–4, 196–7, 205, 241n18
geography. *See* cartography
Gibraltar (Monte Calpe), 113, 198–9
Gil de Biedma, Jaime, 225, 279n10
Gil Herrera, Jorge, 275n33
Gilman, Stephen, 250n5
Gil Polo, Gaspar: *Primera parte de la Diana enamorada*, 18, 20, 120, 139, 268n35
Giménez Caballero, Ernesto, 30–1; *El belén de Salzillo en Murcia (Origen de los Nacimientos en España)*, 223–4; *España nuestra*: *El libro de las juventudes falangistas*, 224–5; and Miguel Hernández, 279n7–n9
gineta, 163–4
glosses, to *Mingo Revulgo*, 46–7, 52
goat/goatherd. See *cabrero*
Godínez, Felipe: *Coloquio primero y segundo de los Pastores de Belén*, 188–9, 227, 262n63, 270n8
Golden Fleece, 130
Gómez Tejada, Cosme: *El filósofo*, 60
Góngora, Luis de: *Soledades*, 209, 277n46
González Alcantud, Antonio, 9
González de Bobadilla, Bernardo: *Ninfas y pastores de Henares*, 18, 256n17, 256n20, 260n47, 268n35
Gonzalo, Arias, 264n12
Good Shepherd. *See* Shepherd, Good
goticismo (Gothicism), 124. *See also* Visigoths
governador christiano, El (Márquez), 57
governmentality, 58–9. *See also* Foucault, Michel
Graeco-Latin. *See* legacy: classical; Romans
Granada (Christian kingdom), 15–17, 51, 103, 115, 127, 133, 158, 189, 197–8, 252n25. *See also* Bermúdez de Pedraza, Francisco; Cueva, Luis de la; Mármol Carvajal, Luis de; Pérez de Hita, Ginés
Granada (Nasrid kingdom): fall of, 5, 10, 79, 94, 109, 125, 152, 243n33, 251n16; Sannazaro, farces on, 280n20; setting of Barrionuevo y Moya, 209–16
Greene, Roland, 243n34
Grell, Chantal, 121
Gruzinski, Serge, 139
Guadalajara y Xavier, Marcos de: *Memorable expulsión y justísimo destierro de los moriscos de España*, 53–4
Guadalaviar (river), 120, 162, 257n23
Guadalevín (river), 198–200, 205–7
Guadalhamar (river), 213
Guadalquivir (river), 117, 119, 129–31, 169, 204, 255n8, 256n14, 257n22, 257n24
Guadalupe, Virgin of. *See under* Mary, Virgin
Guadiana (river), 204, 259n40
Guadix, Diego de: *Recopilación de algunos nombres arábigos que los árabes pusieron a algunas ciudades y otras muchas cosas*, 96, 116, 132–4, 178–9, 257n23, 259n41, 259n43, 273n24
Guidarroman (river), 259n40
Guillén, Claudio, 153
Gutiérrez, Alonso, report on Moriscos, 12, 40–1
Gutiérrez de los Rios, Gaspar: *Noticia general de la estimación de las artes*, 84
Guzmán, Gaspar de, 10, 67
Guzmán de Alfarache (pícaro). *See under* Alemán, Mateo; Luján de Sayavedra, Mateo
Gypsies, 57, 189, 195, 272n15; in Lope de Vega, *Pastores*, 30, 187, 195–7, 217, 272n16
habitus, 13, 76–7, 227; pastoral, 13, 26,

76–8, 104, 208, 220–5, 227–9, 232–4
hagiographies: local, 87; Lope de Vega, *Isidro*, 258n33, 262n63
Ham (son of Noah), 121
Harrison, Faye, 241n16
Harvey, L.P., 246n7, 264n7, 266n25
Heaney, Seamus, 205
Hebrew: language, 130; people, 187–96, 216, 251n9. *See also* Jews
Henry II of Castile, 39–40
Henry III of Castile, 40
Henry IV of Castile, 46–7, 101–2
Hermenegildo, Alfredo, 14, 250n5
hermosura de Raquel, La (Vélez de Guevara), 61–3
Hernández, Gabriel: *Tercera parte de la Diana*, 19
Hernández, Miguel, 30–1, 279n7; *El pastor de la muerte*, 221–5
Hernández Caso, Francisco, 102
Hernández del Portillo, Alonso: *Historia de la muy noble y más leal ciudad de Gibraltar*, 113
Hernández-Pecoraro, Rosilie, 18, 25, 157–8, 244n46
Herod, 193–4, 196–7, 271n11
Herrera, Fernando de, 23
Herrera, Pedro de, 102–3
Herzog, Tamara, 11, 242n24
heterogeneity, 17, 45–7, 253n30; in Lope de Vega, *Pastores*, 187, 193–4; and pastoral, 109, 172, 180. *See also* difference; diversity
heterotopia, 193–4
hidalguía, 117, 128–9, 191, 270n7; of Christ and the Virgin Mary, 190–1, 270n6; of Jews in Lope de Vega, *Pastores*, 193–4. *See also* purity, blood
hilaza, 42–3
Hispalis, 256n15. *See also* Seville
Historia de la insigne ciudad de Segovia (Colmenares), 115
Historia de la muy noble y más leal ciudad de Gibraltar (Hernández del Portillo), 113
Historia del rebelión y castigo de los moriscos del Reino de Granada (Mármol Carvajal), 51, 127, 258n34
Historia verdadera de la conquista de la Nueva España (Díaz del Castillo), 93
historiography:
- early modern, 36, 109, 112–47; and ancient Iberia, 142–7, 261n60; Basque, 128–9; etymology, 129–34; forgeries, 108, 253n38; image of imperial Spain, 158; Lope de Vega, *Batuecas*, 109; Moorish, classified as fiction, 278n53; myths, 23, 214–15; narratives, 131; and pastoral romances, 29, 110–11, 121–9; pastoralization of, 123, 130; place names, 113–21. *See also* Alfonso X of Castile; Bermúdez de Pedraza, Francisco; cartography; Colmenares, Diego de; Díaz del Castillo, Bernal; Echave, Baltasar de; Escolano, Gaspar de; etymology; Hernández del Portillo, Alonso; Luna, Miguel de; Mariana, Juan de; Mármol Carvajal, Luis del; Medina, Pedro de; Morales, Ambrosio de; Nebrija, Antonio de; Ocampo, Florián de
- medieval, 29, 114, 121, 123
- modern, 37; Moriscos, 11; sheep herding, 36–9; and social context, 38–9. *See also* scholarship, modern
Hobsbawm, Eric, 29, 109–10, 254n39
homogeneity, 8, 10, 25, 44, 76, 137, 167–8, 240n10, 242n23, 244n46
homogenization, ethno-religious, 5, 10, 44–5, 50, 100, 123, 134, 137, 168; historical, 110, 123
Huarte de San Juan: *Examen de ingenios para la ciencia*, 272n15
Huerga, Fray Cipriano de la, 49–50, 52–3, 247n19
Hurtado de Toledo, Luis: *Cortes de la muerte*, 50–1, 93, 247n17; *El teatro pastoril en la ribera del Tajo*, 18–19
Hurtado Martínez, Antonio: *Primera parte de Piroeles y Musidoro en Arcadia*, 19

iconography, pastoral, 57, 69, 87, 89, 92, 226
identity:
- African, 230
- aristocratic, 168; and Moorish cultural practices, 30, 151, 161, 184, 241n19, 266n23. *See also* aristocrats; clothing;

cultural practices: Moorish; *pellicos*; *zamarras*
- collective: anxiety, 30; codified by performance of tradition, 92; identification, 11; shepherds as ambivalent figures, 85; shepherds as essentialized model of, 71–5, 134; strategic and situational, 8. *See also* Christians, Old; identity: Spanish
- ethno-religious: *El Abencerraje* in, 151, 153; basis of division of labour, 12; in Cervantes, *Galatea*, 170–3; debates on 17, 21, 29–30, 36, 45, 51–2, 59, 89, 147, 195, 237, 250n3, 262n66, 278n48; in Lope de Vega, *Pastores*, 191–2; and "race," 7–8; in Vélez de Guevara, *Diablo*, 117. *See also* Christians, Old; communities: ethno-religious; Jews; Moors; Moriscos; Muslims
- Galician, 280n19
- Jewish, 76, 85, 193
- local, 88, 168, 241n19
- memory, 11, 92, 158
- Moorish, 153
- Muslim, 229
- Old Christian: 76, 77, 95–6, 103, 104, 108, 193
- Spanish, 10–11, 242n24, 244n46; and Basque, 128; Castile, 4, 108, 110–11, 131, 169, 224–5; essentialized as Christian Roman-Visigothic, 8–9, 147; exclusivist view, 11, 14, 29–30, 39, 76, 102, 147, 154; Gentile, 134–42; *morabito* of Barrionuevo y Moya, *Premio*, 207–8; negotiation post-expulsion of Moriscos, 216–17; oppositional and negative, 76; purity, 109; Tajo and, 169–70; timeless, 37; in Unamuno, 3–4. See *also* Castile; homogeneity; homogenization, ethno-religious; *reconquest*; *sayagués*; Spain: European perception of; Unamuno, Miguel de

ideology, linguistic, 105, 211, 233, 278n50. *See also* sartorial practices
Iglesias, Óscar, 233
Illiberis, 115. *See also* Granada
imaginary, 10, 103, 233; of cultural identity, 73, 169, 175–7; of difference, 4, 9–10, 28, 104, 234; persistence, 220–5, 233–4
immigration: African (post-1975), 31, 227–33. *See also* Africa
inclusions (of Others): figured by the Good Shepherd, 28, 51; in pastoral texts, 12, 51, 154, 166, 187, 195, 199, 210, 215–16, 232; social, 45, 267n30
inclusions and exclusions (system), 4, 13, 134; rhetoric, 127. *See also* exclusions; inclusions (of Others)
Inquisition, 5, 10–11, 53, 54, 89, 94, 188–9, 244n40, 249n30
Isabel de Valois, Queen, 90. *See also* Philip II
Isabel I of Castile, 46, 52, 94, 114. *See also* Catholic monarchs
Isidore of Seville, Saint, 121
Isidro, Saint, 258n33, 262n63
Islamic, as term in the text, 8
Islamic legacy. *See* legacy: Islamic

Jacob: colour of sheep, 66, 68; livestock (Genesis 29–31), 28, 36, 59–71; manipulation of breeding process, 60; prefiguring the Good Shepherd and Christ, 59; representations of, 60–7; sons (Genesis 33–4), 193
- literary representations of, 61–4; in emblem books, 249n30; Monroy y Silva, *El pastor más perseguido*, 62–4; Vélez de Guevara, *La hermosura de Raquel*, 61–2
- pictorial representations of, 64–8; Murillo, *Jacob Laying Peeled Rods before the Flocks of Laban*, 65–6; Murillo, *Laban Searching for His Stolen Household Gods*, 66, 70, 71; Orrente, *Jacob Watering Laban's Sheep before Peeled Branches*, 64, 65–6; Orrente, *Laban Catches up with Jacob*, 66, 68, 69; Ribera, *Jacob among the Sheep*, 72; Ribera, *Jacob with the Herd of Laban*, 73

Jaén, 86
Jameson, Fredric, 21, 26
Japheth (son of Noah), 121
Jaume I of Aragon, 88
Jews: conversions, 5, 41; crypto-, 240n8; defined in Covarrubias, *Tesoro*, 39–40; elided from national origin, 10; expulsion (1492), 5, 44–5, 189, 248n26;

Hebrews, 251n9; Judaizing, punishment, 54, 89; lineages, 55; in Lope de Vega, *Pastores*, 187–92; loss of Hebrew culture, 244n43; marks, 39–41; origin in Tubal, 258n36; outsider, in Barrionuevo y Moya, 215–16; paganism vs. Judaism as model for Old vs. New Christians, 142; scorned, 265n19; Sephardic, 67, 271n10; sheep herding, 215; shepherds in the Adoration, 85, 192–3, 251n10; violence against, 5. See also *conversos*; exclusions; identity: Spanish; inclusions and exclusions; Other, ethno-religious; shepherds
Jiménez de Cisneros, Francisco (Cardinal), 198
Jiménez de Rada, Archbishop Rodrigo, 121
Jiménez de Urrea, Jerónimo: *La famosa Épila*, 19, 236
Jiménez Patón, Bartolomé, 248n26, 262n66
jineta, 163–4
John (Gospel), 14–16, 28, 36, 44–5, 50, 94–5, 101, 180
John the Baptist, Saint: *ʿansara*, 162; festivities, 162–3; in Lope de Vega, *Pastores*, 187; and Moriscos, 265n18
Johnson, Carroll B., 169, 180, 269n40
Josephus: *History of the Jews*, 121
Juan, of Austria, 141
juego de cañas. *See* canes, game of
Julián, Don (Count), 170, 217
Julius Pollux (Iulio Pollux), 130

Kamen, Henry, 240n10
King, Willard, 20, 186, 217
kings (biblical), 193–4, 226–7, 271n12. *See also* Castile: kings *and individual rulers*
Klein, Julius, 36–7
Knapp, Jeffrey, 236
Krauss, Werner, 17, 35–6
Kristeva, Julia, 14–15

Laban, 28, 36, 59–67, 68, 69, 70, 71, 73; model for Philip III and Philip IV, 66. *See also* Jacob
laberinto del fauno, El (Toro), 219–20
Lange, Johannes, 105
languages. *See individual languages*
Lapeyre, Henri, 274n26, 274n31
Lasso de la Vega, Gabriel, 23
Lateran Council (1216), 40
Latin, 27, 37, 183, 246n7, 254n1, 255n5, 256n15, 262n66. *See also* place names: Arabic
Latrás, Lupercio, 227–8
Lavocat, Françoise, 236, 262n63, 270n5
Lea, Henry, 274n31
Lead Books (*Libros de plomo*), 253n38
Ledesma, Juan de, 88
legacy: Basque, 129; classical (Graeco-Latin), 8, 130, 139, 147, 262n66; cultural, 213; Islamic, 4, 9, 16, 23, 25, 29, 36–7, 110–13, 120, 132–4, 147, 225, 254n4; Jewish (Hebrew), 147, 254n4; Moorish, 25, 151–3, 155–6, 158, 164, 168, 184–5; Morisco, 205, 228. *See also* cartography; etymology; paganism; place names
Lemos, Count of, 21
lenguaje pastoril, 108
lenguaje rústico, 108
León, 117–18, 155, 169, 189, 255n11
León, Fray Luis de: *De los nombres de Cristo*, 52–4, 247n19; "La profecía del Tajo," 267n30
León, Valencia de, 84
Lerma, Duke of. *See* Sandoval y Rojas, Francisco de
lexicographers. *See* Aldrete, Bernardo de; Covarrubias, Sebastián de; Guadix, Diego de; Poza, Andrés de; Rosal, Francisco del
ley (law and religion): 22, 50–1, 54, 141, 190, 247n17, 248n26
Libro de la invención de esta Santa Imagen de Guadalupe (Écija), 85
Libro de los secretos de agricultura (Agustín), 42
Libro en que se qüentan los amores de Viraldo y Florindo, 19, 97, 98
Liga deshecha por la expulsión de los Moriscos de España (Méndez de Vasconcelos), 274n29, 276n38
Lily, John: *Galathea*, 236
limpieza de sangre. *See* purity, blood

lineage. *See* genealogy
Linehan, Peter, 86
linguistics, 62, 103, 129; ideology, 105, 233, 278. *See also* etymology; lexicographers; place names
livestock. *See* sheep herding
livestock owners, 4, 25–7, 99, 130
Lobariñas, Gregorio de: *Pastores de Tormes, en prosa y verso*, 19
Lofrasso, Antonio: *Los diez libros de la fortuna de amor*, 18
Longus: *Daphnis and Chloe*, 221, 279n3
Lope. *See* Vega, Lope de
Lopez, Robert Sabatino, 37–9
López de Enciso, Bartolomé: *Desengaño de celos*, 18, 140–2
López de Yanguas, Hernán: *Farsa nuevamente compuesta sobre la felice nueva de la concordia y paz*, 176; *Farsa turquesana*, 93
López de Zárate, Francisco, 58
López Estrada, Francisco, 24, 26, 92, 161, 265nn14–16, 274n27
López García-Berdoy, María Teresa, 265n14, 265n16
"loss of Spain" (*pérdida de España*), 123, 170–1, 267n30. *See also* Rodrigo, Last King of the Visigoths
Low Countries, 55
Luján de Sayavedra, Mateo: *Segunda parte de la vida del pícaro Guzmán de Alfarache*, 75, 77–8, 105, 107, 250n1
Luke (Gospel), 78
Luna, Miguel de: *Historia verdadera del rey don Rodrigo*, 124–5, 259n40, 267n30

Mackay, Angus, 46
MacKay, Ruth, 242n27
Madrid, 118–19, 169, 257n24
Maestro, Jesús, 261n56
Maíllo Salgado, Felipe, 254n1
Majorca, 88–9
Malón de Chaide, Pedro: *La conversión de la Magdalena*, 138
mancha, 60–3, 191. See also *ralea*; *raza*
Mantua Carpetana. *See* Madrid
maquis, 219
Maravall, José Antonio, 243n29
Marek, Margaret, 245n1
Marguerite of Austria, Queen, 55, 84, 90, 105, 107, 162
Marian apparitions, 29, 85–9. *See also* Mary, Virgin
Mariana, Juan de: *Historia general de España*, 116, 131
Marineo, Lucio: *De rebus Hispaniae memorabilibus*, 121
markers, group, 8, 42, 52, 78, 103, 129, 134, 221, 246n7; marks of identity/infamy, 39–41, 247n18. See also *almagre*; Jews; Moriscos; sartorial practices
marlota (Morisco tunic), 102–7, 165–6, 253n34. *See also* clothing
Mármol Carvajal, Luis del: *Historia del rebelión y castigo de los moriscos del Reino de Granada*, 51, 127, 258n34
Márquez, Juan: *El governador christiano*, 57
Márquez Villanueva, Francisco, 83, 244n41, 260n54, 267n27
marranos, 268n38. *See also* pigs
Martínez, Elena María, 7
Martínez de Barros, Juan, gloss to *Mingo Revulgo*, 52
Martínez de Silíceo, Juan, archbishop of Toledo, 48–9, 84, 90
Martire d'Anghiera, Pietro, 115, 120
Marx, Leo, 27
Mary, Virgin, 85–9; of Aránzazu, 251n17; of Guadalupe, 85–8, 251n14; in Lope de Vega, *Pastores*, 190–2, 195; Marguerite of Austria as, in Pantoja de la Cruz, *Birth*, 55; of Nuestra Señora de la Cabeza (Andújar), 85–6; of Our Lady of Lluc (Majorca), 88–9; paintings of, 250nn3–4; pictorial representation, 87, 88; and *reconquest*, 86; in Zújar (Granada), 251n16. *See also* Adoration of the Shepherds; Marian apparitions
Máscara de la expulsión de los moriscos (Mira de Amescua), 93, 110
Massacre of the Innocents, 187, 194
Matthew (Gospel), 133, 194, 278n48
maurophilia, 21, 159, 162, 164, 177, 209, 264n12, 277n43. See also *Abencerraje, El*; ballads: Moorish; canes, game of;

Moorish; Moorishness
Méchoulan, Henri, 7
Medina, Pedro de: *Libro de grandezas y cosas memorables de España*, 116, 257n23, 274n28
Medina-Sidonia, Duke of, 38
Meléndez, Francisco: *Pastores de Tormes*, 19
Memorable expulsión y justísimo destierro de los moriscos de España (Guadalajara y Xavier), 53–4
memory: collective, 86, 158, 193–4, 275n35; memory galleries, 139–40; memory palace in *La Diana*, 157–8. *See also* genealogy
Méndez de Vasconcelos, Juan: *Liga deshecha por la expulsión de los Moriscos de España*, 274n29, 276n38
Mendo, Andrés: *Príncipe perfecto y ministros ajustados*, 249n30
Menéndez Onrubia, Carmen, 188
Menéndez Pelayo, Marcelino, 17–18, 24, 160–1, 265n13
Menocal, María Rosa, 9
Mercader, Gaspar, 167–8, 265n19, 266nn21–3, 267n26; *El prado de Valencia*, 18, 30, 120, 151, 161–8, 184; sonnet in Gaspar de Aguilar, *Expulsión*, 168
Mérida, Fray Diego de: *Viaje a Oriente*, 271n10
Merino sheep, 36–42
Mérique, Cyril, 51
Merivale, Patricia, 262n63
Mesa, Cristóbal de: *Las Navas de Tolosa*, 126
messianism, 16–17; Sephardic, 235
Mesta, 3, 36, 196
Mexía de Ovando, Pedro: *Ovandina*, 255n12
Mexico, 83, 93
Middlebrook, Leah, 24–5
miera, 49
Milhou, Alain, 241n19, 267n30
Mina, La (Ronda), 198–9, 274n28
Mingo Revulgo, Coplas de, 46–7, 49, 52, 101–2, 108, 246n12
Mira de Amescua, Antonio: *Máscara de la expulsión de los moriscos*, 93, 102–3, 110
monarchy: Habsburg, 44; Spanish, 10–11, 44, 92, 235, 258n34. *See also* Castile; identity: Spanish; Spain: regions; Visigoths
Monda. *See* Ronda
Monroy y Silva, Cristóbal: *El pastor más perseguido*, 62–4
Montaña, 117
Montemayor (place name), 155–6
Montemayor, Jorge de, 108, 263n3; *converso*, 244n43, 264n8; *Los siete libros de la Diana*, 14, 17–19, 20, 138–9, 244n43; – and *El Abencerraje*, 29–30, 152–61, 166, 181, 184, 264n11–12; –authorship, 160, 265n13; – and blood purity, 117, 189, 264n8; – and Cervantes, *Don Quixote*, 181, 269n41; – continuations, 20, 159, 161; –editorial history, 152, 155–6, 159–61, 243n36, 244n39, 265n14; – and Lope de Vega, *Pastores*, 189; – modern editions, 30, 160–1; – reception, 158–9; – as roman à clef, 84. See also *Abencerraje, El*; Felismena; pastoral romances
Montero, Juan, 265n14
Montero Reguera, José, 168
Moorish, 9, 38, 101–2, 107. *See also* legacy: Moorish
Moorishness, 9, 15, 21–4, 152, 155; in Mercader, 30, 161. *See also* ballads: Moorish; legacy: Moorish
Moors, 22, 90, 153–4, 166, 198–208. See also *Abencerraje, El*; ballads: Moorish; canes, game of; cultural practices: Moorish; legacy: Islamic; legacy: Moorish; legacy: Morisco; Moorishness; Moriscos
Mora, 132–3
morabito, 198–9, 201, 274n29
Morales, Ambrosio de: *Las antigüedades de las ciudades de España*, 114–16, 122–3, 143–4, 261n60
Morales, Luis de: *Adoration of the Shepherds*, 79, 81
Moreno Báez, Enrique, 161, 265n14
Morínigo, Marcos, 20

Moriscos, 4–5, 11, 51–2, 93, 167–8, 184–5, 189, 196–8, 227–8, 243n34, 263n1, 265n18, 266n25, 266n27, 271n11, 274n26, 275n32, 275n34, 276n37, 276n39, 276n41, 279n5; appearance, 103–4, 165, 240n9; buried treasure, 200–4, 275n35; censuses, 54–5, 59, 198, 248n24; in Cervantes, 180–1, 183, 273n20; clothing, 104–7; *cristianos viejos de moros*, 239n6; crypto-Muslim, 133, 235; debates on, 12–13, 17, 21–4, 40, 124, 134, 152–3, 167–8, 204–6, 240n7, 241n18, 248n23, 248n26, 263n2, 271n12, 276n41; economic activities in historiography, 11–13, 242n25; expulsion of, 5, 13, 19, 21, 24, 28, 30, 44–5, 55, 57, 93, 168, 176, 186–218, 244n46, 248n26, 267n29, 273n20, 274n31, 275n33, 276n41; as *marranos*, 268n38; nostalgia for, 187, 197, 200–4; pastoral, 234–5; phenotypic difference, 228, 276n41; as term in the text, 8; uprisings, 127, 266n26, 268n36. *See also* cultural practices: Morisco; epic; exclusions; *hidalguía*; music; Muslims: conversion; Other; shepherds; Unamuno, Miguel de
moro: as term in the text, 8–9
mudéjares, 5
muladís, 206, 208
Munda. *See* Ronda
Murillo, Bartolomé Esteban, 67–8, 249nn33–4; *The Adoration of the Shepherds*, 79, 82, 99–100; *Jacob Laying Peeled Rods before the Flocks of Laban*, 65–6; *Laban Searching for His Stolen Household Gods*, 66, 70, 71. *See also* Jacob: livestock
music: as ethnographic marker in Lope de Vega, *Pastores*, 194–5; musical instruments and flock in Juan de Ávila, 179–80; pastoral instruments, 91–2, 151, 174–80, 183–4. See also *albogue*; cultural practices: Moorish
Muslims: as African, 103, 222, 257n22, 265n20; under Christian rule (*mudéjares*), 5; conversion, 41, 44–5, 48, 50–1, 213; crypto-, 54, 240n8; frontier wars, 273n22; identifying marks, 39–41; as shepherds, 77, 88; vs. shepherds, 93; in Spanish Civil War, 221–2; as term in the text, 8
mythology. *See* paganism; Pan
myths, 23, 38, 86, 89, 121, 143, 183, 206, 214, 236, 275n35. *See also* "loss of Spain"; Visigoths
Naples, 67, 236
nationalism, 10
Nativity (Christian tradition), 14, 253n29; Blacks in, 226–9; Giménez Caballero on, 223–5
naumachia, 164–6
Navagero, Andrea, 15–17
Navarra, 10–11, 117
Navarrete, Ignacio, 16–17, 243n33
Navarro Gómez, José, 265n13
Navas, Marquis of Las, 84
Navas de Tolosa, Las (battle), 125–7, 214–15, 258n33
Navas de Tolosa, Las (Mesa), 126
Naziism, 7
Nebrija, Antonio de, 262n66; *Dictionarum oppidorum*, 115; *Muestra de la istoria de las antigüedades de España*, 115, 121, 257n21; *Vocabulario de romance en latín*, 42
Nelson, Benjamin, 20, 24–5, 169, 255n10
Nepaulsingh, Colbert I., 244n43, 264n8
Ninfas y pastores de Henares. *See under* González de Bobadilla, Bernardo
Nirenberg, David, 7, 9–10, 58–9
nobles. *See* aristocrats
North Africa. *See* Africa
Noticia general de la estimación de las artes (Gutiérrez de los Ríos), 84
novel: Moorish (see *Abencerraje, El*); picaresque, 24, 77, 198 (*see also* Luján de Sayavedra, Mateo)
Nuestra Señora de la Cabeza (Andújar), 85–6
Numancia, La (Cervantes), 23
Núñez, Miguel, 279n9
Núñez de Reinoso, Alonso, 244n43
Núñez de Velasco, Francisco: *Diálogos de contención entre la milicia y la ciencia*, 114
Núñez Muley, Francisco: *Memorandum*,

271n11

Obabakoak (Atxaga), 232–3
Obra de agricultura (Alonso de Herrera), 41–2
Ocampo, Florián de: *Corónica general de España*, 143
officium pastorum, 79, 91–2
Olivares, Count-Duke of (Gaspar de Guzmán), 10, 67
Oropesa, Alonso de: *Lumen ad revelationem gentium*, 45
Orrente, Pedro, 249n32, 249n34; *Jacob Watering Laban's Sheep before Peeled Branches*, 64–6; *Laban Catches up with Jacob*, 66, 68, 69. *See also* Jacob: livestock
Ortelius, Abraham: *Hispaniae Veteris Descriptio*, 116, 255n7, 257n23, 273n24
orthodoxy, 5–6, 44, 188–9, 240n7, 244n43; in pastoral romances, 137–8, 187–8, 190, 270n5
Osma, José M. de, 262n63
Other, ethno-religious, 4, 93–4, 103, 154, 216, 229, 263n2, 277n46; constructed, 76; marked by sartorial difference, 78, 101; pastoral exclusion of Other and self, 147, 173–4; racial, 232; in Self, 179. *See also* identity: Spanish
Otto, Ton, 110
Our Lady of Cabeza. *See under* Mary, Virgin
Our Lady of Lluc. *See under* Mary, Virgin
ovejero, 245n50. See also *cabrero*; *pastor*; shepherd, figure of

Pacheco, Francisco: *Arte de la pintura*, 253n29
paganism, 134–46; parallel forms of religious thought, 139, 261n55. *See also* Pan
paintings. *See individual artists*
Palafox, Juan de, 83
Pan (god of pastoral), 29, 134–46, 220, 260n47; figure of Christ, 144–6, 262nn63–4; name of Spain, 142–6, 233–4; theology, 135–9
Pan's Labyrinth (Toro), 219–20
Pantoja de la Cruz (Toro), Juan: *The Birth of Christ with the Royal Family* (Adoration of the Shepherds), 55–7, 56, 84, 110
Pascual de Fresno, Gabriel: *El pastor Rusticio*, 19
Passion of Christ, 192
pastor (shepherd), 27, 233, 245n49. See also *cabrero*; shepherd
pastor de Clenarda, El (Botelho de Carvalho), 18
pastor de Fílida, El. See under Gálvez de Montalvo, Luis
pastor de Iberia, El. See under Vega, Bernardo de la
pastor más perseguido, El (Monroy y Silva), 62–4
pastor Quijotiz, El (Camón Aznar), 227–8
pastor Rusticio, El (Pascual de Fresno), 19
pastora de Mançanares y desdichas de Pánfilo, 19, 264n12
pastoral:
– genre: and anti-Islamic propaganda, 280n20; and biopower, 71–2; contesting, Usque, *Consolação*, 235; and denial of history, 172–4; and the expulsion of the Moriscos, 204–6; feminizing genre, 24; global interpretation of, 26; habitus, 13, 26, 76–8, 104, 208, 220–5, 227–9, 232–4; heterogeneous manifestations, 26, 245n48; Italianate form, 17; lack of sentimental fulfilment in, 201; opposed to Moorish in ballads, 22–3; pan-European, 17; power, 58–9, 67–8; satirized in pro-Morisco ballads, 23; shepherd as crucial element, 26–7; terms, 27. *See also* shepherd, figure of
– festivals and festivities, 29, 89–94, 162
– imaginary, 75-7, 104, 227-8, 234
– performances, 89–94; in Cervantes, *Don Quixote*, 173–4. *See also* pastoral festivals and festivities
– power, 58–9, 67–8. *See also* biopower
– romances: Aragonese, 235–6; Basque, 18, 235; in Europe, 17, 236–7; Portuguese, 235; resemanticized in Spain, 15; Sephardic, 235; Spanish: 17–21, 24–6, 84, 113, 117–18, 146–7, 244n40, 244n43,

261n58; – *a lo divino*, 18, 187, 269n2; – contestation of ethnocentricity, 151–85; – decline of, 19–21, 187, 199; – and historiography, 121–9, 130–1, 146–7; – located in immemorial Spain, 139, 146–7; – resemanticized by inclusion of Moorishness, 154–5, 158, 184, 200. *See also under individual authors*
pastoral/pastoril, 27, 245n49
pastoralization of history, 122–3, 132–4; by Moriscos, 131, 259n40. *See also* Guadix, Diego de
pastores de Betis, Los. *See under* Saavedra y Torreblanca, Gonzalo de
Pastores de Tormes (Meléndez), 19
Pastores de Tormes, en prosa y verso (Lobariñas), 19
Pauline interpretation of the Gospel, 45–53
payeses, 225
peasants. *See* rustics
Pedersen, Poul, 110
pellicos (shearling jackets), 77–8, 95–107, 130, 165–6
Peñalosa y Mondragón, Benito: *Cinco excellencias del Español*, 262n64
Perceval, José María, 7, 92, 241n18, 254n1, 267n29, 275n34
pérdida de España. *See* "loss of Spain"
Pérez, Alonso: *Segunda parte de la Diana*, 18, 20, 135, 252n27, 260n47, 260n50
Pérez de Chinchón, Bernardo: *Antialcorano*, 40, 48, 276n37, 276n41
Pérez de Hita, Ginés: *Guerras civiles de Granada*, 21, 104, 158–9, 274n25, 277n43, 278n53
Pérez de Lazarraga, Juan, 18, 235
Petrarchism. *See* poetics: Italianate
phenotype. *See* purity, blood; race: phenotypic difference
Philip II, 53, 83, 89–90, 91–2, 156, 167, 267n30; annexation of Portugal, 10, 140; census information on Moriscos, 55, 248n24; as Good Shepherd, 57; Huerga sermon, 49–50; in López de Enciso, *Desengaño de celos*, 141; and Ortelius, *Hispaniae Veteris Descriptio*, 116; *Relaciones topográficas*, 248n24; statutes of blood purity, 49; suppression of Morisco cultural practices, 5, 16, 54, 176, 254n1, 267n30
Philip III, 53, 55, 57, 84, 90, 162, 193, 266n21; biopower, 66, 69, 194, 248n23; census of 1609, 198; expulsion (1609), 5, 24, 53, 93, 103, 186, 195, 200; as Good Shepherd, 55–9, 67, 194, 248n23; in López de Enciso, *Desengaño de celos*, 141; and Mercader, 167–8; portrait as shepherd, 55–7, 56, 78, 84, 99, 110; as tyrant (Herod) in Lope de Vega, *Pastores*, 194, 197
Philip IV, 66, 248n26
philology. *See* Montemayor, Jorge de
piara, 279n5. *See also* pigs; *porquerizo*
Piçario de Palacios, Álvaro, 53
pícaro. *See* novel: picaresque
pigs, 179–80. See also *piara*; *porquerizos*
Pisuerga (river), 118, 169–70
place names: Arabic, 113–21, 130–4, 213, 257n24, 259n41; and blood purity, 117, 169, 255n11; classicizing against Islamic and Hebrew legacies, 254n4; classicizing ancient Iberian in pastoral romances, 116–21, 169; erasure of original place names by Muslims, 114–16; ideological aspects, 255n8; selective, 120–1; Spain from Pan, 142–6. *See also individual places*
Plato: *Republic*, 43
Pliny, 122, 143
pluralism, ethno-religious, 14–15. See also *convivencia*
Plutarch, 145
Po (river), 169
poetics: ballads, 21–2; Castilian, 197, 254n39; and empire, 243n34; genealogy unknown, 15–16, 243n33; Italianate, 15–17, 158; and political dissidence, 243n34
Pollux, Julius, 130
Ponce, Bartolomé: *Clara Diana a lo divino*, 18, 236, 278n50. *See also* pastoral romances: Spanish
porquerizos (pigherds), 27
Porreño, Baltasar, 247n14
Portugal, 10–11, 139–40, 169, 235

power, pastoral, 58–9
Poza, Andrés de: *Antigua lengua de las Españas*, 101
prado de Valencia, El (pastoral romance). *See under* Mercader, Gaspar
prado de Valencia, El (Tárrega), 266n23
Premio de la constancia y pastores de Sierra Bermeja, El. *See under* Espinel Adorno, Jacinto de
Preto-Rodas, Richard A., 235
Primera crónica general, 240n9. *See also* Alfonso X of Castile
Primera parte de la història de València (Beuter), 120, 257n23
Primera parte de la miscelánea austral (Dávalos y Figueroa), 118, 256n15, 261n60
Primera parte de la soledad entretenida. *See under* Barrionuevo y Moya, Juan de
Primera parte de Piroeles y Musidoro en Arcadia (Hurtado Martínez), 19
Príncipe perfecto y ministros ajustados (Mendo), 249n30
propaganda, anti-Morisco, 12, 22, 40, 168, 204–6, 240n7, 276n41. *See also* Moriscos
prophecy: biblical, 44–5, 50, 278n48; pastoral, 139–42, 195, 198–9, 261n57, 270n5
proverbs, 42–3, 220, 246n7, 254n39
Ptolemy: *Geography*, 115–16
Puente, Juan de la: *Conveniencia de las dos Monarquías Católicas*, 145, 262nn63–4, 276n41
Pulgar, Fernando del, 45–6, 52
Pulido Serrano, Juan Ignacio, 269n1
purity, blood (*limpieza de sangre*), 8, 16, 25, 28–9, 45–59, 62, 79–83, 142, 147, 191–2, 242n27, 244n46, 245n47; and aristocrats, 245n47; in Barrionuevo y Moya, 213; in Espinel Adorno, *Premio*, 206; and genealogy as ethnicity, 5–8, 206, 240n10; and *hidalguía*, 191; historiography of, 6–7, 240n10, 240n13; invisible and ritualized, 94; in Lope de Vega, *Pastores*, 191–2, 196–7; and medieval marks of infamy, 247n18; in Montemayor, *Diana*, 159–61, 189; mystical, 43; mythical geography of, 117–18; narrative trope, 147; racial, 208; and rustics in early drama, 83; and shepherd, 84; statutes (*estatutos de limpieza de sangre*), 5–6, 8, 21, 45, 48–9, 52, 55, 67, 147, 192, 240n7, 247n18, 251n9, 269n42; terms, 6, 43; and white sheep, 270n8. *See also* Christians, New; *conversos*; *hilaza*; *mancha*; markers, group; *ralea*; *raza*; rustics: speech
purity, cultural, 8, 101, 109, 129, 159–61, 186–7, 206, 234. *See also* cleansing

Querol Gavaldá, Miguel, 175
Quinault, Philippe: *La généreuse ingratitude*, 278n47
Quinta parte del Flos Sanctorum (Villegas), 100–1, 145

race, 6–8, 216, 241nn14–16; etymology, 246n7; phenotypic difference, 6–7, 60, 76, 225–33, 240nn9–10, 272n14; social racialization (perception), 232–3. *See also* difference: biological; ethnocentrism; purity, blood; racism
racism, 6–8, 58–9, 147, 227–8, 241n16; early modern, 10, 28–9, 240n10, 240n13; modern, 30, 241n17
ralea, 42–3
Rallo Grus, Asunción, 161, 265n15
raza (*raça*), 4, 5, 42–3, 53, 59, 60–7, 70–1, 99, 222–3, 239n3, 246n7
readers, early modern, 117, 139, 142, 159–60, 188, 194, 244n43, 264n4
Real Academia (Española), 175
Real Academia de la Historia, 273n24
reconquest, 10, 22, 114–15, 172, 241n19, 242n21, 254n3; and *El Abencerraje*, 263n2; in Montemayor, *Diana*, 155, 157–8; and Spanish Civil War, 222, 279n4. *See also* restoration of primeval Spanish identity
Reina, Casiodoro de la, 246n9
re-Latinization, 10, 254n1. *See also* de-Orientalization
religion: Amerindian, 139; freedom of, 5. See also *convivencia*; pluralism, ethno-religious; *and individual religions*
Rennert, Hugo, 24, 274n30
Resentidos, Os, “O belén da caixa,” 228–9,

233, 280n16
Restauración de la abundancia de España (Caxa de Leruela), 258n27, 262n64
restoration of primeval Spanish identity, 29, 114, 118, 122, 159, 183, 205. See also *reconquest*
retratos a lo divino, 84
Ribeiro, Bernardim de, 244n43
Ribera, Jusepe, 67; *Jacob among the Sheep*, 72; *Jacob with the Herd of Laban*, 73. *See also* Jacob, livestock
Rodrigo, Last King of the Visigoths, 123–5, 170; in Cervantes, *Galatea*, 170–3; in Luna, *Historia verdadera*, 267n30; in Mira de Amescua, 102; in Suárez de Figueroa, 267n31; in Vives, 267n31
Rojas de Villandrando, Agustín de: *Viaje entretenido*, 94
romance (genre): Byzantine, 18, 20, 199, 277n46; definition, 243n30; heterogeneous, 205, 215; Moorish, 278n53. *See also* pastoral romances
romanceros, 21–2. *See also* ballads
romances. *See* ballads
Romans, 147, 262n66. *See also* legacy: classical
Romans, Hispano-, 207–8
roña (mange), 49, 53, 247n16, 248n26; *roñoso*, 247n20
Ronda (Munda), 197–8, 273n24, 274n26
Rosal, Francisco del: *Alfabeto primero*, 42–3, 104
Rose, Constance, 235–6, 244n40, 260n50
Rosselló i Verger, Vicenç, 120
Rueda, Lope de, 94; "Paso de Polo y Olalla negra," 226, 279n11; "Paso de Ysacaro y la negra," 280n12
Rufo, Juan: *La Austríada*, 268n36
Ruiz, Juan: *Libro de buen amor*, 42, 177
rulers: Christian (*see individual names*); Muslim, 93, 167; Almansur, 155
rustics, 76, 82–4, 94–104; Muslim, 214; nominalist opposition to Moorish, 107; speech, 108–11. *See also* culture: popular; pastoral clothing; *pellico*; *sayo*; shepherds; *zamarra*
Saavedra y Guzmán, Martín de, 20, 138
Saavedra y Torreblanca, Gonzalo de: *Los pastores de Betis*, 18, 20, 257n22, 265n20
Said, Abou, el Nisaburi, 177
Salazar, Adolfo, 177
Salazar de Mendoza, Pedro, 57–8
Salcedo Olid, Manuel de: *Panegírico historial de Nuestra Señora de la Cabeza*, 88
Salfo, 213
Salomon, Noël, 100, 254n39
Salucio, Fray Agustín de: *Discursos sobre los estatutos de limpieza de sangre*, 43, 53
Salvatierra, Martín de (bishop of Segorbe), 12, 51
sambenitos, 54
San Antonio de Padua (Alemán), 270n6
Sánchez Albornoz, Claudio, 6, 9
Sánchez Belda, Antonio, and María Sánchez Trujillo, 37–8
Sánchez Cantón, Francisco Javier, 78–9
Sánchez de Badajoz, Diego, 79; *Farsa de la Yglesia*, 51, 93
Sandoval y Rojas, Francisco de (Marquis of Denia; Duke of Lerma), 58, 93, 102, 163–4, 195, 274n31
San Germán, Marquis of, 57
sangre, limpieza de. *See* purity, blood
Sannazaro, Jacopo: *Arcadia*, 14, 17, 134, 181, 187, 236; farces, 280n20
Santa Cruz, Marquis of, 189
Sarmiento, Fray Martín, 38
Sarmiento, Pero, 45
sartorial practices, 28, 103, 107; as *almagre*, 39–41; constructed oppositions of rustics and Moriscos, 102–3; difference, and Other, 101, 104; homogenization, ethno-religious, 41, 46–7, 54, 195; homogenization, regional, 100, 253n30; in Luján de Sayavedra, 77–8, 105, 107; as Old Christian, 104; practices indistinguishable, 103–4; primeval Basque, 128; rustics, 94–104; *sambenitos*, 54; seen by foreigners, 104–5, 253n30; shared across classes, 109; shepherd, bond with herd, 95, 101, 252n25; uniformity, 95, 101. *See also* clothing; identity; *pellicos*; rustics; shepherds; *zamarras*
Sayago (Salamanca), 108

sayagués, 108, 253n37. *See also* ballads; *fabla antigua*; rustics: speech
sayo (rustic tunic), 101, 104–7, 253n34. See also *marlota*
scholarship, modern, 5–7, 9, 15, 29–30, 159–61, 217–18, 236, 240n10; historiography, 11–12, 126–7
Sebastião of Portugal, 10, 140
Sebeto (river), 169
Segorbe, bishop of, 12, 51
Segovia, 91–2
segregation, 5–8, 26, 39, 42, 47, 49, 53–4, 57
Segunda parte de la década primera de la historia ... de Valencia (Escolano), 116, 257n23
Segunda parte de la Diana. See under Pérez, Alonso
Segunda parte de la enamorada Elisea. See under Covarrubias Herrera, Jerónimo de
Segunda parte de la vida del pícaro Guzmán de Alfarache (Luján de Sayavedra), 75, 77–8, 105, 107, 250n1
Seidenspinner-Núñez, Dayle, 244n43
selection, breeding. *See* culling; sheep: Merino
self-representation, collective, 90–1
sermons, 14, 45, 48–50, 95, 179–80, 276n41
serrano, 97, 253n35
Serrano Brochero, Juan, 197
Serranos, portal de los (Valencia), 105, 107, 253n35
Servet, Miguel, 115–16, 255n5
Sessa, Dukes of, 89
Seville, 90, 119, 256nn14–15
shearling. *See* clothing: pastoral; *pellicos*; *zamarras*
sheep, 31, 156–7; Christian community as flock, 28, 38–54, 66–7, 95, 101, 194, 247n15; converted Moriscos as sheep, 51; genetic instability and logic of social segregation and culling, 41–2; interbreeding and genetic selection, 37, 41; Matthew (Gospel) on sheep and goats, 278n48; Merino, 36–42; in pastoral imaginary, 75–6; vs. pigs, in Cervantes, *Don Quixote*, 179–80; shepherd and, 76, 94–5, 101, 278n50. See also *almagre*; sheep herding; Shepherd, Good; shepherds; wool
sheep herding: and Africanness, 225; cultural continuity, 123–5, 134; cultural prestige, 4, 225; for ethnic cleansing, 67; historiography, 36–9, 122–9; history, 3–4, 25, 35–6, 86, 251n13; in modern editions of pastoral romances, 39; Moorish, 38; and rhetoric for race and ethno-religious difference, 36, 39; and Spanish pastoral romances, 245n1; terms, 36, 39–43, 129–34, 246n8, 248n23. *See also* culling; sheep: Merino; Shepherd, Good; shepherds
Shem (son of Noah), 121
shepherd, figure of, 14, 243n29; in Balkans, 262n65; biblical, 14, 28, 30, 43–74, 78–89; and blood purity, 79–83; collapses time before and after Islamic presence, 113, 127, 147–8; dignification of commoners, 84, 255n12; grounds popular culture, 112; in Lope de Vega, *Pastores*, 187–95; model of community, 75–111; model of monarch, 75; model of social process, 58–9, 67–71, 75, 248n23, 248n28; Otherness to Jews, Moors, Moriscos, Turks, 76–8, 93, 102–3, 153, 208; potential introducer of heresy, 247n21; refused by aristocracy, 163; representing elites, to legitimize power, 84, 92; and Spanish ethnocentrism, 24, 28, 71–2, 89, 129. *See also* aristocrats; exclusions; homogenization, ethno-religious; inclusions and exclusions; identity: collective; identity: Spanish; pastoral: power; rustics; sartorial practices; Shepherd, Good; shepherds, community of
Shepherd, Good, 14, 28, 36, 43–59, 246n9, 428n28; in art, 55–9; basis for pastoral clothing as Christian marker, 95; and biopower, 248n23; and blood purity, 45–59; in classical thought, 43; culling undesirable sheep, 58–9, 247n22; in discourses in defense of blood purity, 46–7; in discourses in defense of *conversos*, 45; in ecclesiastical reform, 44; Jacob as prefiguration, 59; Pauline interpretation in defense of those

converted, 48–53, 247n15; priests implicitly as, 247n22
shepherds: in contemporary Spain, 30–1, 220–5; in festivals and festivities, 89–94; as Gentile Spaniards, in pastoral romances, 142; as key figures in historical events, 123–7; as Jews in early Spanish drama, 270n9; of Las Navas de Tolosa, 125–7, 258nn32–3; in Marian apparitions, 85–9, 127; Moorish, in Barrionuevo y Moya, 209–11, 278n47; Muslim in North Africa, in Luján de Sayavedra, 77; as Old Christian, 24, 26, 192–3, 264n8; opposed to Moors in ballads, 21–4, 86, 88; in Pauline interpretation of Good Shepherd, 46; and race, 225–34; rustic vs. literary, 112–13; in stories of Rodrigo, 123–5; term as translation of *pastor*, 27, 245n49; in theatrical representations, 79–83; as victims of Muslim raids, 86–7. See also *cabrero*; sheep herding
shepherds, community of, 112–47; in Cervantes, 184; in Cervantes, *Galatea*, 170–3; Hebrew, in Lope de Vega, *Pastores*, 196; invalidated by Moors in Espinel Adorno, *Premio*, 197, 200; in Mercader, *Prado*, 166; in Montemayor, *Diana*, 154, 158; and paganism, 134. *See also* communities: pastoral; Tubal
Sicroff, Albert, 7
Sidney, Philip: *Arcadia*, 17
Sierra Bermeja (mountains), 197–8
siete libros de la Diana, Los. See under Montemayor, Jorge de
Siete partidas (Alfonso X), 40, 108
Siglo de Oro en las selvas de Erífile. See under Balbuena, Bernardo de
Sil (river), 139–40
Silva, Feliciano de, 244n43
sirgueros de la Virgen sin original pecado, Los. See under Bramón, Francisco
Smith, Paul Julian, 9
Soldina. *See* Écija
soledad entretenida, Primera parte de la. See under Barrionuevo y Moya
Solé-Leris, Amadeu, 24
Solórzano Pereira, Juan de: *Emblemata Centum, regio politica*, 249n30
Souviron López, Begoña, 216, 263n3
Spain: European perception of, 104–6, 158–9; name, 142–7; origins, 121; regions, 4, 11, 131, 225, 253n30; as term: 11; transition to democracy (post-Franco), 31, 225, 227, 233, 280n17; unification under Catholic monarchs, 10. *See also individual rulers of Spain*; "loss of Spain"
Spanish (language), 108, 254n1, 255n5, 262n66
Spanish-American War, 3
Spanish Civil War, 9, 30–1, 219–22, 225, 279n4
Spanish empire, 3–4, 20, 44–5, 141, 158, 184, 278n47
Stallaert, Christiane, 7
state, 10–11; church and, 44, 170; role in managing population, 55, 57, 194, 248n25. *See also* Council of State
Stephanus of Byzantium (Stephano): *Ethnica*, 130, 144, 262n61
stockbreeding. *See* breeding, sheep
Strabo, 122
Suárez de Figueroa, Cristóbal: *La constante Amarilis*, 18, 267n31
supersession: cultural, 15–31; Jewish shepherds in Adoration of the Shepherds, 88; in Lope de Vega, *Pastores*, 190, 270n5; of paganism by Christianity, 145; of Ptolemaic map of Iberia, 115; of Visigoths by Islam, 172
Surtz, Ronald, 14, 83, 191
Sweet, James, 240n13

Tablas poéticas (Cascales), 23
Taggard, Mindy Nancarrow, 64–5, 67, 249n31, 249n33
Tajo (river), 273n23; in Aguilar, *Expulsión*, 204–5; in Cervantes, *Galatea*, 169–70; in Cervantes, *Trabajos*, 182–3; Fray Luis de León, "Profecía," 267n30; prophecy in López de Enciso, *Desengaño de celos*, 140–2
Talavera, Hernando de, 51; *Breve suma*, 276n37; *Tratado de la misa*, 192
tale, Moorish: resemanticized in pastoral,

154–5, 158, 184, 200. See also *Abencerraje, El*
Tárrega, Francisco Agustín: *El prado de Valencia*, 266n23
teatro pastoril en la ribera del Tajo, El (Hurtado de Toledo), 18–19
Teijeiro Fuentes, Miguel, 161, 265n15
Tejeda, Jerónimo de: *Tercera parte de la Diana*, 18–20, 257n24, 264n12
teleology, 10, 94
Tendilla, Count of, 115
theatre. *See* drama
Tiber (river), 169–70
time: of completion, 44, 205; and ethnogenesis, 100; halted in Garcilaso de la Vega, 182; immemorial, 3, 29, 37, 101, 109, 146, 191; passage without effect on genealogy, 207; past and present not differentiated, 126; and pastoral romances, 24, 113; pastoralization, 123
Timoneda, Juan de: *Aucto de la oveja perdida*, 51; *Coloquios pastoriles*, 108
Toledo, 119, 170, 256n19; anti-Semitic riots, 27; as Carpentania, in Cervantes, *Trabajos*, 182–3; and pastoral, 49, 84, 90, 100; and Visigoths, 170, 269n42. *See also* clothing: pastoral; Madrid; Martínez de Silíceo, Juan;
Toledo Cathedral, 45, 48–9
Toro, Guillermo del, 219–20
Torquemada, Antonio de: *Coloquios satíricos*, 44, 96, 100, 135, 137, 247n21
Torres Naharro, Bartolomé, 79
tradition, 92, 109–10, 254n39. *See also* culture: popular; folklorism; Hobsbawm, Eric
Tragedias de amor. *See under* Arce Solorzeno, Juan de
transcultural competence, 211
transubstantiation, 145. *See also* Pan
trashumancia, 39, 251n13
Tratado de los dos caminos, 268n36
tratado de Uceda contra los estatutos de limpieza de sangre, El (Uceda), 142, 247n18
trope, cultural: Moorishness, 9; shepherd, 109
Tubal, 101, 121, 128, 258nn35–6
Turia (river). *See* Guadalaviar
Turks, Ottoman, 40–1, 93, 157, 167, 252n23
Twiddy, Ian, 205
tyranny, 92, 194, 197

Uceda, Gaspar de: *El tratado de Uceda contra los estatutos de limpieza de sangre*, 142, 247n18
Unamuno, Miguel de: *En torno al casticismo*, 3–4, 11, 13, 25, 30–1, 220, 228
Urdemalas, Pedro de: *Viaje de Turquía*, 40
Urfé, Honoré d', *L'Astrée*, 17, 236
Ursaria. *See* Madrid
Usque, Samuel: *Consolação às Tribulações de Israel*, 235–6
Uztarroz, Andrés de: *Diálogo de la verdadera honra militar*, 19

Valencia, 51, 54, 77, 90, 116, 162, 267n27, 276n40, 278n50; aristocratic identity and Moorishness, 30, 161, 266n23; in Cervantes, *Galatea*, 183–4; in Luján de Sayavedra, *Segunda parte*, 105, 107, 253n35; in Mercader, *Prado*, 161–8, 266n22; and Moriscos, 267n27. *See also* Guadalaviar
Valencia, Pedro de: moral concerns about expulsion, 274n31; *Tratado acerca de los Moriscos de España*, 12–13, 51, 276n41
Valencia de León, 84
valiente negro en Flandes, El (Claramonte), 226
Valiñas López, Francisco Manuel, 251n10
Valladolid, 40, 90
Vallés, Francisco del: *De sacra philosophia*, 60
Valverde, 91–2
Vandalia. *See* Andalusia
Vega, Bernardo de la: *El pastor de Iberia*, 18, 97, 252n28, 253n37, 257n22, 264n12
Vega, Garcilaso de la, 108, 138, 152–3, 170, 180, 236; eclogues, 14, 182; in Cervantes, *Trabajos*, 182–3; in Vélez

de Guevara, *Diablo Cojuelo*, 117–18
Vega, Garcilaso de la, El Inca: *Comentarios reales*, 261n55
Vega, Lope de, 67, 244n41; Alemán, *San Antonio de Padua*, 270n6; *Arcadia*, 18, 21, 120, 138–40, 187, 257n22, 257n24, 260n53–4, 261n58; *Las Batuecas del Duque de Alba*, 109; "Las fortunas de Diana," 187; "Guzmán el Bravo," 187; *La hermosura de Angélica*, 211; *Isidro*, 258n33, 262n63; *El pastor de Galatea*, 19, 187; *Los pastores de Belén*, 18, 20, 21, 30, 85, 187–97, 216–17, 226, 261n58, 270nn4–5, 271n11, 271n13; *El prodigio de Etiopía*, 226; *La villana de Getafe*, 273n21
Vegas, Damián de: *Poesía cristiana, moral y divina*, 247n22
Vélez de Guevara, Luis: *El Diablo Cojuelo*, 117–18; *La hermosura de Raquel*, 61–3. *See also* Jacob: livestock
verdadero Dios Pan, El (Calderón de la Barca), 262n63
Verdú, Blas: *Discurso de la expulsión de los moriscos de España*, 265n18, 271n12
Viaje a Oriente (Mérida), 271n10
Viaje de Turquía, 40
Viaje entretenido (Rojas de Villandrando), 94
Vicente, Gil, 79: *Auto da fé*, 101
Vida de Marcos de Obregón (Espinel), 198, 210–11
Vigilia y octavario de San Juan Baptista (Abarca de Bolea), 18, 20
Villegas, Alonso de: *Quinta parte del Flos Sanctorum*, 100–1, 145
Villegas, Antonio de: *Inventario*, 263n1
Vincent, Bernard, 11, 274n31
Virgil, 180; *Eclogue*, 10, 134; Encina translation, 278n47
Virgin Mary. *See* Mary, Virgin
Visigoths, 9, 109, 124, 126, 170–2, 182–4, 206, 208, 269n42. See also *goticismo*; "loss of Spain"; Rodrigo, Last King of the Visigoths
Visweswaran, Kamala, 241n16
Viterbo, Annio de, pseudo-Beroso, 115; *Commentaria super opera auctorum diversorum de antiquitatibus loquentium*, 121
Vives, Juan Luis: *Instrucción de la mujer cristiana*, 267n31
Vizcaya, 117

Walter, Monika, 181, 269n40
weddings, royal. *See* pastoral festivals and festivities
Weiditz, Christoph: *Trachtenbuch*, 105–6, 253n30
Wertheimer, Elaine, 14, 270n9
wool: colour, 36–42; Marian apparitions, 85; trade, 3–4, 11–12, 35–6, 42, 122, 130

Xenophon: *Memorabilia*, 43
Ximénez Dembún, Hierónimo, 263n1

zamarras (sheepskin jackets), 95–7, 100–6; *çamarrón*, 101–2; *çamarros*, 96, 101, 124–5, 130, 255n12
zambra (Morisco dance), 24, 103, 176–9, 244n42
Zaragoza, 90
zootechnics, 36, 39, 57, 191
Zorrilla, Laura, 277n45
Zújar (Granada), 251n16

Toronto Iberic

1 Anthony J. Cascardi, *Cervantes, Literature, and the Discourse of Politics*
2 Jessica A. Boon, *The Mystical Science of the Soul: Medieval Cognition in Bernardino de Laredo's Recollection Method*
3 Susan Byrne, *Law and History in Cervantes' Don Quixote*
4 Mary E. Barnard and Frederick A. de Armas (eds), *Objects of Culture in the Literature of Imperial Spain*
5 Nil Santiáñez, *Topographies of Fascism: Habitus, Space, and Writing in Twentieth-Century Spain*
6 Nelson R. Orringer, *Lorca in Tune with Falla: Literary and Musical Interludes*
7 Ana M. Gómez-Bravo, *Textual Agency: Writing Culture and Social Networks in Fifteenth-Century Spain*
8 Javier Irigoyen-García, *The Spanish Arcadia: Sheep Herding, Pastoral Discourse, and Ethnicity in Early Modern Spain*
9 Stephanie Sieburth, *Survival Songs: Conchita Piquer's* Coplas *and Franco's Regime of Terror*
10 Christine Arkinstall, *Spanish Female Writers and the Freethinking Press, 1879–1926*